Evolutionary Cell Processes in Primates

Evolutionary Cell Biology

PUBLISHED TITLES

Deferred Development: Setting Aside Cells for Future Use in Development in Evolution
Edited by Cory Douglas Bishop and Brian K. Hall

Cellular Processes in Segmentation
Edited by Ariel Chipman

Cellular Dialogues in the Holobiont
Edited by Thomas Bosch and Michael G. Hadfield

Evolving Neural Crest Cells
Edited by Daniel Meulemans Medeiros, Brian Frank Eames, Igor Adameyko

Development of Sensory and Neurosecretory Cell Types: Vertebrate Cranial Placodes, Volume 1
Gerhard Schlosser

Evolutionary Origin of Sensory and Neurosecretory Cell Types: Vertebrate Cranial Placodes, Volume 2
Gerhard Schlosser

Evolutionary Cell Processes in Primates: Bones, Brains, and Muscle, Volume I
Edited by M. Kathleen Pitirri and Joan T. Richtsmeier

Evolutionary Cell Processes in Primates: Genes, Skin, Energetics, Breathing, and Feeding, Volume II
Edited by M. Kathleen Pitirri and Joan T. Richtsmeier

For more information about this series, please visit: www.crcpress.com/Evolutionary-Cell-Biology/book-series/CRCEVOCELBIO

Evolutionary Cell Processes in Primates

Genes, Skin, Energetics, Breathing, and Feeding

Volume II

Edited by

M. Kathleen Pitirri
Joan T. Richtsmeier

CRC Press
Taylor & Francis Group
Boca Raton London New York

CRC Press is an imprint of the
Taylor & Francis Group, an **informa** business

First edition published 2022
by CRC Press
6000 Broken Sound Parkway NW, Suite 300, Boca Raton, FL 33487–2742

and by CRC Press
2 Park Square, Milton Park, Abingdon, Oxon, OX14 4RN

Library of Congress Cataloging-in-Publication Data
Names: Pitirri, Kathleen M., editor. | Richtsmeier, Joan T., editor.
Title: Evolutionary cell processes in primates / edited by Kathleen M. Pitirri and Joan T. Richtsmeier.
Description: First edition. | Boca Raton : CRC Press, 2022. | Includes bibliographical references and index. | Contents: Volume 1. Bone, brains, and muscle — Volume 2. Genes, skin, energetics, breathing, and feeding | Summary: "This volume demonstrates the role of cellular mechanisms in the production of specializations defining primates with respect to locomotion, energetics and diet. We highlight how genetic analysis, visualization of cells and tissues, and merging Evo-Devo with cell biology combine to answer questions central to understanding primate evolution"— Provided by publisher.
Identifiers: LCCN 2021023379 | ISBN 9780367437688 (v. 1 ; hardback) | ISBN 9781032072715 (v. 1 ; paperback) | ISBN 9780367437671 (v. 2 ; hardback) | ISBN 9781032072784 (v. 2 ; paperback) | ISBN 9781003206231 (v. 1 ; ebook) | ISBN 9781003206293 (v. 2 ; ebook)
Subjects: LCSH: Primates—Cytology. | Primates—Evolution. | Cells—Mechanical properties. | Primates—Anatomy.
Classification: LCC QL737.P9 E97 2022 | DDC 599.8—dc23
LC record available at https://lccn.loc.gov/2021023379

ISBN: 978-0-367-43767-1 (hbk)
ISBN: 978-1-032-07278-4 (pbk)
ISBN: 978-1-003-20629-3 (ebk)

Typeset in Times
by Apex CoVantage, LLC

Dedication

We dedicate these volumes to our
families, whose love, support, and
patience throughout this process
and during a worldwide lockdown
made these volumes possible.

Ben and Russ – thanks for making this
past year in lockdown so much fun, you
made it one of the best times of my life.

Thrill, Hannah, Lute, and Faith – you'll
probably never know that these books
exist, much less read them. Still, I like
to imagine you finding them one day
and reading a page aloud to one another
Wizard People, Dear Reader-style

Contents

Preface for Volume II ix
Acknowledgments xi
Editors xiii
Contributors xv

Chapter 1 Molecular and Cellular Processes of Pelage Growth and Pigmentation in Primate Evolution 1

Elizabeth Tapanes, Jason M. Kamilar, and Brenda J. Bradley

Chapter 2 Cell Processes and Key Genes in the Evolution of Pigmentation Variation in Humans 35

Heather L. Norton

Chapter 3 Cell Processes Underpinning the Evolution of Primate Dental Form and Formula 55

Cassy M. Appelt, Elsa M. Van Ankum, Denver F. Marchiori, and Julia C. Boughner

Chapter 4 Gene Regulatory Processes in the Development and Evolution of Primate Skeletal Traits 83

Genevieve Housman

Chapter 5 Processes That Generate Modularity in the Mammalian Skull: Implications for Primate Evolution 109

Nandini Singh

Chapter 6 Brown Adipose Tissue, Nonshivering Thermogenesis, and Energy Availability 131

Maureen J. Devlin

Chapter 7 Interaction between Environmental Temperature and Craniofacial Morphology in Human Evolution: A Focus on Upper Airways 161

Laura Maréchal and Yann Heuzé

Chapter 8 Evolution and Development of the Nasal Airways in Primates: The Influence of Eye Size and Position on Chondrogenesis and Ossification of the Nasal Skeleton 189

Timothy D. Smith and Valerie B. DeLeon

Chapter 9 Stem Cells in Primate Evolution 213

Emily L. Durham and M. Kathleen Pitirri

Index 237

Preface for Volume II

This is the second book in a two-volume series focused on cell processes in the evolution of primate characteristics. Volume II resumes the theme of the preceding volume, aiming to: (1) present current relevant work in a manner of interest to specialists who study specific complex traits and to those more generally interested in a cellular biology approach in anthropology, (2) provide a broad body of work enabling a general evaluation of this approach for study of fundamental questions of primate evolution, (3) assess modes of analyses and current knowledge of cellular biology underlying the evolution of primate-specific complex traits, and (4) promote further interest and investigation of cell biology as the basis for understanding variation in primate characteristics and their evolution.

Chronicling patterns of morphological change in primate evolution is the backbone of biological anthropology and is a required, though preliminary, step in understanding those patterns. We recruited contributors to write chapters that would inform anthropologists about work being done to uncover cellular behaviors and processes that build primate phenotypes. Our intent is to introduce and explain the essential role of cellular processes in the production of some of the primates' most celebrated complex traits and highlight this approach as a promising area of anthropological research.

Contributors of chapters in the first section of this volume focus on cellular processes underlying the evolution of hair, skin, and teeth in primates. Elizabeth Tapanes and colleagues (Chapter 1) review the cellular processes involved in pelage growth and pigmentation that have likely contributed to the evolution of primate hair and fur. These authors suggest that combining molecular and cellular techniques with hair phenotyping will allow anthropologists to answer fundamental questions about primate evolution that span behavior, ecology, signaling, and biodiversity. In Chapter 2, Heather L. Norton reviews the physiological factors and key genes that influence melanin production and how this can be used to study the evolution of hair, skin, and eye pigmentation in primates. Due in part to their preservation in the fossil record, the evolution of teeth has been heavily studied across all primate clades. In Chapter 3, Cassy M. Appelt and colleagues review the cellular behaviors involved in dental development and tooth loss and how their study can inform us about changes in dental morphology, function, and formulae that occurred during primate evolution.

The second section of Volume II focuses on energetics, aging, life history, and growth patterns in primate evolution. Variation in skeletal structure has played a ubiquitous and unique role in anthropological research. This section begins with Genevieve Housman's (Chapter 4) review of the contribution of gene regulation in skeletal cells to primate evolution. Housman's chapter shows how an appreciation for the collective impact of genetic variation, environmental disturbances, and regulatory processes on shaping the skeleton is crucial for proper interpretations of skeletal variation. Chapter 4 concludes by proposing future in vitro models that will allow the evaluation of gene-by-environment interactions that contribute to skeletal traits in human and nonhuman primates. Nandini Singh (Chapter 5) continues the discussion of evolutionarily conserved signaling pathways involved in the development of skeletal traits by evaluating the effects of a change in regulation of Sonic Hedgehog signaling on the skull. Singh's work underscores the potential of studies in laboratory animals for clarifying the evolution of integration and modularity in the primate skull.

Selective advantages of differing energetic regimes are the cornerstone of many important theories in anthropology, including those that concern dietary profiles, locomotor strategies, reproductive success, and climatic adaptations. However, the majority of the hypotheses that drive these studies do not consider what happens at the cellular level. Maureen J. Devlin (Chapter 6) addresses this by reviewing the role of specialized fat cells, brown adipocytes, in brown and beige adipose tissue (BAT) in energy expenditure through nonshivering thermogenesis. She assesses how BAT abundance and metabolic activity change across the life course and discusses the role of BAT in hominin expansion to cold climates. Devlin concludes by providing suggestions for future research directions to better understand the importance of BAT in hominin evolution, the framework of which could be applied widely across the study of the evolution of energetics in primates. Environmental influences on cell activity are examined in Chapter 7, where Laura Maréchal and Yann Heuzé review temperature-related morphological variation of nasal and paranasal structures. The authors outline the role that temperature-sensitive developmental pathways might have in producing morphological variation by directly altering the activity of the major types of bone cells: preosteoblasts, osteoblasts, and osteoclasts, and indirectly through protein and hormone production involved in the activity of these cells. The authors close by suggesting novel approaches to improve our understanding of the role of these mechanisms in human evolution. Morphological variation in nasal structures across primates is further examined by Timothy Smith and Valerie B. DeLeon (Chapter 8). These authors examine how two hallmarks of primate evolution, increased reliance on vision (orbital convergence) and decreased reliance on olfaction (reduction of ethmoturbinals), influence each other at the cellular level during development.

Volume II concludes with Emily L. Durham and M. Kathleen Pitirri's discussion of how stem cell research can be applied in the study of mechanisms underlying significant events in human and primate evolution. The authors introduce the various types of stem cells, their role in development, and their maintenance throughout life. They summarize what is currently known about stem cells in different primate species and present hypotheses of how comparative analyses of stem cells in primates might inform us about the evolution of slowed life histories in primates.

To date, the field of biological anthropology has informed us about the who, what, where, and when of important events in primate evolution. The chapters in these volumes demonstrate that studying cellular processes can provide information pertaining to the question of *how*—and perhaps even *why*—these evolutionary events occurred. It is our hope that these chapters inspire new research within biological anthropology that looks to cellular processes as an avenue for explanation and discovery of the mechanistic basis for traits that have been studied since the dawn of the discipline.

Acknowledgments

We greatly appreciate the efforts of our chapter contributors and of the reviewers of each of these chapters who persevered in the face of challenging topics that cross disciplines and overcame obstacles created by the Covid-19 pandemic. We thank the editorial staff of CRC Press/Taylor & Francis for guiding us through the final stages of manuscript preparation and editing. Individuals who provided informative and supportive reviews of the chapters in Volume II of this series are listed in the following.

Dana Al-Hindi
PhD Candidate
Department of Anthropology
University of California, Davis

Christina Bergey
Assistant Professor
Department of Genetics
Rutgers University

Terence Capellini
Richard B. Wolf Associate Professor
Department of Human Evolutionary Biology
Harvard University

Yang Chai
University Professor
George and MaryLou Boone Chair in Craniofacial Molecular Biology
Ostrow School of Dentistry
University of Southern California

Lynn Copes
Associate Professor of Medical Sciences
School of Medicine
Quinnipiac University

Maureen J. Devlin
Associate Professor
Department of Anthropology
University of Michigan

Eva Garrett
Assistant Professor
Department of Anthropology
Boston University

Brenna Henn
Associate Professor
Department of Anthropology
University of California, Davis

Nina Jablonski
Evan Pugh Professor
Department of Anthropology
Pennsylvania State University

R. Amanda C. Larue
Professor
Department of Pathology and Laboratory Medicine
Medical University of South Carolina

Stephanie Levy
Assistant Professor
Department of Anthropology
City University of New York, Hunter College

Scott Maddux
Assistant Professor
Center for Anatomical Sciences
Graduate School of Biomedical Sciences
University of North Texas Health Science Center

Neus Martínez-Abadías
Associate Professor
Department of Evolutionary Biology, Ecology and Environmental Sciences
Universitat de Barcelona

Tesla Monson
Assistant Professor
Department of Anthropology
Western Washington University

Susan Motch-Perrine
Assistant Research Professor
Department of Anthropology
Pennsylvania State University

Cara Ocobock
Assistant Professor
Department of Anthropology
University of Notre Dame

Matthew Ravosa
Professor
Department of Biological Sciences
University of Notre Dame

Roger Reeves
Professor
Departments of Physiology and Genetic Medicine
Johns Hopkins University School of Medicine

Mark Shriver
Professor
Department of Anthropology
Pennsylvania State University

Abigail Tucker
Dean for Research
Faculty of Dentistry, Oral & Craniofacial Sciences
and Professor of Development & Evolution
King's College London

Editors

M. Kathleen Pitirri, PhD, is a postdoctoral scholar in the Department of Anthropology at Pennsylvania State University. She received her PhD in 2019 from the University of Toronto, where she studied primate evolution, focusing specifically on the taxonomic, ontogenetic, and functional basis of mandibular shape variation in living and fossil primates. During her PhD research, Dr. Pitirri developed a novel methodology for studying shape variation of mandibular fragments that are part of the primate fossil record. She found a strong relationship between the shape of the mandibular corpus and molar crypt formation in great apes, suggesting that mandibular shape is linked to an extended period of development in great apes, representing an important evolutionary shift in primates. Upon joining the Richtsmeier Lab, Dr. Pitirri began using mouse models to study the cellular mechanisms involved in transferring information from the genotype to the phenotype. The changes observed in mouse models can be used to interpret the cellular basis for changes observed in skull shape in primates because mechanisms that build the craniofacial skeleton during development also drive variation in disease and evolution. Dr. Pitirri is particularly interested in the evolutionary consequences of change in developmental processes driving the patterning of cellular activities involved in embryogenesis of skull bones, the role of the chondrocranium in skull development, and the genetic pathways regulating the relationship between tooth and bone formation during embryonic development.

Joan T. Richtsmeier, PhD, is Distinguished Professor of Anthropology at the Pennsylvania State University. She received her PhD from Northwestern University in 1985 and joined the faculty of the Department of Cell Biology and Anatomy, Johns Hopkins University School of Medicine, in 1986. There, she focused on establishing new quantitative methods for studying change in biological shape through time, especially in primates, with Professor Subhash Lele. In 1999, she became the 55th woman to achieve the rank of professor at Johns Hopkins University School of Medicine since the school opened in 1893. In 2000, Dr. Richtsmeier moved her lab to the Pennsylvania State University. There, her focus turned to joining developmental biology with evolutionary biology, and with collaborators and students, she has worked to integrate the study of mouse models carrying known genetic variants with understanding the biological basis of patterns of craniofacial disease and evolutionary change. She is particularly interested in early formation of the chondrocranium and how and why cells decide to become osteoblasts and make bone. Dr. Richtsmeier was elected Fellow of the American Association of Anatomists (AAA) in 2018, received the Henry Gray Scientific Achievement Award of the AAA in 2019, and received the David Bixler Excellence in Craniofacial Research Award of the Society for Craniofacial Genetics and Developmental Biology in 2019. She was elected Fellow of the AAAS (Section on Biological Sciences) in 2020. Her work is supported by grants from the National Science Foundation, the National Institutes of Health, and the Wellcome Trust.

Contributors

Cassy M. Appelt
Department of Anatomy, Physiology & Pharmacology
College of Medicine
University of Saskatchewan
Saskatoon, SK, Canada

Julia C. Boughner
Department of Anatomy, Physiology & Pharmacology
College of Medicine
University of Saskatchewan
Saskatoon, SK, Canada

Brenda J. Bradley
Center for the Advanced Study of Human Paleobiology
Department of Anthropology
The George Washington University
Washington, DC

Valerie B. DeLeon
Department of Anthropology
University of Florida
Gainesville, Florida

Maureen J. Devlin
Department of Anthropology
University of Michigan
Ann Arbor, Michigan

Emily L. Durham
Department of Anthropology
Pennsylvania State University
University Park, Pennsylvania

Yann Heuzé
Université Bordeaux, CNRS, MC, PACEA
Pessac, France

Genevieve Housman
Section of Genetic Medicine
University of Chicago
Chicago, Illinois

Jason M. Kamilar
Graduate Program in Organismic and Evolutionary Biology
Department of Anthropology
University of Massachusetts Amherst
Amherst, Massachusetts

Denver F. Marchiori
Department of Anatomy, Physiology & Pharmacology
College of Medicine
University of Saskatchewan
Saskatoon, SK, Canada

Laura Maréchal
Université Bordeaux, CNRS, MC, PACEA
Pessac, France

Heather L. Norton
Department of Anthropology
University of Cincinnati
Cincinnati, Ohio

M. Kathleen Pitirri
Department of Anthropology
Pennsylvania State University
University Park, Pennsylvania

Nandini Singh
Department of Anthropology
California State University Sacramento
Sacramento, California

Timothy D. Smith
School of Physical Therapy
Department of Biology
Slippery Rock University
Slippery Rock, Pennsylvania

Elizabeth Tapanes
Center for the Advanced Study of
Human Paleobiology
Department of Anthropology
The George Washington University
Washington, DC

Elsa M. Van Ankum
Department of Anatomy, Physiology &
Pharmacology
College of Medicine
University of Saskatchewan
Saskatoon, SK, Canada

1 Molecular and Cellular Processes of Pelage Growth and Pigmentation in Primate Evolution

Elizabeth Tapanes, Jason M. Kamilar, and Brenda J. Bradley

CONTENTS

1.1 Mammalian Hair Evolution ... 1
1.1.1 Primate Hair Biology ... 2
1.1.2 The Last Century of Hair Evolution Research ... 3
1.2 The Cellular World of the Hair Keratinization Pathway ... 4
1.2.1 The Hair Growth Cycle ... 6
1.2.2 Genes and Mechanisms Potentially Driving Primate Hair Growth ... 6
1.3 The Cellular World of the Hair Pigmentation Pathway ... 7
1.3.1 The Hair Melanogenesis Pathway ... 7
1.3.2 Genes and Mechanisms Potentially Underlying Variation in Primate Hair Pigmentation ... 14
1.4 The Promise of Genomics for Untangling Our Hairy Evolution ... 19
1.4.1 Using Genomics to Study Primate Hair ... 20
1.4.2 Beyond Genomics ... 21
1.4.3 Incorporating Comparative Vertebrate Taxa ... 22
1.5 Conclusions ... 23
1.6 Acknowledgments ... 24
1.7 References ... 24

1.1 MAMMALIAN HAIR EVOLUTION

Pelage (i.e., hair/fur) is the first barrier providing mammals protection against their external environments. It is one of the defining characteristics of mammals and a likely key to their evolutionary success. Hair is so essential that basal mammals were likely unable to capitalize on terrestrial niches before the evolution of hair—whose primary functions were possible protection from abrasion and insulation (Maderson 2003). However, pelage has multiple additional functions, including, but not limited to: individual/kin/species visual recognition (Allen and Higham 2013), camouflage (Hoekstra et al. 2005; Barrett et al. 2019), protection against pathogens (Paus and Cotsarelis 1999), and potentially providing a means for dispersing olfactory communication (Eisenberg and Kleiman 1972).

Variation in pelage is almost certainly a result of—at least to some degree—natural and sexual selection, sometimes acting at odds with each other. For example, darker manes in lions living in the Serengeti function as an honest signal for nutrition and testosterone that influences male-male competition and female mate choice (West and Packer 2008). However, in black fox squirrels (*Sciurus niger*), thicker and darker tail hairs are hypothesized as an adaptation for thermoregulation (Fratto and Davis 2011). Pelage provides potential adaptations to social as well as physical environments (Cuthill et al. 2017). Thus, pelage is highly susceptible to rapid evolutionary change and impacts from climate shifts.

1.1.1 Primate Hair Biology

Pelage is also a central aspect of primate (including human) diversity, as it shows marked intra- and interspecific variation (Bradley and Mundy 2008). For example, in nonhuman primates, examples of pelage variation include ontogenetic hair change (e.g., dusky langurs, *Trachypithecus obscurus*) and facial pattern complexity within and across species (e.g., guenon monkeys, *Cercopithecus* spp.) (Rowe 1996; Anne-Isola Nekaris and Munds 2010; Allen et al. 2014). There is also a high degree of convergence in certain hair phenotypes across the primate clade, such as: facial hair ornamentation, dense body hair, long capes along the torso, yellow/red pigmentation (sometimes in patches), black/brown pigmentation, and black and white pigmentation (Bradley and Mundy 2008).

An earlier theory postulated that hair patterns shift through evolutionary time from agouti-banding to saturation to bleaching based on observed patterns of pigmentation in platyrrhines (i.e., metachromism) (Hershkovitz 1968). Although some early phylogenetic analyses seemed to support this hypothesis (Jacobs et al. 1995), it has since been falsified multiple times based on empirical evidence in distinct clades (Chaplin and Jablonski 1998; Santana et al. 2012). Additionally, patterns of hair pigmentation change across the primate order and generally do not adhere to phylogeny (Kamilar and Bradley 2011).

Overall, key functions for primate hair are similar to other mammals and include camouflage, communication, and thermoregulation. Primate coloration must often balance crypsis with signaling. Across all primate clades, pelage variation conforms to a classic ecogeographical rule where darker-pigmented animals are most likely found in warm and wet habitats such as rainforests or other densely forested environments (i.e., Gloger's rule) (Gloger 1833; Kamilar and Bradley 2011; Santana et al. 2012, 2013). This is because dark pigments may help conceal nonhuman primates in dense foliage from dichromatic predators, such as felids. However, selection could be acting on a distinct phenotype (e.g., immunity, cold tolerance) linked or co-evolving with color polymorphisms (Delhey 2017; Delhey et al. 2019). Yet certain aspects of primate hair color (i.e., red hues, contrast, complexity, hair tufts) may aid in conspecific communication such as kin or species recognition (Chaplin and Jablonski 1998; Sumner and Mollon 2003; Bradley and Mundy 2008; Winters et al. 2020). In fact, while both orange and black hues might be camouflaged from a dichromatic predator in a background of dense foliage, orange hues may be conspicuous to trichromatic conspecifics (Sumner and Mollon 2003). It is worth noting, though, that even in dichromatic species, such as *Eulemur fulvus*, females seem able to differentiate between dull and brightly colored individuals (Cooper and Hosey 2003; Jacobs et al. 2019). Thus, coloration likely aids with communication and camouflage. On the other hand, it is unclear what role primate pigmentation plays in thermoregulation. Primates of distinct colors exhibit no differences in thermoregulatory behaviors (Bicca-Marques and Calegaro-Marques 1998), but data from other organisms suggests black colors may have a thermoregulatory benefit (Fratto and Davis 2011). Nonetheless, the capacity for primate hair to act as thermal insulation is likely dependent mainly on hair density (Tregear 1965). Body hair density is known to similarly vary across populations and across body regions (Schwartz and Rosenblum 1981)—potentially as an adaptation to climate. For example, in Neotropical primates, facial hair length increases in colder areas (Santana et al. 2012). This follows another ecogeographical rule (the 'hair rule'), which posits individuals should have longer and thicker hairs in colder regions (Rensch 1938). However, hair growth hypotheses are less well studied than those focused on color.

Like other primates, humans are covered in hair of various forms (e.g., thin, long, curly) and types (e.g., scalp, nose, axillary, pubic), spanning a swath of color profiles (e.g., blonde, brown, black) (Lasisi et al. 2016). Human hair, like that of nonhuman primates, varies across body regions and between individuals and is also correlated with ancestry (Steggerda and Seibert 1941; Seibert and Steggerda 1999). However, humans are described as 'hairless' because the hair on the torso and limbs tends to be light, thin, and vellus (Pagel and Bodmer 2003). Humans do not necessarily have less hair on their torso, but the hair they have is more sparse, thinner, and shorter than that of other primates (Schwartz and Rosenblum 1981; Sandel 2013). Thus, hair color and growth variation are defining characteristics for humans as much as nonhuman primates and the rest of their mammalian relatives. Unlike other primates, though, human scalp hair graying may be a uniquely human trait (Tapanes et al. 2020). Both the differences and similarities between human and nonhuman primate hair biology can provide signposts for what it means to be human. Yet the molecular and cellular mechanisms underlying human and non-human primate hair evolution remain relatively understudied.

1.1.2 The Last Century of Hair Evolution Research

Comparative genomic analyses indicate that loci associated with keratinization of hair are under selection across major primate taxa (George et al. 2011), but most published studies of primate hair variation to date have focused on specific aspects of pigmentation and one or a few candidate genes (Mundy and Kelly 2003, 2006; Bradley et al. 2013). Few studies have attempted to identify genomic signatures of selection underlying human hair phenotypes. Here, selection is defined as the differential reproduction of genetically distinct organisms in one population. Traditionally, signatures of positive selection (which are synonymous with adaptation) are examined via evidence of accelerated evolution or extended haplotype analyses (Hancock and Di Rienzo 2008; Graur 2016). For example, accumulated variations associated with hair shape in humans have been identified as potential targets of selection (Adhikari et al. 2016). Most of what we know about the molecular basis of human hair variation, though, comes from disease cases in the dermatological literature (Seibert and Steggerda 1999; Shimomura et al. 2009; Fujimoto et al. 2012)—not from an evolutionary framework. While we know a lot about the genetics of skin color (Norton et al. 2007; Quillen et al. 2019), more work on the cellular and genomic basis of primate hair evolution is needed.

Unlike many phenotypes, we have a good understanding about the physiology and genetics of hair variation in some non-primate mammals, mainly from work done on laboratory or domestic species (Drögemüller et al. 2007; Sponenberg 2009; Lamoreaux et al. 2010; Harel and Christiano 2012). Many hair-related genes are highly conserved—meaning the amino acid sequence of hair genes (those involved in growth and pigmentation) are minimally changed among highly diverged species (Protas and Patel 2008). We can thus generate hypotheses about the evolution of primate hair diversity based on studies of hair in other mammals.

In this chapter, we will discuss what is known about the evolution, cellular biology, and genomics of hair phenotypes common in primates. First, we provide a brief overview of some of the major cells and pathways involved in hair keratinization (i.e., growth) and pigmentation. Based on what is known about other mammals (mostly laboratory or domestic), we review key candidate loci potentially involved in generating distinct hair phenotypes. Second, we highlight that genomics can help scientists study the evolution of hair biology through the use of robust demographic models and genome scans. We comment briefly on functional studies that can be used as supplements. Last, we argue that nonhuman primates provide an ideal comparative framework to established model systems (i.e., beyond *Mus*) for studying hair evolution. Given most of what we know originates in the laboratory rodent, we follow gene/protein nomenclature of *Mus* and *Rattus*, unless specifically talking about primates.

1.2 THE CELLULAR WORLD OF THE HAIR KERATINIZATION PATHWAY

Hair follicles—dynamic 'mini-organs' of skin—form during fetal and perinatal skin development and are responsible for engineering hair shafts (Schneider et al. 2009). The hair follicle is not visible to the naked eye and is composed of the mesenchyme and the dermis. Stem cells in the ectoderm give rise to epithelial components of the follicle: the hair bulb and infundibulum, as well as sebaceous and apocrine glands (Schneider et al. 2009). The lower region contains the hair bulb (or factory for the hair shaft) that houses the dermal papilla (Krause and Foitzik 2006). Within the dermal papilla, hair growth and cycling takes place via keratinocyte (i.e., keratin-producing cell) proliferation and controls morphologies such as hair length and hair width/thickness (Paus and Foitzik 2004). The proximal end of the epithelium contains the infundibulum, which is the opening of the hair canal to the skin surface (Schneider et al. 2009). Last, the upper region—the visible hair—is terminally differentiated dead keratinocytes that are being extended (growing) via the cellular processes in the follicle and dermal papilla.

We know a good deal about cellular biology of hair mostly because of studies on laboratory and fancy mice. Such work highlights that molecular communication between the epidermis and the underlying mesenchyme is the key factor in hair follicle development and subsequent growth cycles (Schneider et al. 2009). This is also known commonly in developmental biology as epithelial-mesenchymal interactions (EMI). The interaction is necessary to develop all appendages on the ectoderm across species and thus is also responsible for the development of scales, feathers, nails, teeth, and exocrine glands (Hardy 1992; Krause and Foitzik 2006)—in addition to pelage. During the development of the hair follicle, interactions between keratinocytes in the epidermis and fibroblasts (connective tissue cells) in the dermal papilla are responsible for generating the hair shaft (Botchkarev and Kishimoto 2003). Fibroblasts in the dermal papilla consist of tightly packaged mesenchymal cells (Figure 1.1A). Two

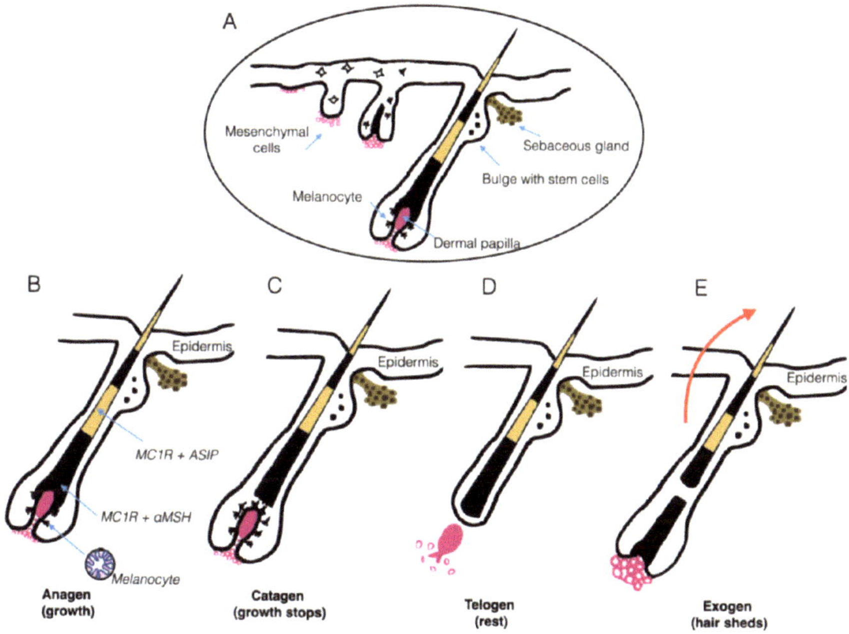

FIGURE 1.1 The hair follicle growth cycle including the (A) main components of the hair follicle, including dermal papilla and melanocytes; (B) the growth phase (or anagen), where melanocytes are deposited into keratinocytes as the hair grows; (C) growth stops (catagen), when the size of the hair follicle decreases and detaches from the dermal papilla; (D) the resting phase (telogen), when the old hair has detached from the dermal papilla and a new hair can begin to grow from; and (E) the resting hair is gradually loosened until it is shed (exogen).

Key: MC1R (*melanocortin-1-receptor*) + *a-MSH* (alpha-melanocyte-stimulating hormone) stimulates the release of eumelanin; MC1R + ASIP (agouti signaling protein) stimulates the release of pheomelanin.

messages are passed between the epidermis and mesenchyme during embryogenesis in mammals: (1) make an appendage in this region, and (2) develop dermal papilla or hair follicle here (Hardy 1992). Tissue recombination experiments on mice have shown that the dermal surface of mice can develop feather buds (from chicken epidermis), but the mouse dermis does not continue to develop scales or feathers beyond this point (Hardy 1992). The latter is hypothesized because hair is the only tissue produced via EMI, which delivers a second message to make a hair follicle.

1.2.1 The Hair Growth Cycle

The hair remains the only mammalian tissue that undergoes life-long tissue regeneration (i.e., "the hair cycle") through rapid growth phases (Paus and Foitzik 2004). Anagen is the first step in the cycle and represents the stage of hair growth. During this stage, keratinocytes proliferate rapidly and differentiate into approximately eight other cell lines (Krause and Foitzik 2006; Schneider et al. 2009). Hair length for any strand is determined by the amount of time spent in anagen. In this growth stage, the connection between mesenchymal cells and keratinocytes in the epidermis intensifies and stimulates signals between growth factors, including *Wnt*, *Bmp*, and *Fgf* (Millar 2002) (Figure 1.1B). During the catagen stage, there is a dramatic decrease in these signaling exchanges and downregulation of factors that typically promote anagen (Figure 1.1C), and this encourages a regression of the lower two-thirds of the shaft (Botchkarev and Kishimoto 2003; Schneider et al. 2009). The hair then enters a period of rest, known as the telogen stage (Figure 1.1D), where there is an increase in the blockage of stimulatory signals via the EMI (Botchkarev and Kishimoto 2003). Specifically, telogen hairs contain inhibitors of hair follicle growth (Paus et al. 1990), such as *Bmp4* (Botchkarev and Kishimoto 2003). The old hair shaft can often remain in the infundibulum for multiple cycles during catagen and telogen stages, and this can contribute to the density of the coat by yielding multiple shafts per follicle (Müller-Röver et al. 2001; Schneider et al. 2009). Last, in the exogen stage (Figure 1.1E), the old hair shafts are shed.

In most mammals, these cycles are hypothesized to play essential roles in responses to seasonal changes in habitat or climate (Green et al. 2010), especially with regard to molting (Ling 1970). In humans, though, it is unclear why the hair follicles continue to cycle. Does this represent an outdated evolutionary relic? Does it promote the regulation of sex hormones? Is it tightly linked to the evolution of sweat glands? Or is there an alternative reason (Kamberov et al. 2013, 2015; Best et al. 2019)? We know little about the evolutionary transitions of human hair and can make few inferences from nonhuman primates due to the sparse comparative research. However, much work on domesticated or laboratory animals has pushed the envelope of our understanding of the molecular controls underlying hair growth evolution.

1.2.2 Genes and Mechanisms Potentially Driving Primate Hair Growth

Since the relationship between the epidermis and mesenchyme is ancient (Schneider et al. 2009), the same genes regulating hair growth/loss in mice or domesticated animals are also hypothesized as involved in the pathway of primate hair growth. While many molecular mechanisms underlying the hair cycle remain obscure, some studies implicate Wnt/β-catenin, bone morphogenetic proteins (*Bmp*) antagonists, and Sonic Hedgehog (*Shh*) as critical players in inducing the anagen stage (Paus and Foitzik 2004; Cotsarelis 2006; Krause and Foitzik 2006). For example, β-catenin in the dermal papilla regulates fibroblast growth factor (*Fgf*) and insulin-like growth factor (*Igf*) pathways, and its activity in the dermal papilla is vital to hair morphogenesis (Enshell-Seijffers et al. 2010). Also, ectodysplasin A receptor (*Edar*) upregulates *Shh*, which leads to follicle proliferation (Schmidt-Ullrich and Paus 2005; Pummila et al. 2007). Unsurprisingly, *EDAR* has strong associations with the degree of hair thickness in humans and mice (Spacek et al. 2010; Adhikari et al. 2016). Ectodysplasin A (*Eda*), a variant of *Edar* (along with Engrailed Homeobox 1 or *EN1*), also seems to underlie the evolution of eccrine glands in mice, in addition to regulating hair growth phenotypes (Kamberov et al. 2013, 2015). Meanwhile, fibroblast growth factor 5 (*Fgf5*) plays a vital role in the transition between anagen and catagen and is the critical gene in the 'angora' phenotype in mice (Hébert et al. 1994). *Fgf5* variation also associates with hair length in humans, felines, and canines (Drögemüller et al. 2007; Cadieu et al. 2009)—often occurring with changes in keratin-associated proteins. It is highly likely this gene may be involved in the production of similarly extravagant hair ornaments in primate hair, such as long capes (e.g., black and white colobus [*Colobus guereza*], geladas [*Theropithecus gelada*]) or mustaches (e.g., emperor tamarin [*Saguinus imperator*]).

Hundreds of molecules have been detected in distinct hair growth cycles in the epithelium and mesenchyme (Botchkarev and Kishimoto 2003). We provide a list of key candidate genes involved in primate hair growth (Table 1.1) based on their prevalence on hair growth phenotypes in other mammals. The list we provide contains a variety of keratin-associated proteins, where simple deletions and insertions may have played a role in the evolution of primate hair variation (Khan et al. 2014).

1.3 THE CELLULAR WORLD OF THE HAIR PIGMENTATION PATHWAY

The hair growth cycle is also linked to the production of melanin pigments in the visible shaft, either eumelanin (black/brown) or phaeomelanin (red/yellow). Neural crest cells (NCCs)—a transient population of cells unique to vertebrates that arise from the embryonic ectoderm during neurogenesis—gives rise to diverse cell types, including melanocytes (pigment cells), craniofacial cartilage and bone, smooth muscle, peripheral and enteric neurons, and glia (Rawles 1947). Melanoblasts, precursor cells of melanocytes, are 'fated' to migrate during embryogenesis dorsolaterally and ventrally (Erickson and Goins 1995; Slominski et al. 2005; Adameyko and Lallemend 2010)—from the dermis to the epidermis, until they finally settle as part of the developing hair follicle.

1.3.1 The Hair Melanogenesis Pathway

In early stages of melanocyte differentiation, cells express SRY-Box transcription factor 10 (*Sox10*) and KIT-Proto oncogene, receptor tyrosinase kinase (*Kit*) (Mort

TABLE 1.1
Top Candidate Genes Likely Involved in Keratin Production and Hair Growth

Gene	Name	HGNC ID	Example Cases	Example References
ALOX15B	Arachidonate 15-Lipoxygenase, Type B	434		1
DPP7	Dipeptidyl Peptidase 7	14892		1
*KRTAP6**	Keratin Associated Protein 6–1	18931		2, 3
*KRTAP7**	Keratin Associated Protein 7–1	18934		2
*KRTAP8**	Keratin Associated Protein 8–1	18935		2
LIPH	Lipase H	18483		4
FGFR2	Fibroblast Growth Factor Receptor 2	3696		5

EDAR	Ectodysplasin A Receptor	2895		6
KRT71	Keratin 71	28927		7, 8, 9
KRT27	Keratin 27	30841		10
KRT25	Keratin 25	30839		10
KRT74	Keratin 74	28929		8, 11
LPAR6	Lysophosphatidic Acid Receptor 6	15520		4
LNX1	Ligand of Numb-Protein X 1	6657		6
TCHH	Trichohyalin	11791		6

(Continued)

TABLE 1.1
(Continued)

Gene	Name	HGNC ID	Example Cases	Example References
GATA3	GATA Binding Protein 3	4172		6
PRSS53	Protease, Serine 53	34407		6
PRSS8	Protease, Serine 8	9491		12
PREP	Prolyl Endopeptidase	9358		6
FOXP2	Forkhead Box P2	13875		6
FOXL2	Forkhead Box L2	1092		6
PAX3	Paired Box 3	8617		6
FGF5	Fibroblast Growth Factor 5	3683		7, 13

RSPO2	R-Spondin 2	28583		7
KRT41P	Keratin 41 Pseudogene	6457		14
*DKK1**	Dickkopf WNT Signaling Pathway Inhibitor 1	2851		15
JAK2	Janus Kinase 2	6192		16
STAT3	Signal Transducer and Activator of Transcription 3	11363		16
SOX2	SRY-Box 2	11195		17
EDN1	Engrailed Homeobox 1	3342		18

Note: Example cases indicated with a ram (sheep), monkey (nonhuman primates), human (human), dog (domestic canines), cat (domestic and/or wild felines), and mouse (laboratory and/or natural rodent populations).

Key: * Other genes in subfamily may also be involved (i.e., KRTAP6–2, DKK2–4, etc.). References: 1. (Adelson et al. 2004), 2. (Khan et al. 2014), 3. (Li et al. 2017), 4. (Khan et al. 2011), 5. (Akihiro Fujimoto et al. 2009), 6. (Adhikari et al. 2016), 7. (Cadieu et al. 2009), 8. (Fujimoto et al. 2012), 9. (Harel and Christiano 2012), 10. (Tanaka et al. 2007), 11. (Shimomura et al. 2010), 12. (Spacek et al. 2010), 13. (Drögemüller et al. 2007), 14. (Winter et al. 2001), 15. (Sick et al. 2006), 16. (Harel et al. 2015), 17. (Driskell et al. 2009), 18. (Kamberov et al. 2015).

et al. 2015). In later stages, microphthalmia-associated transcription factor (*Mitf*) is the master regulator that drives final melanocyte differentiation (Baxter and Pavan 2003; Levy et al. 2006) that produces a neuron-shaped cell (Figure 1.2). Melanocytes contain melanosome organelles, which house pigment granules (Schneider et al. 2009). In melanosomes, the pigment granules are stored and later transported to the tip of the melanocyte protrusions, where they are subsequently transferred into neighboring keratinocytes (Mort et al. 2015).

Melanin pigment patterns are thus the result of a complex process that begins during embryogenesis. In most laboratory and domesticated animals, melanocyte distribution and survival are central to pigment patterning (Lamoreaux et al. 2010). Thus, white hair patches were historically assumed to be regions that lacked melanocytes. Nevertheless, work on natural rodent populations indicates many natural/wild populations have melanocytes in white hair regions or "patches" (i.e., areas interspersed with pigmented hairs) (Manceau et al. 2010; Haupaix and Manceau 2019). White hair then likely arises from spatially constrained processes during development that underlie non-random patterns (Chaplin and Jablonski 1998; Haupaix et al. 2018; Caro and Mallarino 2020).

Several molecules directly bind to the melanocytes and regulate the production of pigment (Caro and Mallarino 2020). This process begins with proopiomelanocortin (coding by the *Pomc* gene), which produces four melanocortin peptides, and

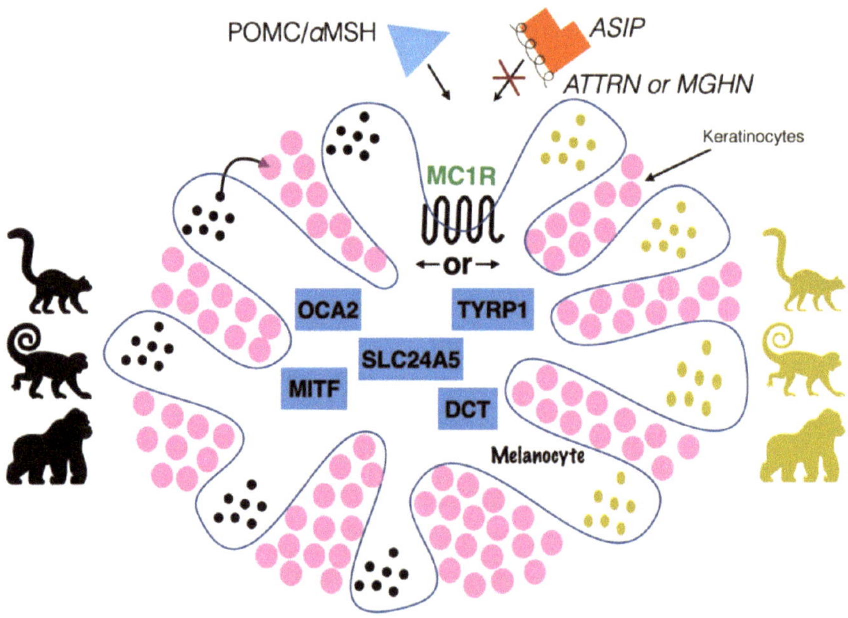

FIGURE 1.2 Melanocyte containing pigment granules, depicting the simplified biochemical pathway that is responsible for the production of eumelanin pigmentation versus phaeomelanin pigmentation. Pigments (eumelanin or pheomelanin) are deposited into keratinocytes and pushed up the growing hair follicle. Primate schematics in yellow and black represent the two pigments that color the primate coat (eumelanin (black) or phaeomelanin (yellow)).

the production of eumelanin or phaeomelanin is determined by interactions between alpha-melanocyte-stimulating hormone (*aMSH* derived from Pomc), *melanocortin-1-receptor* (*Mc1r*), and agouti-signaling protein antagonist (*Asip*) (Brash 1996; Emaresi et al. 2013). In non-human mammals, in the absence of *Asip*, *aMSH* stimulates *Mc1r* (bound on the membrane of the melanocyte) and causes an increase in intracellular cyclic AMP (*cAMP*) (Figure 1.1B) (Hunt et al. 1995; Emaresi et al. 2013). This stimulates the enzymatic activity of tyrosinase (*Tyr*), which oxidizes to dopaquinone (Caro and Mallarino 2020). Cyclic AMP also affects the enzymatic activity of other genes and transcription factors, such as: tyrosinase-related protein 1 (*Tyrp1*), solute carrier family 24 member 5 (*Slc24a5*), and oculocutaneous albinism II melanosomal transmembrane protein (*Oca2*) (Buscà and Ballotti 2000; Newton et al. 2007). This process results in the production of eumelanin within the melanosomes. However, when *Asip* is present, it acts as an antagonist to *Mc1r* and thus decreases intracellular *cAMP* and *Tyr* (Figure 1.1B). This causes a cascade of chemical reactions within the melanosome that switches over to the production of phaeomelanin (orange/yellow color) (Figure 1.2) (Suzuki et al. 1997). Recent work in mice shows that attractin (*Attrn*) attaches to *Asip* and can be a responsible precursor for antagonizing *Mc1r* (Barrett et al. 2019).

The keratinocytes containing pigment granules are then deposited into the hair shaft during anagen (Krause and Foitzik 2006). Thus, as the hair shaft grows, pigments are also continually being deposited. The protein β-catenin in the dermal papilla plays a key role in pigment-type switching responsible for agouti banded pelage (i.e., hairs with eumelanin and phaeomelanin bands [Figure 1.1])—and also contributes to hair morphogenesis (Enshell-Seijffers et al. 2010). This creates two independent processes in the hair papilla and likely creates tradeoffs between hair growth and coloration. In other words, pigmentation and growth (i.e., hair density, width, follicle density) are primarily decoupled cellular processes. For example, in eastern fox squirrels, black individuals have both thinner hairs on their dorsal surface (where solar absorption is greater) and longer tail hairs (which they use to warm themselves at night)—in comparison to gray morphs (Fratto and Davis 2011). Thus, a range of hair growth phenotypes can co-occur with the same pigmentation pattern due to the cellular processes being disassociated from each other.

Melanosomes control the production of pigment slowly through molecular and cellular pathways involving amino acid sequence and/or regulatory changes. Often, the transfer of melanin into keratinocytes occurs as the epidermis responds to the external environment (i.e., fetal environment, climate, and social environment). To illustrate, snowshoe hares typically switch into a white coat during winter months (cued by seasonal light), and this helps individuals camouflage against snow (Zimova et al. 2016). However, with recent warming climates, the frequency of hares maintaining their brown spring coats has increased in areas with less snowfall due to admixture (Jones et al. 2018). In essence, there is an ongoing transfer of phaeomelanin into keratinocytes as pelage is growing in the hares. The allele underlying this change (at the *Asip* locus) is projected to reach fixation should warming climates continue (Jones et al. 2019). Similarly, keratin genes present in modern humans may trace back to admixture events between *Homo sapiens* and *Homo neanderthalensis* that potentially led to humans being able to adapt to colder environments (Sankararaman et al. 2014). Pelage, in many ways, provides adaptations to distinct environments.

The maintenance of genomic diversity underlying pelage pigmentation may provide significant plasticity for evolution to act on during periods of climate change (Millien et al. 2006). In fact, during the climate warming that has occurred over the past 30 years, the frequency of light-colored Soay sheep (*Ovis aries*) has decreased in some populations (Maloney et al. 2009). Similar responses have likely occurred in primate hair, since populations have evolved against a backdrop of changing climates while living in forests largely modified by anthropogenic pressures (Fleagle and Gilbert 2006).

1.3.2 Genes and Mechanisms Potentially Underlying Variation in Primate Hair Pigmentation

The mechanisms behind primate hair pigmentation remain an enigma, though a combination of coding and regulatory changes may be responsible for pigmentation differences across primates. As with other phenotypes, there is debate about whether adaptive evolution more commonly targets coding or regulatory regions (King and Wilson 1975; Hoekstra and Coyne 2007; Carroll 2008; Harris et al. 2019). For example, in deer mice, both regulatory and coding changes are responsible for the resulting hair phenotype in the same population of mice (Barrett et al. 2019). However, new evidence suggests that alternative splicing diverges quicker than differential expression and thus has a potentially important role in the rapid evolutionary change of an adaptive trait (Merkin et al. 2012). Despite the lack of associations between primate coat color and *MC1R* or *ASIP* (Mundy and Kelly 2003, 2006), we cannot rule out coding changes as an important source of variation, namely because such coding changes commonly lead to hair color changes in a variety of other taxa (i.e., *Mus* (Nachman et al. 2003; Hoekstra et al. 2005; Barrett et al. 2019)).

However, it is important to note that mechanisms underlying hair evolution between humans and mice can be distinct, though complementary. For example, in humans, *MC1R* variants are strongly associated with red hair phenotypes (as well as fair skin and freckles) (Valverde et al. 1995; Raimondi et al. 2008; Zorina-Lichtenwalter et al. 2019). Yet in most non-human mammals, *Mc1r* variants are most strongly associated with black hair color (Nachman et al. 2003; Hofreiter and Schöneberg 2010; Da Silva et al. 2017). Interestingly, human *MC1R* and *Mus Mc1r* exhibit important differences that may help explain these contrasting results (Healy et al. 2001). Human melanocytes have a lower density of MC1R receptors in comparison to mouse melanocytes (Montjoy 1994). Human MC1R is also more sensitive to hormone stimulation and less sensitive to antagonism by agouti (Suzuki et al. 1997). Thus, while work on laboratory mice provides important insights into the cellular and genomic pathways underlying human hair variation, it is an imperfect model (Ito and Wakamatsu 2011). This is why studies of nonhuman primate hair would serve as an important comparison.

A variety of genes are associated with pigmentation phenotypes broadly and make good candidates for further exploration in nonhuman primates. *SLC24A5* is a protein-coding gene that is strongly associated with pigmentation phenotypes in a variety of human populations as well as in zebrafish (Lamason et al. 2005; Mallick et al. 2013; Adhikari et al. 2016). Another protein-coding gene, *Tyrp1*, plays a significant role in generating the brown coat color of felines (Lyons et al. 2005) and 'chestnut' pigmentation in horses (Li et al. 2014), as well as a range of black/brown coats in dogs (Schmutz et al. 2002). Recent studies in African striped mice (*Rhabdomys pumilio*) indicate the transcription factor ALX Homeobox 3 (*Alx3*) can directly repress *Mitf* to produce lightly colored stripes. The same study showed that *Alx3* is also responsible for generating light stripes in Eastern chipmunks (*Tamias striatus*) (Mallarino et al. 2016a)—showing the same mechanism is responsible for convergent pelage phenotypes in distantly related taxa. We hypothesize this newly discovered pigmentation gene likely also underlies some primate phenotypes (e.g., stripe patterns in *Callithrix flaviceps*). Due to similar evidence in other mammalian systems, a variety of other protein-coding genes or transcription factors may be involved in generating primate coat color diversity (Table 1.2)—either via simple mutations and/or regulatory changes on or nearby only one of a handful of genes.

While all primates are born with hair, many nonhuman primates are born with natal coats that are either un-pigmented and only develop pigment post-partum (e.g., *Propithecus*, *Colobus*) or are born with pigmented coats that later develop patterns (e.g., *Indri*). This is likely because the first anagen cycle in humans (and nonhuman primates) does not occur until approximately four weeks after birth (Müller-Röver et al. 2001). From studies in quail, we know the position of patterns can be species specific and established in early development, but the size of patterns is directed by spatially restricted gene regulation (Haupaix et al. 2018; Caro and Mallarino 2020). Considering this knowledge, it is fascinating to consider the development of nonhuman primate hair color patterns since they are among the most complex of all mammals.

In primates, we hypothesize the pigmentation cycle is broadly divided into at least two stages (Figure 1.3). First, the migration and distribution of embryonic neural crest cells migrate throughout the body and likely settle in a species-specific predetermined pattern along epithelium (Dupin and Le Douarin 2003; Schneider 2018). These patterns are determined by a broad range of transcription factors, such as *MITF* (Levy et al. 2006). Of all pigmentation genes, *MITF* is most likely to contain simple coding changes—potentially under positive selection—across distinctly patterned species (e.g., sifakas, Asian colobines) that could provide key insight into primate pattern development. Second, the extent that those patterns are present (i.e., size) is then likely regulated throughout life up until sexual maturity (Caro and Mallarino 2020). However, little to nothing is known about what genes govern developmental hair pigmentation patterns in primates. Yet this developmental process in primates is likely more complex than any other organism previously studied—and cannot even compare to development of stripes in a chipmunk (Chaplin and Jablonski 1998). Genes involved in this process are likely a huge key to understanding how pigmentation differences arise within and between primate species.

We note that many genes involved in the hair melanogenesis pathway are likely also implicated in skin and eye pigmentation, and it may be difficult to tie an association solely to hair pigmentation (Quillen et al. 2019). For example, *SLC24A5*, *ASIP*, and *OCA2* are also associated with human skin and/or eye color phenotypes (Norton et al. 2007; Beleza et al. 2013). However, in relation to human skin, nonhuman primate hair is vastly understudied, and it is hair—not skin—that is the defining

TABLE 1.2
Top Candidate Genes Possibly Involved in Melanin Production of Hair Color and Patterning

Gene	Long Name	HGNC ID	Example Cases	Example References
MC1R	Melanocortin 1 Receptor	4286		1, 2
EXOC2	Exocyst Complex Component 2	24968		3
IRF4	Interferon Regulatory Factor 4	6119		3, 4
TYRP1	Tyrosinase Related Protein 1	12450		5
ASIP	Agouti Signaling Protein	745		6, 7, 8
ATRN	Attractin	885		6, 9
ALX3	ALX Homeobox 3	449		10
LVRN	Laeverin	26904		11, 12
KITLG	KIT Ligand	6343		5, 10
KIT	KIT Proto-Oncogene Receptor Tyrosine Kinas	6342		5, 10

MITF	Melanocyte Inducing Transcription Factor	7105		10, 13
TYR	Tyrosinase	449		4, 10
TYRP1	Tyrosinase Related Protein 1	7146		8, 10
OCA2	Oculocutaneous Albinism II Melanosomal Transmembrane Protein	8101		10
GPR143	G Protein-Coupled Receptor 143	20145		10
EDNRB	Endothelin Receptor Type B	3180		5
EDN3	Endothelin 3			5
TRPM1	Transient Receptor Potential Cation Channel Subfamily M Member 1	7146		10, 14
POMC	Proopiomelanocortin	9201		15, 16
SLC45A2	Solute Carrier Family 45 Member 2	16472		4

(Continued)

TABLE 1.2
(Continued)

Gene	Long Name	HGNC ID	Example Cases	Example References
MGRN1	Mahogunin Ring Finger 1	20254		5
SERPINB2*	Serpin Family B Member 2	8584		17, 18
TCF7	Transcription Factor 7	11639		18
IRF4	Interferon Regulatory Factor 4	6119		4
SLC24A5	Solute Carrier Family 24 Member 5	20611		4
WNT10B	Wnt Family Member 10B	12775		19, 20

Note: Example cases indicated with a mouse (laboratory and/or natural rodent populations), human (human), cat (domestic and/or wild felines), horse (horses), ox (cattle), and monkey (nonhuman primates).

Key: * These genes are not generally thought of as involved in the melanogenesis pathway but have been found in association studies as potential players in melanin production. References: 1. (Hoekstra et al. 2005), 2. (Nachman et al. 2003), 3. (Branicki et al. 2008), 4. (Adhikari et al. 2016), 5. (Lamoreaux et al. 2010), 6. (Barrett et al. 2019), 7. (Gershony et al. 2014), 8. (Sponenberg 2009), 9. (Gunn et al. 2001), 10. (Mallarino et al. 2016a), 11. (Eizirik et al. 2010), 12. (Kaelin et al. 2013), 13. (Levy et al. 2006), 14. (Bellone et al. 2008), 15. (Krude et al. 1998), 16. (Wang et al. 2016), 17. (Ganesan et al. 2008), 18. (Meyer et al. 2015), 19. (Sick et al. 2006), 20. (Ye et al. 2013)

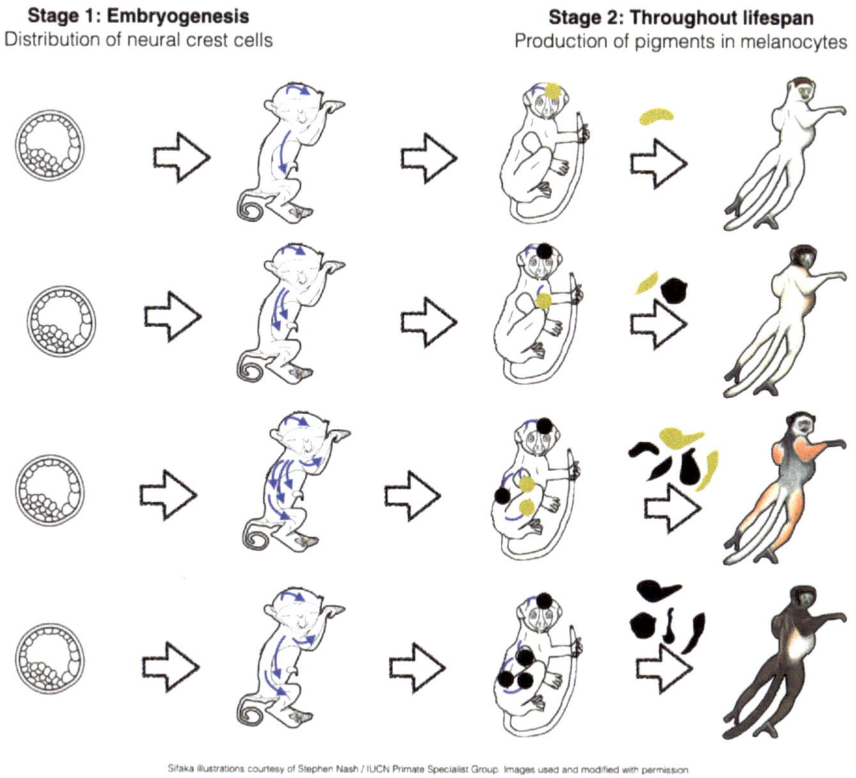

FIGURE 1.3 The development of pigmentation as a two-stage process. Blue arrows on the sifaka infant during embryogenesis represents the migration of neural crest cells, and circles represent the distribution of melanocytes occurring during the first true anagen cycle. The colors of the circles represent the type of melanin for each body region: yellow (phaeomelanin) or black (eumelanin). This process is subsequently responsible for the production of pigmented patches on distinct body regions (here, shapes above the final arrows) across species.

characteristic trait of mammals. While pleiotropic effects are likely, we argue the genomics and cellular biology of primate hair needs more attention.

1.4 THE PROMISE OF GENOMICS FOR UNTANGLING OUR HAIRY EVOLUTION

Hair is an excellent system by which to study adaptation and to identify signatures of selection in the genome. Identifying candidate loci responsible for distinct phenotypes via genome scans is often invoked to argue for genomic evidence of selection (Barrett and Hoekstra 2011). However, the power to detect selection within species is significantly limited by population demography (Harris et al. 2019). Changes to hair that are advantageous to a population can occur and 'sweep' through a population rapidly (Sturm and Duffy 2012; Szpak et al. 2019) and can potentially sort based on micro-habitat (Nachman et al. 2003; Hoekstra et al. 2005). For example, the frequency of coat color in a wild population of deer mice has been altered in just a few generations—in response to predation pressures that impact *Agouti/Asip* variant frequency (Barrett et al. 2019). Thus, rapid (or detectable) changes in coat color can occur for mammals in just a handful of generations.

However, the impact of such a 'sweep' is mostly dependent on demographic parameters such as effective population size (Nam et al. 2017). Selection also needs to be larger than the rate of migration in order for the allele to reach fixation (Lenormand 2002), and, when migration is high, adaptive loci are more likely found in close genomic proximity (Pfeifer et al. 2018). The later is why, in naturally occurring rodent populations, only a few mutations near or at one locus (typically *Mc1r* or *Asip*) are responsible for marked coat color changes (Nachman et al. 2003; Hoekstra et al. 2005; Barrett et al. 2019).

In many natural mammalian populations, some of the most persuasive examples of adaptive evolution have involved identification of hair loci under selection via population genomics and other methods (e.g., rock pocket mice [*Chaetodipus intermedius*], beach mice [*Peromyscus polionotus*], deer mice [*Peromyscus maniculatus*], and snowshoe hares [*Lepus americanus*]). For example, work on beach mice indicates that in populations that have undergone substantial bottlenecks, it might be difficult (at best) to impossible (at worst) to identify signatures of selection in the genome (Poh et al. 2014). This is particularly important if hypothesized selection on hair genes is potentially due to crypsis/camouflage (as is the case for many primates), because this often follows the alteration of a current habitat or colonization of a new habitat (Harris et al. 2019). Such environmental shifts often occur in conjunction with major population bottlenecks, changes in migration rate, and/or admixture (Lande and Shannon 1996; Lenormand 2002). Without such data, any genomic signature of selection on pigmentation or keratin genes may be inconclusive.

1.4.1 Using Genomics to Study Primate Hair

We argue that studying the interplay between selection and demography on hair traits in primate populations is well within reach through the use of whole-genome sequencing or whole-exome sequencing. Whole-genome sequencing screens both neutral and functional regions. Alternatively, exome-sequencing favorably sequences protein-coding regions of the genome; however, a large percentage of exome sequence reads (40–60%) recover neutral intronic and intergenic sequences (Samuels et al. 2013; Webster 2015; Fuentes-Pardo and Ruzzante 2017). In comparison to whole-exome sequencing, whole-genome sequencing produces 100× the amount of data, a larger number of variants, and a higher probability of detecting structural variants (Warr et al. 2015). Whole genome scans are powerful and have been used to elucidate adaptations and natural selection in snub-nosed monkeys (*Rhinopithecus* spp.) and chimpanzees (*Pan spp.*) (Zhou et al. 2014; Cagan et al. 2016; Yu et al. 2016). Yet there are numerous limitations to whole-genome sequencing that likely make whole-exome sequencing a more attractive alternative for many research questions in the field of primatology.

Whole-genome sequencing is 2–3× the cost of whole-exome sequencing per sample. For example, for a short-read sequence on Illumina platforms, a 30× genome can cost approximately $1,500. In contrast, similar coverage for an exome sequence would cost approximately $300 (Schwarze et al. 2018). For phenotype-genotype questions, there is a clear advantage to sequencing a greater number of individuals for greater statistical power over fewer individuals with more genomic information (Warr et al. 2015). Importantly, whole-genome sequencing produces greater numbers of variants, partly because non-functional regions are not well conserved. These regions are less well understood than the coding regions, making it difficult to predict when a non-exonic variant would be significantly associated with a given trait, especially given a likely lower average minor allele frequency for amino-acid substitutions compared to non-exonic variants (Mu et al. 2011; Ward and Kellis 2012).

Exome sequencing of nonhuman primates using human baits (i.e., sequences used to perform genomic capture) has proven effective for apes, African monkeys, and strepsirrhines (Vallender 2011; Webster et al. 2018). Specifically, exome-sequencing can recover a high percentage of neutral (as well as coding) regions. Thus, studies using exome sequencing in nonhuman primates can get a robust picture of population structure and other demographic parameters (Tennessen et al. 2011). Moreover, since exome sequencing provides great coverage of protein coding regions, this same data can facilitate tests for selection on coding regions (Tennessen et al. 2011). Similar sequencing methods for targeting protein-coding regions have been used in other mammals to identify potential genes under selection (Roffler et al. 2016; Jones et al. 2018; Phifer-Rixey et al. 2018).

1.4.2 Beyond Genomics

Still, studying hair biology requires an integrated approach, similar to the study of morphological evolution from an evolutionary developmental (*evo-devo*) perspective, including: (1) quantitation of variation in hair pigmentation and/or growth morphologies, (2) identification of potential genes and pathways driving variation (while accounting for population demography), and last, and most critical of all, (3) performance of functional tests (Mallarino and Abzhanov 2012) confirming the causal relationship between the genetic and phenotypic variation.

A variety of tools and methods exist to quantify pelage pigmentation phenotypes in primates, including spectrophotometry (Zuk and Decruyenaere 1993) and digital photography (Gerald et al. 2001; Stevens et al. 2007, 2009), which includes, for example, the gray-scale method (Allen et al. 2014) as well as various more subjective methodologies (Santana et al. 2012, 2013; Rakotonirina et al. 2017). To quantify hair growth (density, width, follicle clustering, and length), light microscopy has proven robust for primate pelage—while also producing consistent results between living and museum specimens and showing no effect of storage time (Bradley et al. 2014).

Identification of key loci and pathways can be made by combining demographic models with a candidate gene approach (Nachman et al. 2003; Hoekstra et al. 2005), genome scans (Adhikari et al. 2016), or even transcriptome profiling (i.e., RNAseq). The last would hold immense promise if sequencing skin biopsies with hair or hair tufts containing follicles pulled from live animals. Specifically, previous work shows that sufficient quantities of RNA can be extracted from hair tufts with only ~5–10 hairs (Bradley et al. 2005). Plucking from animals with an unknown age and/or opportunistically in the wild means that hair follicles can be at distinct stages of the growth cycle (Figure 1.1), and this may weaken indicators of differential gene expression. Thus, since sampled hair tufts (usually consisting of ~50–100 hairs for primates) subjected to gene expression analyses may contain a mixture of hairs at different stages, standardizing for RNA quantity is especially important when comparing gene expression in this way.

For genomic identification of hair-related polymorphisms to be of value, they must be followed up by functional tests. Such studies could entail the manipulation of animal genomes (e.g., CRISPR-Cas9 method), including gene disruption and gene introduction—and these have already begun in nonhuman primates (Kang et al. 2019). Specifically, the *PPRAG* and *RAG1* genes of long-tailed macaques (*Macaca fasicularis*) and marmosets (*Callithrix jacchus*) were recently altered (Sasaki et al. 2009; Niu et al. 2014). The causal molecular mechanisms of hair color in mice are often studied in this way (Crawford et al. 2017; Barrett et al. 2019). However, despite primates making ideal comparators to the study of pelage biology, creating transgenic primates to study hair variation is highly unlikely (Rogers 2018) due to slow life histories, greater costs, and ethical considerations (Warren et al. 2018).

1.4.3 Incorporating Comparative Vertebrate Taxa

It would be more feasible to assess primate hair data comparatively against functional tests on model rodent species (i.e., *Mus musculus* and *Rattus norvegicus*). To accomplish this, forward genetics (e.g., whole-exome sequencing or RNAseq on a natural *Propithecus* population) could be used to establish hypotheses about causal molecular mechanisms underlying hair traits (Mallarino et al. 2016b). Then these hypotheses could be tested using functional studies on transgenic mice. This approach has been successfully adopted to study limb loss in snakes (Infante et al. 2015), bat limb formation (Cretekos et al. 2008), and skeletal impacts to hominin hybridization (Warren et al. 2018) and to understand human hair pigmentation and growth (Kamberov et al. 2013; Guenther et al. 2016). However, this work should not be considered in isolation from comparative studies to other natural rodent populations. Organisms such as deer mice, African striped mice, spiny mice (*Acomys*), and chipmunks are offering a window into the diversity of pelage and skin across the rodent clade that may help comparatively characterize primate diversity. Even the threespine stickleback (*Gasterosteus aculeatus*) would make an interesting comparative vertebrate model for primate hair evolution and development, especially for those interested in parallel evolution. Many of the same genes that likely control traits such as primate hair length (*EDA*) and hair pigmentation patterning (*KITLG*) also contribute to generating diversity in threespine sticklebacks (Colosimo et al. 2005; Miller et al. 2007).

Using other vertebrate taxa as comparative systems would help clarify the cause and effect underlying primate pelage patterns. In these systems, it is also easier and more cost effective to perform genome manipulations. Other avenues that would add substance to a study on hair biology may include histological examination of skin punches or even the study of hair stem cells (Blanpain et al. 2007; Grogan and Perry 2020) across primate and rodent models.

1.5 CONCLUSIONS

There is still so much we do not know about the cellular and genomic biology of hair, and we know even less about the mechanisms and pathways underlying normal hair diversity in primates. Most of what we know about the cellular biology of the hair originates from the dermatological literature (Seibert and Steggerda 1999; Shimomura et al. 2009; Fujimoto et al. 2012). From an evolutionary perspective, most studies have focused on understanding hair pigmentation loci, pathways, and causes (Mundy and Kelly 2003; Mundy et al. 2003; Hoekstra et al. 2005; Mundy and Kelly 2006; Bradley et al. 2013; Mallarino et al. 2016a; Barrett et al. 2019). Studies aiming to uncover the molecular mechanisms behind hair growth have been largely limited to studies on domestic or livestock animals (Drögemüller et al. 2007; Cadieu et al. 2009; Li et al. 2017). However, it has been shown that β-catenin in the dermal papilla is responsible for generating two independent pathways (one leading to hair growth and the other to hair pigmentation during anagen) (Enshell-Seijffers et al. 2010). This leaves us with new questions: Are distinct or similar evolutionary pressures acting on hair growth and pigmentation? How do those evolutionary factors translate across micro- and macro-evolutionary scales among humans, nonhuman primates, and the larger mammalian clade? Are the proximate mechanisms underlying hair phenotypes between sifakas and rodents (or humans) the same? What can a comparative perspective on hair evolution tell us about why and how humans evolved to appear so 'hairless'?

Under the scope of this topic, nonhuman primates provide a useful comparative framework situated between humans and model rodent systems. Primates are the most variably pigmented mammalian order and exhibit convergence to other mammals in their pelage patterns. For example, lemur hair pigmentation can often vary to similar extents inter- and intra-specifically (within a population and across a whole genus). Yet they converge with many other animals in exhibiting variegated black and white coat patterns (e.g., *Varecia varigata*), stripes (e.g., *Lemur catta*), and melanic individuals that occur naturally in populations (e.g., *Propithecus diadema* in Tsinjoarivo, Madagascar). A variety of primate taxa may help address distinct questions about the evolution of hair biology. For example, sifakas make an excellent contemporary proxy for asking questions about which ecological and environmental pressures potentially drove the evolution of hair variation in humans—since they live dispersed throughout all of Madagascar's eco-regions (Irwin 2006; Mittermeier et al. 2010). They vary in pelage, body size, activity cycle, vision, and even behavioral traits; however, they are all bipedal hoppers, meaning that similar to humans, their heads always directly face the sun. Alternatively, insights into *Cercopithecus* spp. pelage biology can offer direct insights into how and why hair is altered during recurrent gene flow, since a number of populations have experienced ancient and/or recent hybridization leading to novel hair phenotypes (Detwiler 2019; van der Valk et al. 2020). In fact, only small regions of the genome that pertain to coloration often distinguish admixed populations of birds and rabbits (Jones et al. 2018) and alone can confer new ecological adaptations.

Quantifying and studying the molecular mechanisms in natural populations (i.e., captive, wild) and linking them back to genotypes, fitness, and environmental conditions will be essential—as will functional testing on vertebrates. We have proposed in this chapter a substantive list of candidate loci (Tables 1.1, 1.2), which has been massively reduced (there are ~500 loci associated with hair biology in the Gene Ontology Database). Most of the proposed candidates have either been linked to phenotypes in natural populations and/or in multiple taxa. We hope these lists can be a robust starting place for candidate gene studies or used in genome scans in combination with demographic modeling and measures of genetic diversity.

On that note, primate population genomics provides a robust tool to test evolutionary hypotheses (Perry 2014)—especially with regard to the evolution of hair. Scientists working with primates already have an advantage of the availability of multiple long-term field sites (10+ years of data collection) where population genetic data already exists or can be easily sequenced. The data primatologists *already* collect at these long-term field sites (usually including foods eaten, seasonal patterns in climate or grouping, reproductive success, pedigree information, parasitic load, and predation rates) would give many studies originating with primates the advantage of more easily linking phenotype to genotype to fitness. Many studies in the non-primate mammalian literature struggle with this and usually are limited to linking genotype to fitness or phenotype to genotype (Harris et al. 2019). The work on natural populations of mice illustrates that well-designed research can effectively identify links among phenotypes-genotypes-fitness-function (see more at Barrett et al. 2019). Hair is an ideal system in which to ask (and answer) major questions about evolutionary biology and adaptation. Work towards understanding the evolution of hair in nonhuman primates, evolving against a backdrop of changing climates, gene flow, and selective pressures, will undoubtedly prove rewarding for evolutionary biology and anthropology well beyond the next century.

1.6 ACKNOWLEDGMENTS

The authors thank two anonymous reviewers who provided helpful feedback to an earlier version of this draft. We also thank Stephen Nash for providing the sifaka illustrations—modified and used here with permission. For helpful conversations about hair evolution and genomics, we thank members of The George Washington University's Primate Genomics Lab (PGL), the Comparative Primatology Lab at UMass Amherst, and University of Utah's Primate Evolution and Genomics Lab (PEGL). JMK and BJB's work on primate hair evolution has been supported by the National Science Foundation (BCS #1546730, BCS #1606360), the Leakey Foundation, and the Wenner-Gren Foundation. ET's work on primate hair evolution has been supported by the Leakey Foundation, the Explorer's Club, and the International Primatological Society.

1.7 REFERENCES

Adameyko, I., and F. Lallemend. 2010. Glial versus melanocyte cell fate choice: Schwann cell precursors as a cellular origin of melanocytes. *Cellular and Molecular Life Sciences* 67:3037–55.

Adelson, D. L., G. R. Cam, U. DeSilva, and I. R. Franklin. 2004. Gene expression in sheep skin and wool (hair). *Genomics* 83:95–105.

Adhikari, K., T. Fontanil, S. Cal, J. Mendoza-Revilla, et al. 2016. A genome-wide association scan in admixed Latin Americans identifies loci influencing facial and scalp hair features. *Nature Communications* 7:10815.

Allen, W. L., and J. P. Higham. 2013. Analyzing visual signals as visual scenes. *American Journal of Primatology* 75:664–82.

Allen, W. L., M. Stevens, and J. P. Higham. 2014. Character displacement of cercopithecini primate visual signals. *Nature Communications* 5:4266.

Barrett, R. D. H., and H. E. Hoekstra. 2011. Molecular spandrels: tests of adaptation at the genetic level. *Nature Reviews Genetics* 12:767–80.

Barrett, R. D. H., S. Laurent, R. Mallarino, S. P. Pfeifer, et al. 2019. Linking a mutation to survival in wild mice. *Science* 363:499–504.

Baxter, L. L., and W. J. Pavan. 2003. Pmel17 expression is Mitf-dependent and reveals cranial melanoblast migration during murine development. *Gene Expression Patterns* 3:703–7.

Beleza, S., N. A. Johnson, S. I. Candille, D. M. Absher, et al. 2013. Genetic architecture of skin and eye color in an African-European admixed population. *PLoS Genetics* 9:1003372.

Bellone, R. R., S. A. Brooks, L. Sandmeyer, B. A. Murphy, et al. 2008. Differential gene expression of Trpm1, the potential cause of congenital stationary night blindness and coat spotting patterns (LP) in the appaloosa horse (*Equus caballus*). *Genetics* 179:1861–70.

Best, A., D. Lieberman, and J. Kamilar. 2019. Diversity and evolution of human eccrine sweat gland density. *Journal of Thermal Biology* 84:331–8.

Bicca-Marques, J. C., and C. Calegaro-Marques. 1998. Behavioral thermoregulation in a sexually and developmentally dichromatic neotropical primate, the black-and-gold howling monkey (*Alouatta caraya*). *American Journal of Physical Anthropology* 106:533–46.

Blanpain, C., V. Horsley, and E. Fuchs. 2007. Epithelial stem cells: turning over new leaves. *Cell* 128:445–58.

Botchkarev, V. A., and J. Kishimoto. 2003. Molecular control of epithelial-mesenchymal interactions during hair follicle cycling. *Journal of Investigative Dermatology Symposium Proceedings* 8:46–55.

Bradley, B. J., M. S. Gerald, A. Widdig, and N. I. Mundy. 2013. Coat color variation and pigmentation gene expression in rhesus macaques (*Macaca mulatta*). *Journal of Mammalian Evolution* 20:263–70.

Bradley, B. J., and N. I. Mundy. 2008. The primate palette: the evolution of primate coloration. *Evolutionary Anthropology* 17:97–111.

Bradley, B. J., J. Pastorini, and N. I. Mundy. 2005. Successful retrieval of mRNA from hair follicles stored at room temperature: implications for studying gene expression in wild mammals. *Molecular Ecology Notes* 5:961–4.

Bradley, B. J., S. Walsh, A. Nishimura, J. Karlsson, et al. 2014. The evolution of primate pelage: morphological analysis of museum research skins. *American Journal of Physical Anthropology* 153:84.

Branicki, W., U. Brudnik, J. Draus-Barini, T. Kupiec, et al. 2008. Association of the *SLC45A2* gene with physiological human hair colour variation. *Journal of Human Genetics* 5:966–71.

Brash, G. S. 1996. The genetics of pigmentation: from fancy genes to complex traits. *Trends in Genetics* 12:299–305.

Buscà, R., and R. Ballotti. 2000. Cyclic AMP a key messenger in the regulation of skin pigmentation. *Pigment Cell Research* 13:60–9.

Cadieu, E., M. W. Neff, P. Quignon, K. Walsh, et al. 2009. Coat variation in the domestic dog is goverened by variants in three genes. *Science* 326:150–3.

Cagan, A., C. Theunert, H. Laayouni, G. Santpere, et al. 2016. Natural selection in the great apes. *Molecular Biology and Evolution* 33:3268–83.

Caro, T., and R. Mallarino. 2020. Coloration in mammals. *Trends in Ecology and Evolution* 35:357–66.

Carroll, S. B. 2008. Evo-devo and an expanding evolutionary synthesis: a genetic theory of morphological evolution. *Cell* 134:25–36.

Chaplin, G., and N. G. Jablonski. 1998. The integument of the odd-nosed colobines. In *The Natural History of the Doucs and Snub-Nosed Monkeys*, ed. N. G. Jablonski, 79–104. Singapore: World Scientific Publishing Co.

Colosimo, P. F., K. E. Hosemann, S. Balabhadra, G. Villarreal Jr, et al. 2005. Widespread parallel evolution in sticklebacks by repeated fixation of the Ectodysplasin alleles. *Science* 307:1928–33.

Cooper, V. J., and G. R. Hosey. 2003. Sexual dichromatism and female preference in *Eulemur fulvus* subspecies. *International Journal of Primatology* 24:1177–88.

Cotsarelis, G. 2006. Epithelial stem cells: a folliculocentric view. *Journal of Investigative Dermatology* 126:1459–68.

Crawford, N. G., D. E. Kelly, M. E. B. Hansen, M. H. Beltrame, et al. 2017. Loci associated with skin pigmentation identified in African populations. *Science* 358:eaan8433.

Cretekos, C. J., Y. Wang, E. D. Green, J. F. Martin, et al. 2008. Regulatory divergence modifies limb length between mammals. *Genes and Development* 22:141–51.

Cuthill, I. C., W. L. Allen, K. Arbuckle, B. Caspers, et al. 2017. The biology of color. *Science* 357:aan0221.

Delhey, K. 2017. Gloger's rule. *Current Biology* 27:R681–701.

Delhey, K., J. Dale, M. Valcu, and B. Kempenaers. 2019. Reconciling ecogeographical rules: rainfall and temperature predict global colour variation in the largest bird radiation. *Ecology Letters* 22:726–36.

Detwiler, K. M. 2019. Mitochondrial DNA analyses of *Cercopithecus* monkeys reveal a localized hybrid origin for *C. mitis doggetti* in Gombe National Park, Tanzania. *International Journal of Primatology* 40:28–52.

Driskell, R. R., A. Giangreco, K. B. Jensen, K. W. Mulder, et al. 2009. Sox2-positive dermal papilla cells specify hair follicle type in mammalian epidermis. *Development* 136:2815–23.

Drögemüller, C., S. Rüfenacht, B. Wichert, and T. Leeb. 2007. Mutations within the Fgf5 gene are associated with hair length in cats. *Animal Genetics* 38:218–21.

Dupin, E., and L. M. Le Douarin. 2003. Development of melanocyte precursors from the vertebrate neural crest. *Oncogene* 22:3016–23.

Eisenberg, J. F., and D. G. Kleiman. 1972. Olfactory communication in mammals. *Annual Review of Ecology and Systematics* 3:1–32.

Eizirik, E., V. A. David, V. Buckley-Beason, M. E. Roelke, et al. 2010. Defining and mapping mammalian coat pattern genes: multiple genomic regions implicated in domestic cat stripes and spots. *Genetics* 184:267–75.

Emaresi, G., A. L. Ducrest, P. Bize, H. Richter, et al. 2013. Pleiotropy in the melanocortin system: expression levels of this system are associated with melanogenesis and pigmentation in the tawny owl (*Strix Aluco*). *Molecular Ecology* 22:4915–30.

Enshell-Seijffers, D., C. Lindon, E. Wu, M. M. Taketo, et al. 2010. β-catenin activity in the dermal papilla of the hair follicle regulates pigment-type switching. *Proceedings of the National Academy of Sciences of the United States of America* 107:21564–9.

Erickson, C. A., and T. L. Goins.1995. Avian neural crest cells can migrate in the dorsolateral path only if they are specified as melanocytes. *Development* 121:915–24.

Fleagle, J. G., and C. C. Gilbert. 2006. The biogeography of primate evolution: the role of plate tectonics, climate and chance. In *Primate Biogeography*, ed. S. M. Lehman and J. G. Fleagle, 375–418. Boston, MA: Springer.

Fratto, M. A., and A. K. Davis. 2011. Do black-furred animals compensate for high solar absorption with smaller hairs? A test with a polymorphic squirrel species. *Current Zoology* 57:731–6.

Fuentes-Pardo, A. P., and D. E. Ruzzante. 2017. Whole-genome sequencing approaches for conservation biology: advantages, limitations and practical recommendations. *Molecular Ecology* 26:5369–406.

Fujimoto, A., M. Farooq, H. Fujikawa, A. Inoue, et al. 2012. A missense mutation within the helix initiation motif of the keratin K71 gene underlies autosomal dominant woolly hair/hypotrichosis. *Journal of Investigative Dermatology* 132:2342–9.

Fujimoto, A., N. Nishida, R. Kimura, T. Miyagawa, et al. 2009. *FGFR2* is associated with hair thickness in Asian populations. *Journal of Human Genetics* 54:461–5.

Ganesan, A. K., H. Ho, B. Bodemann, S. Petersen, et al. 2008. Genome-wide siRNA-based functional genomics of pigmentation identifies novel genes and pathways that impact melanogenesis in human cells. *PLoS Genetics* 4:e1000298.

George, R. D., G. McVicker, R. Diederich, S. B. Ng, et al. 2011. Trans genomic capture and sequencing of primate exomes reveals new targets of positive selection. *Genome Research* 21:1686–94.

Gerald, M. S., J. Bernstein, R. Hinkson, and R. A. E. Fosbury. 2001. Formal method for objective assessment of primate color. *American Journal of Primatology* 53:79–85.

Gershony, L. C., M. C. T. Penedo, B. W. Davis, W. J. Murphy, et al. 2014. Who's behind that mask and cape? The Asian leopard cat's agouti (Asip) allele likely affects coat colour phenotype in the bengal cat breed. *Animal Genetics* 45:893–7.

Gloger, C. L. 1833. *Das Abandern Der Vögel Durch Einfluss Des Kilma's.* Wroclaw: Germany: August Schulz & Co.

Graur, D. 2016. *Molecular and Genome Evolution*. Sunderland, MA: Sinauer Associates, Inc.

Green, R. E., J. Krause, A. W. Briggs, T. Maricic, et al. 2010. A draft sequence of the Neandertal genome. *Science* 328:710–22.

Grogan, K. E., and G. H. Perry. 2020. Studying human and nonhuman primate evolutionary biology with powerful in vitro and in vivo functional genomics tools. *Evolutionary Anthropology: Issues, News, and Reviews* 29:143–58.

Guenther, C. A., B. Tasic, L. Luo, M. A. Bedell, et al. 2016. A molecular basis for classic blond hair color in Europeans despite thousands of years of interest in hair. *Nature Genetics* 46:748–52.

Gunn, T. M., T. Inui, K. Kitada, S. Ito, et al. 2001. Molecular and phenotypic analysis of attractin mutant mice. *Genetics* 158:1683–95.

Hancock, A. M., and A. Di Rienzo. 2008. Detecting the genetic signature of natural selection in human populations: models, methods, and data. *Annual Review of Anthropology* 37:197–217.

Hardy, M. H. 1992. The secret life of the hair follicle. *Trends in Genetics* 8:55–61.

Harel, S., and A. M. Christiano. 2012. Keratin 71 mutations: from water dogs to woolly hair. *Journal of Investigative Dermatology* 132:2315–17.

Harel, S., C. A. Higgins, J. E. Cerise, Z. Dai, et al. 2015. Pharmacologic inhibition of JAK-STAT signaling promotes hair growth. *Science Advances* 1:1–13.

Harris, R. B., K. Irwin, M. R. Jones, S. Laurent, et al. 2019. The population genetics of crypsis in vertebrates: recent insights from mice, hares, and lizards. *Heredity* 124:1–14.

Haupaix, N., C. Curantz, R. Bailleul, S. Beck, et al. 2018. The periodic coloration in birds forms through a prepattern of somite origin. *Science* 361:eaar4777.

Haupaix, N., and M. Manceau. 2019. The embryonic origin of periodic color patterns. *Developmental Biology* 460:70–6.

Healy, E., S. A. Jordan, P. S. Budd, R. Suffolk, et al. 2001. Functional variation of *MC1R* alleles from red-haired individuals. *Human Molecular Genetics* 10:2397–402.

Hébert, J. M., T. Rosenquist, J. Götz, and G. R. Martin. 1994. Fgf5 as a regulator of the hair growth cycle: evidence from targeted and spontaneous mutations. *Cell* 78:1017–25.

Hershkovitz, P. 1968. Metachromism or the principle of evolutionary change in mammalian tegumentary colors. *Evolution* 22:556–75.

Hoekstra, H. E., and J. A. Coyne. 2007. The locus of evolution: evo devo and the genetics of adaptation. *Evolution* 61:995–1016.

Hoekstra, H. E., J. G. Krenz, and M. W. Nachman. 2005. Local adaptation in the rock pocket mouse (*Chaetodipus intermedius*): natural selection and phylogenetic history of populations. *Heredity* 94:217–28.

Hofreiter, M., and T. Schöneberg. 2010. The genetic and evolutionary basis of colour variation in vertebrates. *Cellular and Molecular Life Sciences* 67:2591–603.

Hunt, G., S. K. Kyne, K. Wakamatsu, S. Ito, et al. 1995. Nle^4DPhe^7 α-melanocyte-stimulating hormone increases the eumelanin: phaeomelanin ratio in cultured human melanocytes. *Journal of Investigative Dermatology* 104:83–5.

Infante, C. R., A. G. Mihala, S. Park, J. S. Wang, et al. 2015. Shared enhancer activity in the limbs and phallus and functional divergence of a limb-genital cis-regulatory element in snakes. *Developmental Cell* 35:107–19.

Irwin, M. T. 2006. Ecologically enigmatic lemurs: the sifakas of the eastern forests (*Propithecus candidus, P. diadema, P. edwardsi, P. perrieri*, and *P. tattersalli*). In *Lemurs: Ecology and Adaptation*, ed. L. Gould and M. L. Sauther, 305–26. New York: Springer.

Ito, S., and K. Wakamatsu. 2011. Human hair melanins: what we have learned and have not learned from mouse coat color pigmentation. *Pigment Cell and Melanoma Research* 24:63–74.

Jacobs, R. L., C. C. Veilleux, E. E. Louis Jr., J. P. Herrera, et al. 2019. Less is more: lemurs (*Eulemur* spp.) may benefit from loss of trichromatic vision. *Behavioral Ecology and Sociobiology* 73:22.

Jacobs, S. C., A. Larson, and J. M. Cheverud. 1995. Phylogenetic relationships and orthogenetic evolution of coat color among tamarins (genus *Saguinus*). *Systematic Biology* 44:515–32.

Jones, M. R., L. S. Mills, P. C. Alves, C. M. Callahan, et al. 2018. Adaptive introgression underlies polymorphic seasonal camouflage in snowshoe hares. *Science* 360:1355–8.

Jones, M. R., L. S. Mills, J. D. Jensen, and J. M. Good. 2019. Convergence and gene flow shape the evolution of seasonal camouflage in snowshoe hares. *Evolution* 74:2033–45.

Kaelin, C. B., X. Xu, L. Z. Hong, V. A. David, et al. 2013. Specifying and sustaining pigmentation patterns in domestic and wild cats. *Science* 337:1536–41.

Kamberov, Y. G., E. K. Karlsson, G. L. Kamberova, D. E. Lieberman, et al. 2015. A genetic basis of variation in eccrine sweat gland and hair follicle density. *Proceedings of the National Academy of Sciences of the United States of America* 112:9932–7.

Kamberov, Y. G., S. Wang, J. Tan, P. Gerbault, et al. 2013. Modeling recent human evolution in mice by expression of a selected EDAR variant. *Cell* 152:691–702.

Kamilar, J. M., and B. J. Bradley. 2011. Interspecific variation in primate coat colour supports Gloger's Rule. *Journal of Biogeography* 38:2270–7.

Kang, Y., C. Chu, F. Wang, and Y. Niu. 2019. CRISPR/Cas9-mediated genome editing in nonhuman primates. *Disease Models and Mechanisms* 12:dmm039982.

Khan, I., E. Maldonado, V. Vasconcelos, S. J. O'Brien, et al. 2014. Mammalian keratin associated proteins (KRTAPs) subgenomes: disentangling hair diversity and adaptation to terrestrial and aquatic environments. *BMC Genomics* 15:779.

Khan, S., R. Habib, H. Mir, Umm-E-Kalsoom, et al. 2011. Mutations in the *LPAR6* and *LIPH* genes underlie autosomal recessive hypotrichosis/woolly hair in 17 consanguineous families from Pakistan. *Clinical and Experimental Dermatology* 36:652–4.

King, M.-C., and A. C. Wilson.1975. Evolution at two levels in humans and chimpanzees. *Science* 188:107–16.

Krause, K., and K. Foitzik. 2006. Biology of the hair follicle: the basics. *Seminars in Cutaneous Medicine and Surgery* 25:2–10.

Krude, H., H. Biebermann, W. Luck, R. Horn, et al. 1998. Severe early-onset obesity, adrenal insufficiency and red hair pigmentation caused by *POMC* mutations in humans. *Nature Genetics* 19:155–7.

Lamason, R. L., M. P. K. Mohideen, J. R. Mest, A. C. Wong, et al. 2005. *SLC24A5*, a putative cation exchanger, affects pigmentation in zebrafish and humans. *Science* 310:1782–7.

Lamoreaux, M. L., V. Delmas, L. Larue, and D. C. Bennett, eds. 2010. *The Colors of Mice: A Model Genetic Network*, 1st ed. West Sussex, UK: Wiley-Blackwell.

Lande, R., and S. Shannon. 1996. The role of genetic variation in adaptation and population persistence in a changing environment. *Evolution* 50:434–7.

Lasisi, T., S. Ito, K. Wakamatsu, and C. N. Shaw. 2016. Quantifying variation in human scalp hair fiber shape and pigmentation. *American Journal of Physical Anthropology* 160:341–52.

Lenormand, T. 2002. Gene flow and the limits to natural selection. *Trends in Ecology and Evolution* 17:183–9.

Levy, C., M. Khaled, and D. E. Fisher. 2006. *MITF*: master regulator of melanocyte development and melanoma oncogene. *Trends in Molecular Medicine* 12:406–14.

Li, B., X. He, Y. Zhao, Q. Zhao, et al. 2014. Tyrosinase-related protein 1 (Tyrp1) gene polymorphism and skin differential expression related to coat color in mongolian horse. *Livestock Science*:58–64.

Li, S., H. Zhou, H. Gong, F. Zhao, et al. 2017. Variation in the ovine KAP6–3 gene (KRTAP6–3) is associated with variation in mean fibre diameter-associated wool traits. *Genes* 8:204.

Ling, J. K. 1970. Pelage and molting in wild mammals with special reference to aquatic forms. *The Quarterly Review of Biology* 45:16–54

Lyons, L. A., I. T. Foe, H. C. Rah, and R. A. Grahn. 2005. Chocolate coated cats: TYRP1 mutations for brown color in domestic cats. *Mammalian Genome* 16:356–66.

Maderson, P. F. A. 2003. Mammalian skin evolution: a reevaluation. *Experimental Dermatology* 12:233–6.

Mallarino, R., and A. Abzhanov. 2012. Paths less traveled: evo-devo approaches to investigating animal morphological evolution. *Annual Review of Cell and Developmental Biology* 28:743–63.

Mallarino, R., C. Henegar, M. Mirasierra, M. Manceau, et al. 2016a. Developmental mechanisms of stripe patterns in rodents. *Nature* 539:518–23.

Mallarino, R., H. E. Hoekstra, and M. Manceau. 2016b. Developmental genetics in emerging rodent models: case studies and perspectives. *Current Opinion in Genetics and Development* 39:182–6.

Mallick, B. C., F. M. Iliescu, M. Möls, S. Hill, et al. 2013. The light skin allele of *SLC24A5* in South Asians and Europeans shares identity by descent. *PLoS Genetics* 9:1003912.

Maloney, S. K., A. Fuller, and D. Mitchell. 2009. Climate change: is the dark soay sheep endangered? *Biology Letters* 5:826–9.

Manceau, M., V. S. Domingues, C. R. Linnen, E. B. Rosenblum, et al. 2010. Convergence in pigmentation at multiple levels: mutations, genes and function. *Philosophical Transactions of the Royal Society B: Biological Sciences* 365:2439–50.

Merkin, J., C. Russell, P. Chen, and C. B. Burge. 2012. Evolutionary dynamics of gene and isoform regulation in mammalian tissues. *Science* 338:1593–9.

Meyer, W. K., A. Venkat, A. R. Kermany, B. V. D. Geijn, et al. 2015. Evolutionary history inferred from the de novo assembly of a nonmodel organism, the blue-eyed black lemur. *Molecular Ecology* 24:4392–405.

Millar, S. E. 2002. Molecular mechanisms regulating hair follicle development. *Journal of Investigative Dermatology* 118:216–25.

Miller, C. T., S. Beleza, A. A. Pollen, D. Schluter, et al. 2007. Cis-regulatory changes in kit ligand expression and parallel evolution of pigmentation in sticklebacks and humans. *Cell* 131:1179–89.

Millien, L., L. Olson, F. A. Smith, A. B. Wilson, et al. 2006. Ecotypic variation in the context of global climate change: revisiting the rules. *Ecology Letters* 9:853–69.

Mittermeier, R. A., E. E. Louis, Jr., M. Richardson, C. Schwitzer, et al. 2010. *Lemurs of Madagascar*, 3rd ed. New York: Conservation International.

Montjoy, K. G. 1994. The human melanocyte stimulating hormone receptor has evolved to become 'super-sensitive' to melanocortin peptides. *Molecular and Cellular Endocrinology* 102:R7–11.

Mort, R. L., I. J. Jackson, E. E. Patton, R. L. Mort, et al. 2015. The melanocyte lineage in development and disease. *Development* 142:620–32.

Mu, X. J., Z. J. Lu, Y. Kong, H. Y. K. Lam, et al. 2011. Analysis of genomic variation in non-coding elements using population-scale sequencing data from the 1000 genomes project. *Nucleic Acids Research* 39:7058–76.

Müller-Röver, S., B. Handjiski, C. V. D. Veen, S. Eichmüller, et al. 2001. A comprehensive guide for the accurate classification of murine hair follicles in distinct hair cycle stages. *Journal of Investigative Dermatology* 117:3–15.

Mundy, N. I., and J. Kelly. 2003. Evolution of a pigmentation gene, the melanocortin-1 receptor, in primates. *American Journal of Physical Anthropology* 121:67–80.

Mundy, N. I., and J. Kelly. 2006. Investigation of the role of the agouti signaling protein gene (*ASIP*) in coat color evolution in primates. *Mammalian Genome*:1205–13.

Mundy, N. I., J. Kelly, E. Theron, and K. Hawkins. 2003. Evolutionary genetics of the melanocortin-1 receptor in vertebrates. *Annals of the New York Academy of Sciences* 994:307–12.

Nachman, M. W., H. E. Hoekstra, and S. L. D'Agostino. 2003. The genetic basis of adaptive melanism in pocket mice. *Proceedings of the National Academy of Sciences of the United States of America* 100:5268–73.

Nam, K., K. Munch, T. Mailund, A. Nater, et al. 2017. Evidence that the rate of strong selective sweeps increases with population size in the great apes. *Proceedings of the National Academy of Sciences of the United States of America* 114:1613–18.

Nekaris, K. A. I., and R. A. Munds. 2010. Using facial markings to unmask diversity: the slow lorises (Primate: Lorisidae: *Nycticebus* spp.) of Indonesia. In *Indonesian Primates*, ed. S. Gursky and J. Supriatna, 383–96. New York: Springer.

Newton, R. A., A. L. Cook, D. W. Roberts, J. H. Leonard, et al. 2007. Post-transcriptional regulation of melanin biosynthetic enzymes by CAMP and resveratrol in human melanocytes. *Journal of Investigative Dermatology* 127:2216–27.

Niu, Y., B. Shen, Y. Cui, Y. Chen, et al. 2014. Generation of gene-modified cynomolgus monkey via Cas9/RNA-mediated gene targeting in one-cell embryos. *Cell* 156:836–43.

Norton, H. L., R. A. Kittles, E. Parra, P. McKeigue, et al. 2007. Genetic evidence for the convergent evolution of light skin in Europeans and East Asians. *Molecular Biology and Evolution* 24:710–22.

Pagel, M., and W. Bodmer. 2003. A naked ape would have fewer parasites. *Proceedings of the Royal Society B: Biological Sciences* 270:117–19.

Paus, R., and G. Cotsarelis. 1999. The biology of hair follicles. *New England Journal of Medicine* 341:491–7.

Paus, R., and K. Foitzik. 2004. In search of the 'hair cycle clock': a guided tour. *Differentiation* 72:489–511.

Paus, R., K. S. Stenn, and R. E. Link. 1990. Telogen skin contains an inhibitor of hair growth. *British Journal of Dermatology* 122:777–84.

Perry, G. H. 2014. The promise and practicality of population genomics research with endangered species. *International Journal of Primatology* 35:55–70.

Pfeifer, S. P., S. Laurent, V. C. Sousa, C. R. Linnen, et al. 2018. The evolutionary history of Nebraska deer mice: local adaptation in the face of strong gene flow. *Molecular Biology and Evolution* 35:792–806.

Phifer-Rixey, M., K. Bi, K. G. Ferris, M. J. Sheehan, et al. 2018. The genomic basis of environmental adaptation in house mice. *PLoS Genetics* 14:1–28.

Poh, Y. P., V. S. Domingues, H. E. Hoekstra, and J. D. Jensen. 2014. On the prospect of identifying adaptive loci in recently bottlenecked populations. *PLoS One* 9:e110579.

Protas, M. E., and N. H. Patel. 2008. Evolution of coloration patterns. *Annual Review of Cell and Developmental Biology* 24:425–46.

Pummila, M., I. Fliniaux, R. Jaatinen, M. J. James, et al. 2007. Ectodysplasin has a dual role in ectodermal organogenesis: inhibition of *Bmp* activity and induction of *Shh* expression. *Development* 134:117–25.

Quillen, E. E., H. L. Norton, E. J. Parra, F. Lona-Durazo, et al. 2019. Shades of complexity: new perspectives on the evolution and genetic architecture of human skin. *American Journal of Physical Anthropology* 168:4–26.

Raimondi, S., F. Sera, S. Gandini, S. Iodice, et al. 2008. MC1R variants, melanoma and red hair color phenotype: a meta-analysis. *International Journal of Cancer* 122:2753–60.

Rakotonirina, H., P. M. Kappeler, and C. Fichtel. 2017. Evolution of facial color pattern complexity in lemurs. *Scientific Reports* 7:1–11.

Rawles, M. E. 1947. Origin of pigment cells from the neural creast in the mouse embryo. *Physiological Zoology* 20:248–66.

Rensch, B. 1938. Some problems of the geographical variation and species-formation. *Proceedings of the Linnean Society London* 150:275–85.

Roffler, G. H., S. J. Amish, S. Smith, T. Cosart, et al. 2016. SNP discovery in candidate adaptive genes using exon capture in a free-ranging alpine ungulate. *Molecular Ecology Resources* 16:1147–64.

Rogers, J. 2018. The behavioral genetics of nonhuman primates: status and prospects. *American Journal of Physical Anthropology* 165:23–36.

Rowe, N. 1996. *The Pictoral Guide to the Living Primates*. East Hampton: Pagonias Press.

Samuels, D. C., L. Han, J. Li, S. Quanghu, et al. 2013. Finding the lost treasures in exome sequencing data. *Trends in Genetics* 29:593–9.

Sandel, A. A. 2013. Brief communication: hair density and body mass in mammals and the evolution of human hairlessness. *American Journal of Physical Anthropology* 152:145–50.

Sankararaman, S., S. Mallick, M. Dannemann, K. Prüfer, et al. 2014. The genomic landscape of Neanderthal ancestry in present-day humans. *Nature* 507:354–7.

Santana, S. E., J. L. Alfaro, and M. E. Alfaro. 2012. Adaptive evolution of facial colour patterns in Neotropical primates. *Proceedings of the Royal Society B: Biological Sciences* 279:2204–11.

Santana, S. E., J. L. Alfaro, A. Noonan, and M. E. Alfaro. 2013. Adaptive response to sociality and ecology drives the diversification of facial colour patterns in catarrhines. *Nature Communications* 4:1–7.

Sasaki, E., H. Suemizu, A. Shimada, K. Hanazawa, et al. 2009. Generation of transgenic non-human primates with germline transmission. *Nature* 459:523–7.

Schmidt-Ullrich, R., and R. Paus. 2005. Molecular principles of hair follicle induction and morphogenesis. *BioEssays* 27:247–61.

Schmutz, S. M., T. G. Berryere, and A. D. Goldfinch. 2002. TYRP1 and MC1R genotypes and their effects on coat color in dogs. *Mammalian Genome* 13:380–7.

Schneider, M. R., R. Schmidt-Ullrich, and R. Paus. 2009. The hair follicle as a dynamic mini-organ. *Current Biology* 19:R132–42.

Schneider, R. A. 2018. Neural crest and the origin of species-specific pattern. *Genesis* 56:1–33.
Schwartz, G. G., and L. A. Rosenblum. 1981. Allometry of primate hair density and the evolution of human hairlessness. *American Journal of Physical Anthropology* 55:9–12.
Schwarze, K., J. Buchanan, J. C. Taylor, and S. Wordsworth. 2018. Are whole-exome and whole-genome sequencing approaches cost-effective? A systematic review of the literature. *Genetics in Medicine* 20:1122–30.
Seibert, H. C., and M. Steggerda. 1999. The size and shape of human head hair—along its shaft. *The Journal of Heredity* 33:302–4.
Shimomura, Y., M. Wajid, L. Petukhova, M. Kurban, et al. 2010. Autosomal-dominant woolly hair resulting from disruption of keratin 74 (KRT74), a potential determinant of human hair texture. *American Journal of Human Genetics* 86:632–8.
Shimomura, Y., M. Wajid, A. Zlotogorski, Y. J. Lee, et al. 2009. Founder mutations in the lipase h gene in families with autosomal recessive woolly hair/hypotrichosis. *Journal of Investigative Dermatology* 129:1927–34.
Sick, S., S. Reinker, J. Timmer, and T. Schlake. 2006. WNT and DKK determine hair follicle spacing through a reaction-diffusion mechanism. *Science* 314:1447–50.
Silva, L. G. D., K. Kawanishi, P. Henschel, A. Kittle, et al. 2017. Mapping black panthers: macroecological modeling of melanism in leopards (*Panthera pardus*). *PLoS One* 12:1–17.
Slominski, A., J. Wortsman, P. M. Plonka, K. U. Schallreuter, et al. 2005. Hair follicle pigmentation. *Journal of Investigative Dermatology* 124:13–21.
Spacek, D. V., A. F. Perez, K. M. Ferranti, L. K. L. Wu, et al. 2010. The mouse frizzy (Fr) and rat 'hairless' (FrCR) mutations are natural variants of protease serine S1 family member 8 (Prss8). *Experimental Dermatology* 19:527–32.
Sponenberg, D. P. 2009. *Equine Color Genetics*, 3rd ed. Ames: Wiley-Blackwell.
Steggerda, M., and H. C. Seibert. 1941. Size and shape of head hair from six racial groups. *Journal of Heredity* 32:315–18.
Stevens, M., C. A. Párraga, I. C. Cuthill, J. C. Partridge, et al. 2007. Using digital photography to study animal coloration. *Biological Journal of the Linnean Society* 90:211–37.
Stevens, M., M. C. Stoddard, and J. P. Higham. 2009. Studying primate color: towards visual system-dependent methods. *International Journal of Primatology* 30:893–917.
Sturm, R. A., and D. L. Duffy. 2012. Human pigmentation genes under environmental selection. *Genome Biology* 13:248.
Sumner, P., and J. D. Mollon. 2003. Colors of primate pelage and skin: objective assessment of conspicuousness. *American Journal of Primatology* 59:67–91.
Suzuki, I., A. Tada, M. M. Ollmann, G. S. Barsh, et al. 1997. Agouti signaling protein inhibits melanogenesis and the response of human melanocytes to α-melanotropin. *Journal of Investigative Dermatology* 108:838–42.
Szpak, M., Y. Xue, Q. Ayub, and C. Tyler-Smith. 2019. How well do we understand the basis of classic selective sweeps in humans? *FEBS Letters* 593:1431–48.
Tanaka, S., I. Miura, A. Yoshiki, Y. Kato, et al. 2007. Mutations in the helix termination motif of mouse type I IRS keratin genes impair the assembly of keratin intermediate filament. *Genomics* 90:703–11.
Tapanes, E., S. Anestis, J. M. Kamilar, and B. J. Bradley. 2020. Does facial hair greying in chimpanzees provide a salient progressive cue of aging? *PLoS One* 15:e0235610.
Tennessen, J. A., T. D. O'Connor, M. J. Bamshad, and J. M. Akey. 2011. The promise and limitations of population exomics for human evolution studies. *Genome Biology* 12:127.
Tregear, R. T. 1965. Hair density, wind speed, and heat loss in mammals. *Journal of Applied Physiology* 20:796–801.

Valk, T. V. D., C. M. Gonda, H. Silegowa, S. Almanza, et al. 2020. The genome of the endangered dryas monkey provides new insights into the evolutionary history of the vervets. *Molecular Biology and Evolution* 37:183–94.

Vallender, E. J. 2011. Expanding whole exome resequencing into non-human primates. *Genome Biology* 12:R87.

Valverde, P., E. Healy, I. Jackson, J. L. Rees, et al. 1995. Variants of the melanocyyte-stimulation hormone receptor gene are associated with red hair and fair skin in humans. *Nature Genetics* 11:328–30.

Wang, H., S. Ma, L. Xue, Y. Li, et al. 2016. MiR-488 determines coat pigmentation by down-regulating the pigment-producing gene pro-opiomelanocortin. *Cellular and Molecular Biology* 62:37–43.

Ward, L. D., and M. Kellis. 2012. Interpreting non-coding variation in complex disease genetics lucas. *Nature Biotechnology* 30:1095–106.

Warr, A., C. Robert, D. Hume, A. Archibald, et al. 2015. Exome sequencing: current and future perspectives. *G3: Genes, Genomes, Genetics* 5:1543–50.

Warren, K. A., T. B. Ritzman, R. A. Humphreys, C. J. Percival, et al. 2018. Craniomandibular form and body size variation of first generation mouse hybrids: a model for hominin hybridization. *Journal of Human Evolution* 116:57–74.

Webster, T. H. 2015. *Genomics of a Primate Radiation: Speciation and Diversification in the Macaques*. Doctoral dissertation. Yale University.

Webster, T. H., E. E. Guevara, R. R. Lawler, and B. J. Bradley. 2018. Successful exome capture and sequencing in lemurs using human baits. *BioRxiv*. doi:10.1101/490839.

West, P. M., and C. Packer. 2008. Sexual selection, temperature, and the lion's mane. *Science* 1339:1339–44.

Winter, H., L. Langbein, M. Krawczak, D. N. Cooper, et al. 2001. Human type I hair keratin pseudogene ΦhHaA has functional orthologs in the chimpanzee and gorilla: evidence for recent inactivation of the human gene after the *Pan-Homo* divergence. *Human Genetics* 108:37–42.

Winters, S., W. Allen, and J. Higham. 2020. The structure of species discrimination signals across a primate radiation. *eLife* 9:e47428.

Ye, J., T. Yang, H. Guo, Y. Tang, et al. 2013. Wnt10b promotes differentiation of mouse hair follicle melanocytes. *International Journal of Medical Sciences* 10:691–8.

Yu, L., G. D. Wang, J. Ruan, Y. B. Chen, et al. 2016. Genomic analysis of snub-nosed monkeys (*Rhinopithecus*) identifies genes and processes related to high-altitude adaptation. *Nature Genetics* 48:947–52.

Zhou, X., B. Wang, Q. Pan, J. Zhang, et al. 2014. Whole-genome sequencing of the snub-nosed monkey provides insights into folivory and evolutionary history. *Nature Genetics* 46:1303–10.

Zimova, M., L. S. Mills, and J. J. Nowak. 2016. High fitness costs of climate change-induced camouflage mismatch. *Ecology Letters* 19:299–307.

Zorina-Lichtenwalter, K., R. N. Lichtenwalter, D. V. Zaykin, M. Parisien, et al. 2019. A study in scarlet: MC1R as the main predictor of red hair and rxemplar of the flip-flop effect. *Human Molecular Genetics* 28:2093–106.

Zuk, M., and J. G. Decruyenaere. 1993. Measuring individual variation in colour: a comparison of two techniques. *Biological Journal of the Linnean Society* 53:165–73.

2 Cell Processes and Key Genes in the Evolution of Pigmentation Variation in Humans

Heather L. Norton

CONTENTS

2.1 Introduction 35
2.2 Pigment Cells across the Body 36
  2.2.1 Epidermis 36
  2.2.2 Hair Cells 39
  2.2.3 Iris 40
2.3 Pigmentation Genetics 41
2.4 The Evolution of Human Skin Pigmentation Variation 43
  2.4.1 Evidence from Ancient DNA 44
  2.4.2 Epistatic Interactions 45
  2.4.3 Convergent Evolution of Pigmentation Phenotypes 46
2.5 Evolution of Human Hair and Iris Variation 46
2.6 Eurocentric Bias in Pigmentation Studies 47
2.7 Summary 47
2.8 Acknowledgments 48
2.9 References 48

2.1 INTRODUCTION

Pigmentation of the skin, hair, and eyes are three phenotypic traits that show extensive variation across the human species. This variation is attributed to the amount and type of melanin produced in specialized neural-crest derived cells known as melanocytes. Variation in human skin pigmentation is the result of long-term adaptation to ultraviolet radiation (UVR) intensity (Jablonski and Chaplin 2000). Although there is no evidence that variation in hair and iris pigmentation also reflects adaptation to UVR, the shared genetic architecture of human pigmentary traits means that variation in these phenotypes may be indirectly shaped by selective pressures on skin pigmentation. Genomic methods, such as genome-wide association studies (GWASs) and selection scans, are relatively new tools that researchers can use to identify loci that influence pigmentary traits. These have contributed to a growing body of knowledge related to pigmentation biology and genetics. However, in order to understand the potential functional significance of loci identified using these methods, it is critical to marry such knowledge with a well-grounded understanding of the cellular processes and mechanisms that underlie the production and distribution of melanin. What follows is a description of pigmentary cellular processes in the skin, hair, and iris; a brief summary of major genes affecting pigmentary phenotypes in humans; and a description of the evolutionary history of human pigmentation variation.

2.2 PIGMENT CELLS ACROSS THE BODY

2.2.1 Epidermis

Skin is composed of three different layers—the external epidermis; middle dermis; and inner hypodermis, a layer of fatty tissue that helps to anchor the dermis to skeletal structures. Of primary importance in understanding pigmentation are the cells found in the basal layer of the epidermis (shown in Figure 2.1), the pigment-producing melanocytes. These cells are found dispersed among basal keratinocyte cells, with a group of cells consisting of 1 melanocyte, ~30–40 keratinocytes, and 1 Langerhans cell forming a functional and structural complex known as the epidermal melanin (EM) unit (Costin and Hearing 2007). Melanocytes are derived from neural crest cells that migrate to their final locations in the body, including the epidermis and dermis, between the 10th and 14th weeks of development (Haake and Holbrook 1999). Once localized to these regions melanocytes begin to produce melanosomes, the membrane-bound organelles where melanin synthesis occurs. As the melanosomes mature (see subsequently), they are transferred to keratinocyte cells in the EM unit. These keratinocyte cells subsequently move upwards through the stratum spinosum and stratum granulosum.

Melanocyte count will vary across different regions of the body, with higher melanocyte counts per mm^2 of skin found in genital regions and lower counts observed in other regions (e.g., back) (Szabó 1954; Whiteman et al. 1999). While

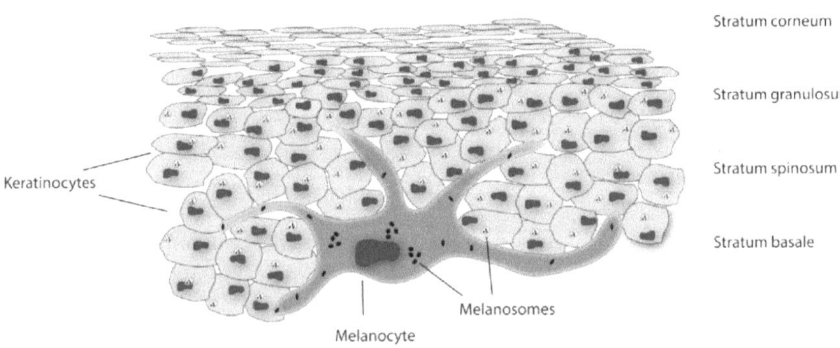

FIGURE 2.1 Layers of the epidermis, highlighting the relationship between melanocyte and keratinocyte cells. Note: Langerhans cells, also part of the epidermal-melanin unit, are not represented here.

melanocyte count does not seem to vary among darker and lighter skin tones, both the melanosomes themselves (including their number, size, and distribution within keratinocytes) and the quantity and type of melanin they contain are critical factors influencing overall skin pigmentation phenotype. Melanosomes are lysosomal-like organelles that originate in the perinuclear region near the Golgi stacks, and move through four developmental stages (I–IV) before being transferred to local keratinocyte cells. Stage I eumelanosomes (melanosomes that will ultimately contain brown-black eumelanin) are spherical organelles that lack a coherent internal structure and do not show evidence of tyrosinase activity (and therefore contain no melanin). In the second stage of development, melanosomal structural proteins such as Pmel17 have transformed the organelle into a more ellipsoidal unit with limited tyrosinase activity and a small amount of melanin deposition (Berson et al. 2001; Kushimoto et al. 2001). As melanin production increases and becomes deposited on internal melanosomal fibrils, the melanosome transitions to stage III. By stage IV, a fully mature eumelanosome is more ellipsoidal in shape, and tyrosinase activity is once again low. Maturation stages for pheomelanosomes are similar, although mature pheomelanosomes (containing red-yellow pheomelanin) tend to be smaller, remain spherical in shape, and lack a strong internal fibrillar structure.

After maturation, melanosomes migrate up through the dendritic arms of the melanocyte. The subsequent transfer of melanosomes to keratinocyte cells has long been a subject of debate—proposed mechanisms include exocytosis, cytophagocytosis, fusion of the plasma membrane, and transfer by membrane vesicles. These mechanisms are reviewed in Tadokoro and Takahashi (2017). Recent work suggests that the melanosomes are released as vesicles from the melanocytes and are incorporated into keratinocyte cells by phagocytosis (Tadokoro et al. 2016). Once in the keratinocyte, pheomelanosomes tend to occur in aggregate groupings of 2–3, while the larger eumelanosomes are found in single units (Toda et al. 1972). As the keratinocyte cells migrate up through the epidermis, the melanosomal membrane degrades and melanin granules localize to a region over the keratinocyte nucleus. This development of a supra-nuclear melanin cap is believed to provide protection from UV-induced DNA damage (Kobayashi et al. 1998). By the time the keratinocyte has completed the process of terminal differentiation and has reached the upper layers of the epidermis, the melanosomes in more lightly pigmented individuals are completely degraded, while eumelanosomes more common in darker skin may still be intact in these cells (Thong et al. 2003).

Within the melanosomes, melanin production (described in a process known as the Raper-Mason pathway) is driven largely by the rate-limiting enzyme tyrosinase (produced by the *TYR* gene) (Oetting 2000). Early in the process, transcription of the gene *TYR* is stimulated by the microphthalmia-associated transcription factor, MITF. Tyrosinase activity is critical for the initial steps in the melanin synthesis pathway, the hydroxylation of tyrosine to 3,4-dihyroxyphenylalanine (DOPA), and the oxidation of DOPA to DOPAquinone. At this point the melanin synthesis pathway diverges; in the presence of melanosomal sulfhydryl compounds, pheomelanogenesis occurs, while in their absence, the enzymes dopachrome tautomerase and tyrosinase-related protein 1 facilitate the production of DHI and DHICA eumelanins (Figure 2.2).

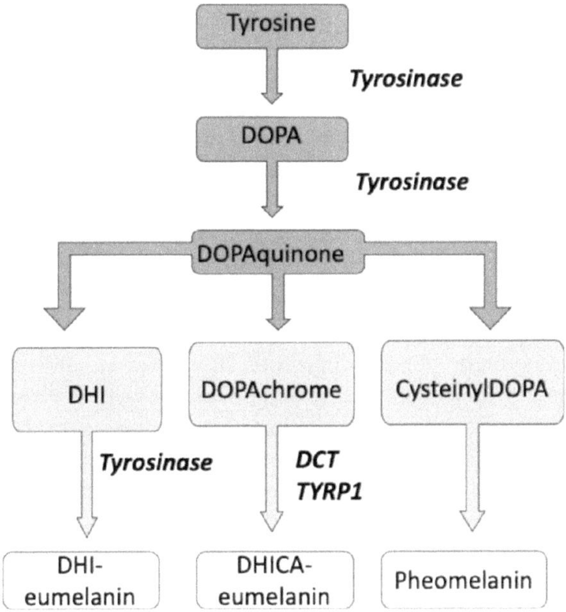

FIGURE 2.2 A simplified depiction of the Raper-Mason pathway, showing the basic steps in the production of eumelanin and pheomelanin.

The switch between the production of eumelanin and pheomelanin is regulated by the melanocortin-1 receptor, a 7-pass transmembrane G-protein-coupled receptor found on the cell surface of the melanocyte. The MC1R responds to both α-melanocyte stimulating hormone (α-MSH) and adrenocorticotropic hormone (ACTH), both of which are produced by proopiomelanocortin (POMC). When the hormone α-MSH binds to the MC1R, it activates adenylyl cyclase, which increases levels of intracellular cAMP and leads to an increase in the expression of the melanocyte master regulator microphthalmia-associated transcription factor (MITF). As described, this stimulates the transcription of *TYR* and initiates the initial steps in the Raper-Mason pathway of melanogenesis, resulting in the production of brown-black eumelanin. However, the MC1R can also bind the antagonist agouti-signaling protein (ASIP). When ASIP is able to successfully compete with α-MSH for MC1R receptor binding, pheomelanin is produced instead of eumelanin. More recently, the protein β-defensin 3 (HBD3) was identified as an additional physiological MC1R antagonist, which may provide a new method for the regulation of MC1R activity.

While tyrosinase (TYR), tyrosinase-related protein (TYRP1), and dopachrome tautomerase (DCT) are all important enzymes for melanin production, other factors also influence this process, including the melanosomal environment. Melanosomes are acidic during early stages of development (when melanin production is initiated), but as they mature, their pH increases until peak tyrosinase activity occurs at a more neutral pH level of 6.8 (Ancans et al. 2001). Consistent with this, the pH of melanosomes obtained from more lightly pigmented individuals is lower than that of melanosomes from more darkly pigmented individuals (Fuller et al. 2001). In addition, individuals with oculocutaneous albinism type II, a pigmentary disorder associated with very low levels of tyrosinase activity, exhibit melanosomes that are highly acidic (Puri et al. 2000; Brilliant 2001).

While an individual's overall skin pigmentation is determined both by the total amount of melanin produced as well as the ratio of eumelanin to pheomelanin, global patterns of human skin pigmentation variation reflect an adaptation to UVR, with darker (more melanogenic) skin being more commonly observed in people with ancestry from high UVR regions and lighter (less melanogenic) skin being more common in people with ancestry from regions were UVR is weaker (Relethford 1997; Jablonski and Chaplin 2000). The strong relationship between melanin and UVR is attributed to the fact that melanin, and in particular eumelanin, has the ability to both scatter and absorb UVR, limiting its ability to penetrate the epidermis (Kaidbey et al. 1979). These photoprotective properties suggest that melanin may act like a natural sunscreen, absorbing 50–70% of UVR (Brenner and Hearing 2008). This has the effect of minimizing the amount of long-wavelength UVA (320–400 nm) and short-wavelength (280–320 nm) UVB radiation that can cause both indirect and direct DNA damage to keratinocyte and melanocyte cells.

2.2.2 Hair Cells

The mechanism of melanin production in hair cells does not significantly differ from its production in skin. A review of these processes can be found in Slominski et al. (2005) and in Tapanes et al. (this volume). The follicular-melanin unit, located in the proximal hair bulb, consists of one melanocyte and ~5 keratinocytes. Compared to epidermal melanocytes, melanocytes of the hair bulb tend to be both larger and more dendritic. They also produce larger melanosomes than are commonly observed in epidermal melanocytes. When mature, these melanosomes are transferred to the precortical keratinocytes, where they do not undergo extensive degradation, unlike melanosomes in epidermal keratinocytes.

Unlike the continuous production of melanin in the epidermis, in hair melanocytes, melanogenesis is tightly linked to the hair growth cycle (Slominski and Paust 1993). Briefly, this process consists of three phases. The first phase, known as anagen, is when hair grows through a shaft formation process, which typically takes place over 3–5 years. Following this, the catagen phase (1–2 weeks) sees the resorption of up to 70% of the hair follicle due to apoptosis. Finally, the telogen phase, lasting ~3 months, is a period of quiescence. Melanin production takes place only during the anagen phase of the cycle. As with skin pigmentation, hair color is determined by the amount and type of melanin produced (Ito and Wakamatsu 2011). For example, darker hair tends to exhibit a higher amount of total melanin and a high eumelanin to pheomelanin ratio. In contrast, blond hair has low total amounts of melanin, and red hair exhibits a low eumelanin to pheomelanin ratio.

Melanin production in hair is age dependent in humans, declining or ceasing altogether in some individuals with increasing age. The age of onset of hair graying is typically in the fourth decade of life for humans (Keogh and Wash 1965). Graying of the hair is the result of a reduction in the number of melanogenically active melanocytes in a follicular melanin unit. However, observations of a dopa oxidation reaction (indicative of tyrosinase activity) in these melanocyte cells suggests that some melanin is still being produced, albeit at lower levels (Commo et al. 2004). The bulbs of white hair do not show evidence of any active melanocytes (Commo et al. 2004).

2.2.3 Iris

There is no analogy to the melanocyte-keratinocyte complex of the skin or the melanocyte-follicular complex of the hair in the iris. Instead, melanosomes remain in the melanocyte cells and do not migrate upon reaching maturity. The iris itself has two layers of pigmented cells. The iris pigmented epithelium (IPE) is the thinner, innermost layer of tissue and is highly pigmented. However, differences in melanin content in this layer have not been observed between individuals with different eye colors, suggesting that it is not a main driver of eye color differences in humans. Instead, the relatively uniform pigmentation observed here may be indicative of the fact that melanin in this layer is important for protecting the retina and absorbing excess light (Wilkerson 1996). Superficial to the IPE are two muscular layers of tissue that do not contain melanocytes. The two most superficial layers of the iris, the anterior-border layer and the stromal layer, are important in determining inter-individual variation in eye color. Melanocytes and fibroblast cells make up the anterior-border layer, while the stroma includes a loose mesh of collagen fibers, melanocytes, fibroblasts, and clump cells (Edwards et al. 2016), as shown in Figure 2.3.

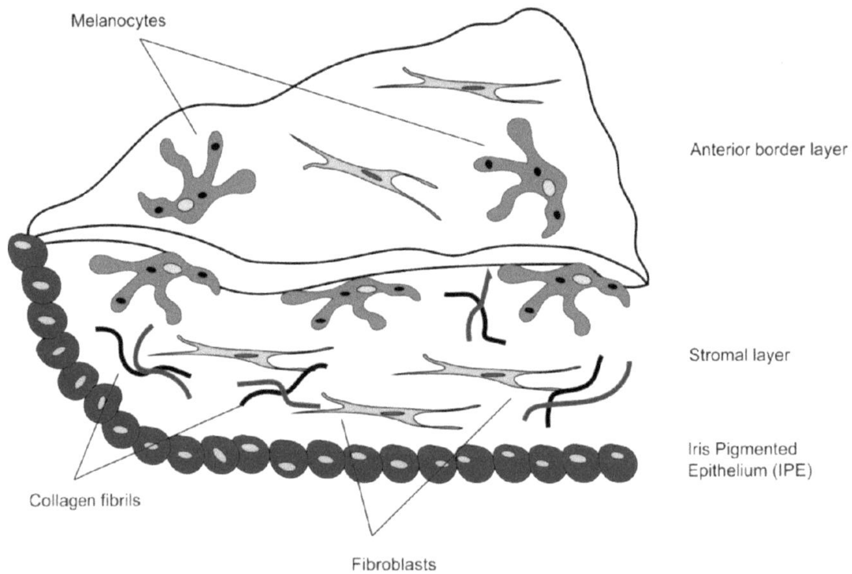

FIGURE 2.3 Three layers of the iris: the iris pigmented epithelium (IPE), stromal layer, and anterior border layer, with relevant cells (melanocytes, fibroblasts) and structures (collagen fibrils) highlighted. Note: Clump cells, located in the stromal layer, are not shown. (Adapted from Stevens and Lowe 2004).

The total amount and type of melanin in the melanosomes found in the anterior-border and stromal layers are thought to be responsible for most of the variation in human eye color. Darker eyes exhibit greater amounts of total melanin and higher eumelanin/pheomelanin ratios in these layers relative to lighter-colored eyes (Peles et al. 2009; Wakamatsu et al. 2007). For example, green eyes show a high pheomelanin content, blue eyes show low levels of both eumelanin and pheomelanin, and brown eyes seem to include a mixture of both eumelanin and pheomelanin (Prota et al. 1998). In addition to melanin content, other structural features of the iris may also affect iris color. These include the depth of the stromal layer, presence of iridial structures (such as Wolfflin nodules), organization of the extracellular components in the anterior-border and stromal layers, corneal thickness and curvature, and the ratio of stromal to anterior-border melanin content (Edwards et al. 2016).

2.3 PIGMENTATION GENETICS

Genes involved in melanin production as well as melanosomal structure and development play critical roles in regulating pigmentation variation. Three genes involved in the melanin synthesis pathway described previously, *TYR*, *TYRP1*, and *DCT*, all have the potential to impact pigmentation phenotype. However, only mutations in *TYR*, the gene that produces the rate-limiting enzyme tyrosinase, have been strongly associated with pigmentation variation. Several nonsense or missense mutations in *TYR* are associated with oculocutaneous albinism type I—a type of albinism in which no melanin is produced (Oetting 2000). However, mutations in *TYR* also have more subtle effects on non-pathological pigmentary variation and contribute to lighter skin pigmentation in European populations (Shriver et al. 2003). While mutations in *TYRP1* cause a much rarer type of albinism (OCA type III) (Boissy et al. 1996), mutations in this gene are also associated with pigmentary variation in other phenotypes, including the blond hair observed in some Melanesian populations (Kenny et al. 2012; Norton et al. 2014). To date, there is no evidence that the gene encoding *DCT* has a major impact on pigmentation phenotypes.

The genes *SLC24A5* and *SLC45A2* (also known as *AIM1* or *MATP*) are melanosomal membrane proteins that both have significant impact on normal variation in pigmentation. The SLC45A2 protein plays a role in the processing and intracellular trafficking of tyrosinase and other melanosomal proteins (Costin et al. 2003). It has been suggested that SLC45A2 may also play a role in modulating melanosomal pH (Bartölke et al. 2014; Bin et al. 2015), which would impact melanin production. Mutations in this gene are associated with oculocutaneous albinism type IV (Newton et al. 2001) as well as lighter skin pigmentation in European (Graf et al. 2005, 2007; Norton et al. 2006) and South Asian populations (Stokowski et al. 2007).

The *SLC24A5* gene is a member of a the SLC24A family of potassium-dependent sodium/calcium exchangers. Early studies of the role this gene played in both human and zebrafish pigmentation hypothesized that the SLC24A5 protein regulated ion transport in and out of the melanosome, providing a mechanism for how *SLC24A5* could have such profound impacts on skin pigmentation (Lamason 2005). Specifically, by influencing pH and/or calcium ion concentration within the melanosome, the rs1426654 mutation in exon 3 of this gene was thought to cause reduced levels of tyrosinase activity and subsequently decreased levels of melanin production. However, later studies could not definitively confirm the localization of SLC24A5 to the melanosomal membrane. More recently, SLC24A5 has been localized to the trans-Golgi-network (Rogasevskaia et al. 2019). This discovery raises new questions about how changes to the SLC24A5 protein may influence pigmentation levels. While the precise mechanism remains unknown, it is clear that the *SLC24A5* gene has one of the largest effects on normal pigmentation in humans observed to date. A single mutation in this gene, rs1426654, is capable of explaining 30% of the variation in skin pigmentation between European and Western African populations (Lamason 2005). More recently, this gene has also been associated with the lighter skin pigmentation observed among the San hunter-gatherer populations of southern Africa (Martin et al. 2017; Lin et al. 2018) and South Asian populations (Basu Mallick et al. 2013; Iliescu et al. 2018). Like several other loci, there is evidence that the derived allele at this mutation has been strongly favored by natural selection in European populations (Norton et al. 2006; Beleza et al. 2013; Lamason 2005; Sabeti et al. 2007). More recently, selection favoring this allele has been observed in Eastern African populations and San populations from southern Africa (Crawford et al. 2017; Lin et al. 2018).

As described, the construction of melanosomes is a complex process that can impact melanin production. One of the key genes involved in this process is the *OCA2*, or *P*, locus. This gene encodes a hydrophobic integral melanosomal membrane protein (Rosemblat et al. 1994) that plays a role in regulating melanosomal pH, and hence melanin production (Brilliant 2001). This protein is part of a chloride ion channel that allows chloride ions to pass through the melanosomal membrane and increase melanosomal pH, thereby facilitating melanin production (Bellono et al. 2014). Mutations in *OCA2* cause oculocutaneous albinism type II, a type of albinism in which small amounts of melanin are produced (Lee et al. 1994). Mutations in this gene have also been associated with lighter skin pigmentation between west African and European populations (Shriver et al. 2003), while still other mutations are associated with lighter skin color in East Asian populations (Akey et al. 2001; Edwards et al. 2010; Rawofi et al. 2017).

For several years, the *OCA2* gene was also implicated as a major locus influencing iris pigmentation. More recently, it has been discovered that this effect is largely mediated by the *HERC2* gene, found immediately upstream of the *OCA2* promoter on chromosome 15 (Eiberg et al. 2008; Kayser et al. 2008; Sturm et al. 2008). Derived alleles in this region regulate the expression of the *OCA2* gene in Europeans and are a major determinant of blue vs. brown eye color in European populations.

Perhaps one of the most well-studied aspects of pigmentation variation is the switch between the production of eumelanin and pheomelanin. Although quite short, the *MC1R* gene exhibits an unusually high level of sequence diversity, particularly in European and East Asian populations (Rana et al. 1999). Several of these mutations are associated with lighter skin pigmentation, freckles, red hair, an increased risk of melanoma, and impaired DNA repair capacity (Flanagan 2000; Palmer et al. 2000; Rees 2000; Valverde 1996; Valverde et al. 1995; van der Velden et al. 2001; Yamaguchi et al. 2012). The elevated melanoma risk associated with some of these mutations is likely attributable to both the production of pheomelanin as well as independent effects influencing DNA repair capacity (Kadekaro et al. 2010). While one might expect diversity at *ASIP* to mirror that of *MC1R*, instead this gene appears to be relatively conserved across humans. A single mutation in the promoter region of this gene has been associated with darker skin pigmentation (Kanetsky et al. 2002; Bonilla et al. 2005).

An additional reason for investigating the genetics of human pigmentation variation is that this may also help to identify genes that regulate pigmentation variation in nonhuman primates. Surprisingly, many of the major genes involved in regulating human pigmentation phenotypes do not play similar roles in other primate species. For example, the highly pheomelanic coat of orangutans does not appear to be due to mutations in the *MC1R* gene, indicating that the genetic cause of this phenotype differs from that observed in European human populations (Haitina et al. 2007). Further, the MC1R antagonist agouti is not responsible for the darker coat color observed in the Sulawesi macaque and black lion tamarin (Mundy and Kelly 2006), and the blue eye color of blue-eyed black lemur is not due to mutations in *HERC2-OCA2* region, as observed in humans. Taken together, these data suggest that we still have much to learn about the genes that regulate nonhuman primate pigmentation variation. Uncovering the genetic architecture of nonhuman primate pigmentation phenotypes may be an important key to understanding the evolution of different coat colors and their roles in camouflage and sexual selection. A more comprehensive review of pigmentation genetics in nonhuman primates can be found in Tapenes et al. (this volume).

2.4 THE EVOLUTION OF HUMAN SKIN PIGMENTATION VARIATION

Human pigmentation has captured the interests of biologists and anthropologists not only because of its extensive phenotypic diversity but also because of the strong role that natural selection has played in shaping the geographic distribution of this variation. The development of tools to quantify skin color has made it possible to formally test hypotheses about the causes of this distribution. These investigations first identified a relationship between skin reflectance and latitude (Relethford 1997), and subsequent work utilizing satellite data identified UVR more specifically as the driving selective force shaping global human skin pigmentation diversity (Jablonski and Chaplin 2000). Today we understand that variation in human skin color reflects long-term adaptation to UVR, driven by the ability of melanin to provide protection to naked human skin from damaging UVA and UVB (Jablonski and Chaplin 2010). Although early hypotheses explaining the prevalence of darker skin pigmentation in high UVR regions focused on the ability of a highly melanized skin to prevent the development of skin cancers (Fitzpatrick 1965), others have argued that this could only have been of secondary or minor evolutionary importance given the late age of onset of most skin cancers (Blum 1961). The prevailing explanations for the global distribution of human skin pigmentation today focus on balancing the need to minimize UV-induced photolysis of folate in geographic regions where UVR is strong with the need to maximize the potential for previtamin D_3 synthesis in regions where UVR is weaker (Jablonski and Chaplin 2000, 2010).

Skin melanin content in humans is shaped by strong natural selection because of the ability of melanin to absorb and scatter UVR. In regions where UVR is intense, a highly melanized skin was hypothesized to be advantageous by preventing or minimizing folic acid photolysis (Branda and Eaton 1978). Folic acid degradation was likely an important selective force in our evolutionary history because of the important role that folate plays in fetal development. Because the photoprotective properties of melanin could limit the destruction of folate in cutaneous blood vessels, Jablonski and Chaplin proposed that darker skin may have been favored in regions of high UVR as a way to limit reductions in fertility associated with neural tube defects (Jablonski and Chaplin 2000, 2010). More recently, the important role of folate in cell division has become apparent, suggesting that the maintenance of folic acid levels would have been important for both males and females in living in high-UVR environments (Jablonski and Chaplin 2010).

The benefits of a highly melanized skin in regions where UVR is strong may become a liability in places where UVR is weak. In regions where UVR is weaker, lower levels of melanin may enable the UV-mediated synthesis of the steroid hormone vitamin D_3 (Loomis 1967; Murray 1934). While humans can obtain vitamin D through dietary means, vitamin D can also be endogenously produced when UVB radiation (290–310 nm) penetrates the skin. 7-dehydrocholestrol (7-DHC) then absorbs this radiation in the epidermis and dermis, leading to the synthesis of previtamin D_3.

Much of the original focus on the evolutionary significance of vitamin D focused on the important role that vitamin D plays in calcium metabolism. Specifically, vitamin D helps to facilitate the absorption of calcium from the gut, allowing it to be distributed to key areas of the body, including the skeletal system and teeth. Vitamin D deficiencies are associated with the skeletal disorders osteomalacia and rickets (Holick 2006). In regions where UVB is weak, therefore, it may be advantageous to have skin with a reduced melanin content to ensure sufficient previtamin D_3 production. Jablonski and Chaplin argue that the selective pressure for sufficient synthesis of previtamin D would have been particularly strong in females, who may have enhanced needs for calcium absorption and processing during pregnancy and breastfeeding (Jablonski and Chaplin 2000).

2.4.1 Evidence from Ancient DNA

The identification of several pigmentation genes targeted by recent directional selection supports the hypothesis that lighter skin pigmentation evolved recently and was favored by strong natural selection (Quillen et al. 2019). For example, efforts to date the onset of selection favoring functional mutations associated with reduced skin melanin content in the genes *SLC24A5*, *SLC45A2*, and *TYRP1* suggest that these alleles were likely swept to high frequency within the past 12–18,000 years in European populations (Beleza et al. 2013). Rapid advancements in aDNA recovery and sequencing technologies have also made it possible to identify pigmentation-associated mutations in fossil remains from Neanderthals (Lalueza-Fox et al. 2007), as well as to document the appearance and shift in the frequencies of these alleles in *Homo sapiens* over time (Olalde et al. 2014, 2018; Ju and Mathieson 2020; Mathieson et al. 2015). Such studies have started to provide valuable insights into the timing and pace of selection-mediated change in European populations.

Early investigations into the Neanderthal genome revealed that although Neanderthals did not share common mutations associated with fair skin and red hair in modern humans, they did harbor Neanderthal-specific variants with predicted functions similar to those of common mutations in European populations associated with lighter skin (Lalueza-Fox et al. 2007). aDNA recovered from the remains of individuals across Eurasia (and in particular northwestern Europe) indicates that the frequencies of several alleles known to play a major role in European pigmentation were still very much in flux through the Neolithic period and that frequencies shifted with the arrival of Neolithic migrants. For example, while derived alleles at some pigmentation loci, such as the HERC2/OCA derived haplotype associated with blue eyes, were already common in European Mesolithic populations (Olalde et al. 2014; Mathieson et al. 2015), derived alleles at *SLC45A2* and *SLC24A5* only increased in frequency in later European populations due to natural selection and introduction by Neolithic migrants from the east (Olalde et al. 2018; Mathieson et al. 2015; Ju and Mathieson 2020).

2.4.2 Epistatic Interactions

It is tempting to jump from documenting the presence or absence of these alleles to making predictions about the appearance of individuals living in these populations (Cerqueira et al. 2012; Brace et al. 2019; Lalueza-Fox et al. 2007) based on the known effects of these alleles in modern populations. However, predicting past phenotypes should be carried out with caution, largely because we cannot fully estimate the effects of these mutations in the context of the full genetic background of these individuals. Combinations of pigmentation alleles not common in modern populations may have been more prevalent in the past, leading to pigmentary phenotypes that differ from what we can observe today. Two examples from modern populations highlight the need for caution in this area. The first comes from a report by Edwards and colleagues, noting that some individuals of South Asian descent carrying two copies of the derived allele at the rs12913832 SNP in the *HERC2/OCA2* complex, commonly associated with blue eye color, displayed a brown-eyed phenotype. This was attributed to possible epistatic interactions with other, currently unknown loci that might mask the effect of the derived allele in these individuals (Edwards et al. 2012), consistent with subsequent reports of gene-gene interactions influencing iris pigmentation (Pośpiech et al. 2014).

More recently, Iliescu et al. demonstrated that gene-gene interactions with *SLC24A5* can affect the impact of this gene on skin pigmentation in certain South Asian populations, where despite the relatively high frequency of the derived allele associated with lighter skin pigmentation, skin melanin index remains high (Iliescu et al. 2018). They argue that the effect of *SLC24A5* is "overprinted" by other, as yet uncharacterized genetic loci. Recognition of the impact that such gene-gene interactions can have on pigmentation phenotype has relevance not just for the reconstruction of ancestral phenotypes but more directly for the construction and application of forensic panels of genetic markers used to predict the phenotype of unknown individuals (Walsh et al. 2011, 2013, 2017). This is particularly important as such panels, typically designed using European reference populations, are applied to other populations.

2.4.3 Convergent Evolution of Pigmentation Phenotypes

As our understanding of pigmentation genetics has improved, so has our comprehension of the evolutionary history of skin pigmentation in humans. One of the more interesting aspects of this evolutionary history has been the convergent evolution of adaptive skin phenotypes in similar UV environments. This is clearly shown in comparisons of European and northern East Asian populations. While both exhibit reduced levels of melanin, as expected under a model where selection has favored lighter skin as a means to maximize previtamin D3 synthesis, the genetic architecture of pigmentation is largely distinct in each region. One of the clearest examples of this can be seen in the gene *SLC24A5*, where the allele and extended haplotype associated with lighter skin color are fixed in European populations but nearly absent in populations from East Asia (Lamason 2005). Similar patterns observed for other loci, including *SLC24A5* and *TYR*, strongly suggest that lighter skin has evolved in European and East Asian populations through largely distinct genetic architectures (Norton et al. 2006; Lao et al. 2007; Myles et al. 2006; Yang et al. 2016). Notably, there has been far less success in identifying a complex of genes affecting skin lightening in East Asian populations. The result is that our understanding of the genetic and evolutionary processes leading to skin lightening across Asia remains largely incomplete.

In regions where UVR is high, we have also had a relatively poor understanding of the genetic architecture and evolutionary history of pigmentary change. A recent paper by Crawford and colleagues has begun to shed light on this issue by identifying derived alleles at two genes, *MFSD12* and *DBB1*, that are associated with darker skin pigmentation in African populations. These alleles are also observed in Australomelanesian and South Asian populations, and the haplotypes shared between these populations and Africans suggests that they are identical by descent. This means that the same haplotypes, shared with a common ancestor, have been favored in multiple high-UVR populations (Crawford et al. 2017). Identifying additional loci associated with darker pigmentation and tracing their evolutionary history across disparate populations inhabiting high-UVR regions is an area of exciting future study.

2.5 EVOLUTION OF HUMAN HAIR AND IRIS VARIATION

While natural selection has clearly played a strong role in shaping human skin pigmentation variation, it is less clear how hair or eye color diversity could have been shaped by similar processes. Given that hair and eye pigmentation are generally more variable in European populations, it is possible that some of this phenotypic diversity is due to the prevalence of alleles influencing lighter skin pigmentation in these populations—the joint role of *MC1R* on both lighter skin and red hair color is a good example of this phenomenon. However, some alleles, such as rs12203592 in the gene *IRF4*, have contrasting effects depending on their cellular environment. In this example, the derived allele at this SNP is associated with both lighter skin and darker hair pigmentation (Han et al. 2008). Sexual selection has also been proposed as an evolutionary factor driving pigmentation phenotype variation in humans (Frost 1988). A comprehensive review of skin reflectance data in humans ultimately did not support a strong model of sexual selection as a driver for human skin pigmentation diversity (Madrigal and Kelly 2007), although it is possible that it has played a role in shaping hair and iris pigmentation (Frost 2014).

2.6 EUROCENTRIC BIAS IN PIGMENTATION STUDIES

Finally, it should be noted that investigations into the genetic and cellular processes of pigmentation variation and evolution have largely focused on European populations. While this has helped to identify and characterize genes that can explain large pigmentation differences between European populations and others (particularly West African populations), and to develop in-depth models exploring the timing and rate of pigmentation change in Europe over the last 40,000 years, the approach has likely created a significant bias in our global understanding of pigmentary genetics. For example, the identification of the genes *MFSD12* and *DBB1* by Crawford and colleagues suggests that we still have much to learn about the genetic and cellular processes regulating global human pigmentary diversity. While many of the genes known to impact pigmentation in European populations influence eumelanin vs. pheomelanin production, the melanin production pathway, or the melanosomal membrane, the identified genes affect pigmentation through novel mechanisms. Specifically, because *MFSD12* localizes to lysosomes and not melanosomes, its effect on pigmentation may be related to a modification of lysosomal function (Crawford et al. 2017). This study also reported that pigmentary phenotypes may be influenced by pleiotropic genes such as *DBB1*, which affects pigmentation, cellular response to UVR, and female fertility (Crawford et al. 2017).

An additional large-scale investigation into African pigmentation genetic diversity also confirmed the association of known as well as novel loci associated with pigmentary phenotypes in the region, highlighting the need for an expansion of pigmentation genetic studies beyond Europe (Martin et al. 2017). This is echoed in a recent GWAS in admixed populations from the Americas, which identified *MFSD12* as an important regulator of pigmentation variation in East Asian and Native American populations (Adhikari et al. 2019). This work also supports the argument that epistatic interactions among pigmentary genes can have significant impacts on pigmentation phenotypes (Adhikari et al. 2019).

2.7 SUMMARY

Pigmentation of the skin, hair, and eyes in humans is the result of a complex series of cellular processes that are centered around the production of melanin within melanocyte cells. The general mechanisms of melanin production are consistent across the skin, hair, and eyes, although some key differences can be found. First, in the skin, melanin production is more or less continuous and can be stimulated (at least in some people) by exposure to UVR, in what may be a short-term physiological adaptation to minimize subsequent UVR damage. In contrast, melanin production in hair cells is tied tightly to the hair growth cycle and ultimately declines or ceases with increasing age. Cellular processes in the iris differ from those in both the skin and hair in that iridial melanocytes do not form complexes with keratinocyte or other cells. Instead, mature melanosomes remain within iridial melanocytes. The shared cellular processes involved in pigmentation mean that many of the same genes influence pigmentation in the skin, hair, and eyes, although some of these genes may exhibit different patterns of expression in different cell types. Increasing research into the nature of how key pigmentary proteins interact with each other, particularly in more diverse populations, should help to improve our understanding of the complex nature of pigmentation variation. This will be particularly important as genome-wide surveys of variation identify novel genes and mutations that impact pigmentation. In addition to identifying such mutations, a key goal of these studies should be to functionally validate how these variants affect the cellular processes that regulate melanin production and melanosomal environment. Finally, a better understanding of the cellular and genetic processes regulating pigmentation diversity in globally diverse samples will improve our understanding of the evolutionary history of pigmentary phenotypes across the human species.

2.8 ACKNOWLEDGMENTS

I'm grateful to the editors for the invitation to participate in this volume. I also extend my gratitude to two anonymous reviewers for their valuable comments and suggestions, which have substantially improved this manuscript.

2.9 REFERENCES

Adhikari, K., J. Mendoza-Revilla, A. Sohail, M. Fuentes-Guajardo, et al. 2019. A GWAS in Latin Americans highlights the convergent evolution of lighter skin pigmentation in Eurasia. *Nat Comm* 10:358. doi:10.1038/s41467-018-08147-0.

Akey, J. M., H. Wang, M. Xiong, H. Wu, et al. 2001. Interaction between the Melanocortin-1 receptor and *P* genes contributes to inter-individual variation in skin pigmentation phenotypes in a Tibetan population. *Hum Gen* 108:516–20. doi:10.1007/s004390100524.

Ancans, J., D. J. Tobin, M. J. Hoogduijn, N. P. Smit, et al. 2001. Melanosomal PH controls rate of melanogenesis, eumelanin/phaeomelanin ratio and melanosome maturation in melanocytes and melanoma cells. *Exp Cell Res* 268:26–35. doi:10.1006/excr.2001.5251.

Bartölke, R., J. J. Heinisch, H. Wieczorek, and O. Vitavska. 2014. Proton-associated sucrose transport of mammalian solute carrier family 45: an analysis in *Saccharomyces cerevisiae*. *Biochem J* 464:193–201. doi:10.1042/BJ20140572.

Basu Mallick, C., F. M. Iliescu, M. Möls, S. Hill, et al. 2013. The light skin allele of SLC24A5 in South Asians and Europeans shares identity by descent. *PLoS Genetics* 9:e1003912. doi:10.1371/journal.pgen.1003912.

Beleza, S., A. M. Santos, B. McEvoy, I. Alves, et al. 2013. The timing of pigmentation lightening in Europeans. *Mol Biol Evol* 30:24–35. doi:10.1093/molbev/mss207.

Bellono, N. W., I. E. Escobar, A. J. Lefkovith, M. S. Marks, et al. 2014. An intracellular anion channel critical for pigmentation. *eLife* 3:e04543. doi:10.7554/eLife.04543.

Berson, J. F., D. C. Harper, D. Tenza, G. Raposo, et al. 2001. PMEL17 initiates premelanosome morphogenesis within multivesicular bodies. *Mol Biol of the Cell* 12:3451–64. doi:10.1091/mbc.12.11.3451.

Bin, B.-H., J. Bhin, S. H. Yang, M. Shin, et al. 2015. Membrane-associated transporter protein (MATP) regulates melanosomal PH and influences tyrosinase activity. *PLoS One* 10:e0129273. doi:10.1371/journal.pone.0129273.

Blum, H. F. 1961. Does the melanin pigment of human skin have adaptive value? An essay in human ecology and the evolution of race. *Quart Rev Biol* 36:50–63.

Boissy, R. E., Z. Huiquan, W. S. Oetting, L. M. Austin, et al. 1996. Mutation in and lack of expression of tyrosinase-related protein-1 (TRP-1) in melanocytes from an individual with brown oculocutaneous albinism: a new subtype of albinism classified as 'OCA3'. *Am J Hum Genet* 58:1145–56.

Bonilla, C., L.-A. Boxill, S. A. Mc Donald, T. Williams, et al. 2005. The 8818G allele of the agouti signaling protein (ASIP) gene is ancestral and is associated with darker skin color in African Americans. *Hum Gen* 116:402–6. doi:10.1007/s00439-004-1251-2.

Brace, S., Y. Diekmann, T. J. Booth, L. van Dorp, et al. 2019. Ancient genomes indicate population replacement in early Neolithic Britain. *Nat Ecol & Evol* 3:765–71. doi:10.1038/s41559-019-0871-9.

Branda, R. F., and J. W. Eaton. 1978. Skin color and nutrient photolysis: an evolutionary hypothesis. *Science* 201:625–6.

Brenner, M., and V. J. Hearing. 2008. The protective role of melanin against UV damage in human skin. *Photochem Photobiol* 84:539–49. doi:10.1111/j.1751-1097.2007.00226.x.

Brilliant, M. H. 2001. The mouse *p* (pink-eyed dilution) and human *P* genes, oculocutaneous albinism type 2 (OCA2) and melanosomal PH. *Pigment Cell Research* 14:86–93.

Cerqueira, C. C. S., V. R. Paixão-Côrtes, F. M. B. Zambra, F. M. Salzano, et al. 2012. Predicting *Homo* pigmentation phenotype through genomic data: from Neanderthal to James Watson. *Am J Hum Biol* 24:705–9. doi:10.1002/ajhb.22263.

Commo, S., O. Gaillard, and B. A. Bernard. 2004. Human hair greying is linked to a specific depletion of hair follicle melanocytes affecting both the bulb and the outer root sheath. *Brit J Dermatol* 150:435–43. doi:10.1046/j.1365-2133.2004.05787.x.

Costin, G., and V. J. Hearing. 2007. Human skin pigmentation: melanocytes modulate skin color in response to stress. *The FASEB Journal* 21:976–94. doi:10.1096/fj.06-6649rev.

Costin, G., J. C. Valencia, W. D. Vieira, L. Lamoreux, et al. 2003. Tyrosinase processing and intracellular trafficking is disrupted in mouse primary melanocytes carrying the Underwhite (Uw) mutation. A model for oculocutaneous albinism (OCA) Type 4. *J Cell Sci* 116:3203–12.

Crawford, N. G., D. E. Kelly, M. E. B. Hansen, M. H. Beltrame, et al. 2017. Loci associated with skin pigmentation identified in African populations. *Science* 358:eaan8433. doi:10.1126/science.aan8433.

Edwards, M., A. Bigham, J. Tan, S. Li, et al. 2010. Association of the OCA2 polymorphism His615Arg with melanin content in East Asian populations: further evidence of convergent evolution of skin pigmentation. *PLoS Genetics* 6:e1000867. doi:10.1371/journal.pgen.1000867.

Edwards, M., D. Cha, S. Krithika, M. Johnson, et al. 2016. Analysis of iris surface features in populations of diverse ancestry. *Roy Soc Open Sci* 3:150424. doi:10.1098/rsos.150424.

Edwards, M., A. Gozdzik, K. Ross, J. Miles, et al. 2012. Technical note: quantitative measures of iris color using high resolution photographs. *Am J Phys Anthropol* 147:141–9. doi:10.1002/ajpa.21637.

Eiberg, H., J. Troelsen, M. Nielsen, A. Mikkelsen, et al. 2008. Blue eye color in humans may be caused by a perfectly associated founder mutation in a regulatory element located within the HERC2 gene inhibiting OCA2 expression. *Hum Gen* 123:177–87. doi:10.1007/s00439-007-0460-x.

Fitzpatrick, T. B. 1965. Introductory lecture. In *Recent Progress in Photobiology*, ed. E. J. Bower, 365–73. New York: Academic Press.

Flanagan, N. 2000. Pleiotropic effects of the melanocortin 1 receptor (MC1R) gene on human pigmentation. *Hum Mol Gen* 9:2531–7. doi:10.1093/hmg/9.17.2531.

Frost, P. 1988. Human skin color: a possible relationship between its sexual dimorphism and its social perception. *Perspec Biol Med* 32:38–58. doi:10.1353/pbm.1988.0010.

Frost, P. 2014. The puzzle of European hair, eye, and skin color. *Advan Anthropol* 4:78–88. doi:10.4236/aa.2014.42011.

Fuller, B. B., D. T. Spaulding, and D. R. Smith. 2001. Regulation of the catalytic activity of preexisting tyrosinase in Black and Caucasian human melanocyte cell cultures. *Exp Cell Res* 262:197–208. doi:10.1006/excr.2000.5092.

Graf, J., R. Hodgson, and A. van Daal. 2005. Single nucleotide polymorphisms in the MATP gene are associated with normal human pigmentation variation. *Hum Mutation* 25:278–84. doi:10.1002/humu.20143.

Graf, J., J. Voisey, I. Hughes, and A. van Daal. 2007. Promoter polymorphisms in the *MATP* (*SLC45A2*) gene are associated with normal human skin color variation. *Hum Mutation* 28:710–17. doi:10.1002/humu.20504.

Haake, A., and K. Holbrook. 1999. The structure and development of skin. In *Fitzpatrick's Dermatology in General Medicine*, ed. I. Freedberg et al., 70–114. New York: McGraw-Hill.

Haitina, T., A. Ringholm, J. Kelly, N. I. Mundy, et al. 2007. High diversity in functional properties of melanocortin 1 receptor (MC1R) in divergent primate species is more strongly associated with phylogeny than coat color. *Mol Biol Evol* 24:2001–8. doi:10.1093/molbev/msm134.

Han, J., P. Kraft, H. Nan, Q. Guo, et al. 2008. A genome-wide association study identifies novel alleles associated with hair color and skin pigmentation. *PLoS Genetics* 4:e1000074. doi:10.1371/journal.pgen.1000074.

Holick, M. F. 2006. Resurrection of vitamin D deficiency and rickets. *J Clin Invest* 116:2062–72. doi:10.1172/JCI29449.

Iliescu, F. M., G. Chaplin, N. Rai, G. S. Jacobs, et al. 2018. The influences of genes, the environment, and social factors on the evolution of skin color diversity in India. *Am J Hum Biol* 30:e23170. doi:10.1002/ajhb.23170.

Ito, S., and K. Wakamatsu. 2011. Diversity of human hair pigmentation as studied by chemical analysis of eumelanin and pheomelanin: human hair pigmentation. *J Euro Acad Dermatol Venereol* 25:1369–80. doi:10.1111/j.1468-3083.2011.04278.x.

Jablonski, N., and G. Chaplin. 2000. The evolution of human skin coloration. *J Hum Evol* 39:57–106. doi:10.1006/jhev.2000.0403.

Jablonski, N., and G. Chaplin. 2010. Human skin pigmentation as an adaptation to UV radiation. *PNAS* 107:8962–8. doi:10.1073/pnas.0914628107.

Ju, D., and I. Mathieson. 2020. The evolution of skin pigmentation associated variation in West Eurasia. *PNAS* 118:e2009227118. doi.org/10.1073/pnas.2009227118

Kadekaro, A. L., S. Leachman, R. J. Kavanagh, V. Swope, et al. 2010. Melanocortin 1 receptor genotype: an important determinant of the damage response of melanocytes to ultraviolet radiation. *The FASEB Journal* 24:3850–60. doi:10.1096/fj.10-158485.

Kaidbey, K. H., P. P. Agin, R. M. Sayre, and A. M. Kligman. 1979. Photoprotection by melanin—a comparison of Black and Caucasian skin. *J Am Acad Dermatol* 1:249–60. doi:10.1016/S0190-9622(79)70018-1.

Kanetsky, P. A., J. Swoyer, S. Panossian, R. Holmes, et al. 2002. A polymorphism in the agouti signaling protein gene is associated with human pigmentation. *Am J Hum Gen* 70:770–5. doi:10.1086/339076.

Kayser, M., F. Liu, A. C. J. W. Janssens, F. Rivadeneira, et al. 2008. Three genome-wide association studies and a linkage analysis identify HERC2 as a human iris color gene. *Am J Hum Gen* 82:411–23. doi:10.1016/j.ajhg.2007.10.003.

Kenny, E. E., N. J. Timpson, M. Sikora, M.-C. Yee, et al. 2012. Melanesian blond hair is caused by an amino acid change in TYRP1. *Science* 336:554. doi:10.1126/science.1217849.

Keogh, E. V., and R. J. Wash. 1965. Rate of greying of human hair. *Nature* 207:877–8.

Kobayashi, N., A. Nakagawa, T. Muramatsu, Y. Yamashina, et al. 1998. Supranuclear melanin caps reduce ultraviolet induced DNA photoproducts in human epidermis. *J Investigative Dermatol* 110:806–10. doi:10.1046/j.1523-1747.1998.00178.x.

Kushimoto, T., V. Basrur, J. Valencia, J. Matsunaga, et al. 2001. A model for melanosome biogenesis based on the purification and analysis of early melanosomes. *PNAS* 98:10698–703. doi:10.1073/pnas.191184798.

Lalueza-Fox, C., H. Rompler, D. Caramelli, C. Staubert, et al. 2007. A melanocortin 1 receptor allele suggests varying pigmentation among Neanderthals. *Science* 318:1453–5. doi:10.1126/science.1147417.

Lamason, R. L. 2005. SLC24A5, a putative cation exchanger, affects pigmentation in zebrafish and humans. *Science* 310:1782–6. doi:10.1126/science.1116238.

Lao, O., J. M. de Gruijter, K. van Duijn, A. Navarro, et al. 2007. Signatures of positive selection in genes associated with human skin pigmentation as revealed from analyses of single nucleotide polymorphisms: signatures of positive selection at human pigmentation genes. *Ann Hum Gen* 71:354–69. doi:10.1111/j.1469-1809.2006.00341.x.

Lee, S. T., R. D. Nicholls, S. Bundey, R Laxova, et al. 1994. Mutations of the *P* gene in oculocutaneous albinism, ocular albinism, and Prader-Willi syndrome plus albinism. *New Eng J Med* 330:529–34.

Lin, M., R. L. Siford, A. R. Martin, S. Nakagome, et al. 2018. Rapid evolution of a skin-lightening allele in Southern African KhoeSan. *PNAS* 115:13324–9. doi:10.1073/pnas.1801948115.

Loomis, W. F. 1967. Skin-pigment regulation of vitamin-D biosynthesis in man. *Science* 157:501–6.

Madrigal, L., and W. Kelly. 2007. Human skin-color sexual dimorphism: a test of the sexual selection hypothesis. *Am J Phys Anthropol* 132:470–82. doi:10.1002/ajpa.20453.

Martin, A. R., M. Lin, J. M. Granka, J. W. Myrick, et al. 2017. An unexpectedly complex architecture for skin pigmentation in Africans. *Cell* 171:1340–53.e14. doi:10.1016/j.cell.2017.11.015.

Mathieson, I., I. Lazaridis, N. Rohland, S. Mallick, et al. 2015. Genome-wide patterns of selection in 230 ancient Eurasians. *Nature* 528:499–503. doi:10.1038/nature16152.

Mundy, N. I., and J. Kelly. 2006. Investigation of the role of the agouti signaling protein gene (ASIP) in coat color evolution in primates. *Mammalian Genome* 17:1205–13. doi:10.1007/s00335-006-0056-0.

Murray, F. G. 1934. Pigmentation, sunlight, and nutritional disease. *Am Anthropol* 36:438–45.

Myles, S., M. Somel, K. Tang, J. Kelso, et al. 2006. Identifying genes underlying skin pigmentation differences among human populations. *Hum Gen* 120:613–21. doi:10.1007/s00439-006-0256-4.

Newton, J. M., O. Cohen-Barak, N. Hagiwara, J. M. Gardner, et al. 2001. Mutations in the human orthologue of the mouse Underwhite gene (Uw) underlie a new form of oculocutaneous albinism, OCA4. *Am J Hum Gen* 69:981–8.

Norton, H. L., E. A. Correa, G. Koki, and J. S. Friedlaender. 2014. Distribution of an allele associated with blond hair color across Northern Island Melanesia. *Am J Phys Anthropol* 153:653–62. doi:10.1002/ajpa.22466.

Norton, H. L., R. A. Kittles, E. Parra, P. McKeigue, et al. 2006. Genetic evidence for the convergent evolution of light skin in Europeans and East Asians. *Mol Biol Evol* 24:710–22. doi:10.1093/molbev/msl203.

Oetting, W. S. 2000. The tyrosinase gene and oculocutaneous albinism type 1 (OCA1): a model for understanding the molecular biology of melanin formation. *Pigment Cell Research* 13:320–5. doi:10.1034/j.1600-0749.2000.130503.x.

Olalde, I., M. E. Allentoft, F. Sánchez-Quinto, G. Santpere, et al. 2014. Derived immune and ancestral pigmentation alleles in a 7,000-year-old Mesolithic European. *Nature* 507:225–8. doi:10.1038/nature12960.

Olalde, I, Selina Brace, Morten E. Allentoft, Ian Armit, Kristian Kristiansen, Thomas Booth, Nadin Rohland, et al. 2018. The Beaker phenomenon and the genomic transformation of northwest Europe. *Nature* 555(7695):190–6. doi:10.1038/nature25738.

Palmer, J. S., D. L. Duffy, N. F. Box, J. F. Aitken, et al. 2000. Melanocortin-1 receptor polymorphisms and risk of melanoma: is the association explained solely by pigmentation phenotype? *Am J Hum Gen* 66:176–86. doi:10.1086/302711.

Peles, D. N., L. Hong, D.-N. Hu, S. Ito, et al. 2009. Human iridal stroma melanosomes of varying pheomelanin contents possess a common eumelanic outer surface. *J Phys Chem B* 113:11346–51. doi:10.1021/jp904138n.

Pośpiech, E., A. Wojas-Pelc, S. Walsh, F. Liu, et al. 2014. The common occurrence of epistasis in the determination of human pigmentation and its impact on DNA-based pigmentation phenotype prediction. *Forensic Sci Int Genet* 11:64–72. doi:10.1016/j.fsigen.2014.01.012.

Prota, G., D.-N. Hu, M. R. Vincensi, S. A. McCormick, et al. 1998. Characterization of melanins in human irides and cultured uveal melanocytes from eyes of different colors. *Exp Eye Res* 67:293–9. doi:10.1006/exer.1998.0518.

Puri, N., J. M. Gardner, and M. H. Brilliant. 2000. Aberrant PH of melanosomes in pink-eyed dilution (p) mutant melanocytes. *J Invest Dermatol* 115:607–13. doi:10.1046/j.1523-1747.2000.00108.x.

Quillen, E. E., H. L. Norton, E. J. Parra, F. Lona-Durazo, et al. 2019. Shades of complexity: new perspectives on the evolution and genetic architecture of human skin. *Am J Phys Anthropol* 168:4–26. doi:10.1002/ajpa.23737.

Rana, B. K., D. Hewett-Emmett, L. Jin, N. Sambuughin, et al. 1999. High polymorphism at the human melanocortin 1 receptor locus. *Genetics* 151:1547–57.

Rawofi, L., M. Edwards, S. Krithika, P. Le, et al. 2017. Genome-wide association study of pigmentary traits (skin and iris color) in individuals of East Asian ancestry. *PeerJ* 5:e3951. doi:10.7717/peerj.3951.

Rees, J. L. 2000. The melanocortin 1 receptor (MC1R): more than just red hair. *Pigment Cell Res* 13:135–40. doi:10.1034/j.1600-0749.2000.130303.x.

Relethford, J. H. 1997. Hemispheric difference in human skin color. *Am J Phys Anthropol* 104:449–57.

Rogasevskaia, T. P., R. T. Szerencsei, A. H. Jalloul, F. Visser, et al. 2019. Cellular localization of the K^+-dependent Na^+–Ca^{2+} exchanger NCKX 5 and the role of the cytoplasmic loop in its distribution in pigmented cells. *Pigment Cell Melanoma Res* 32:55–67. doi:10.1111/pcmr.12723.

Rosemblat, S., D. Durham-Pierre, J. M. Gardner, Y. Nakatsu, et al. 1994. Identification of a melanosomal membrane protein encoded by the pink-eyed dilution (type II oculocutaneous albinism) gene. *PNAS* 91:12071–5. doi:10.1073/pnas.91.25.12071.

Sabeti, P. C., P. Varilly, B. Fry, J. Lohmueller, et al. 2007. Genome-wide detection and characterization of positive selection in human populations. *Nature* 449:913–18. doi:10.1038/nature06250.

Shriver, M. D., E. J. Parra, S. Dios, C. Bonilla, et al. 2003. Skin pigmentation, biogeographical ancestry and admixture mapping. *Hum Gen* 112:387–99. doi:10.1007/s00439-002-0896-y.

Slominski, A., and R. Paust. 1993. Melanogenesis is coupled to murine anagen: toward new concepts for the role of melanocytes and the regulation of melanogenesis in hair growth. *J Invest Dermatol* 101:8.

Slominski, A., J. Wortsman, P. M. Plonka, K. U. Schallreuter, et al. 2005. Hair follicle pigmentation. *J Invest Dermatol* 124:13–21. doi:10.1111/j.0022-202X.2004.23528.x.

Stevens, A., and J. Lowe. 2004. *Human Histology*, 3rd ed. St. Louis: Mosbey.

Stokowski, R. P., P. V. K. Pant, T. Dadd, A. Fereday, et al. 2007. A genomewide association study of skin pigmentation in a South Asian population. *Am J Hum Gen* 81:1119–32. doi:10.1086/522235.

Sturm, R. A., D. L. Duffy, Z. Z. Zhao, F. P. N. Leite, et al. 2008. A single SNP in an evolutionary conserved region within Intron 86 of the HERC2 gene determines human blue-brown eye color. *Am J Hum Gen* 82:424–31. doi:10.1016/j.ajhg.2007.11.005.

Szabó, G. 1954. The number of melanocytes in human epidermis. *Brit Med J* 1:1016.

Tadokoro, R., H. Murai, K.-I. Sakai, T. Okui, et al. 2016. Melanosome transfer to keratinocyte in the chicken embryonic skin is mediated by vesicle release associated with rho-regulated membrane blebbing. *Sci Rep* 6:38277. doi:10.1038/srep38277.

Tadokoro, R., and Y. Takahashi. 2017. Intercellular transfer of organelles during body pigmentation. *Curr Opin Gen Dev* 45:132–8. doi:10.1016/j.gde.2017.05.001.

Thong, H.-Y., S.-H. Jee, C.-C. Sun, and R. E. Boissy. 2003. The patterns of melanosome distribution in keratinocytes of human skin as one determining factor of skin colour. *Brit J Dermatol* 149:498–505. doi:10.1046/j.1365-2133.2003.05473.x.

Toda, K., M. A. Pathak, J. A. Parrish, and T. B. Fitzpatrick. 1972. Alterations of racial differences in melanosome distribution in human epidermis after exposure to ultraviolet light. *Nat New Biol* 236:143–5.

Valverde, P. 1996. The Asp84Glu variant of the melanocortin 1 receptor (MC1R) is associated with melanoma. *Hum Mol Gen* 5:1663–6. doi:10.1093/hmg/5.10.1663.

Valverde, P., E. Healy, I. Jackson, J. L. Rees, et al. 1995. Variants of the melanocyte-stimulating hormone receptor gene are associated with red hair and fair skin in humans. *Nat Genet* 11:328–30.

Velden, P., A. van der, L. A. Sandkuijl, W. Bergman, S. Pavel, et al. 2001. Melanocortin-1 receptor variant R151C modifies melanoma risk in Dutch families with melanoma. *Am J Hum Gen* 69:774–9. doi:10.1086/323411.

Wakamatsu, K., D.-N. Hu, S. A. McCormick, and S. Ito. 2007. Characterization of melanin in human iridal and choroidal melanocytes from eyes with various colored irides: melanin in human uveal melanocytes. *Pigment Cell & Melanoma Research* 21:97–105. doi:10.1111/j.1755-148X.2007.00415.x.

Walsh, S., L. Chaitanya, K. Breslin, Ch. Muralidharan, et al. 2017. Global skin colour prediction from DNA. *Hum Gen* 136:847–63. doi:10.1007/s00439-017-1808-5.

Walsh, S., A. Lindenbergh, S. B. Zuniga, T. Sijen, et al. 2011. Developmental validation of the IrisPlex system: determination of blue and brown iris colour for forensic intelligence. *Foren Sci Int Gen* 5:464–71. doi:10.1016/j.fsigen.2010.09.008.

Walsh, S., F. Liu, A. Wollstein, L. Kovatsi, et al. 2013. The IrisPlex system for simultaneous prediction of hair and eye colour from DNA. *Foren Sci Int Gen* 7:98–115. doi:10.1016/j.fsigen.2012.07.005.

Whiteman, D. C., P. G. Parsons, and A. C. Green. 1999. Determinants of melanocyte density in adult human skin. *Arch Dermatol Res* 291:511–16. doi:10.1007/s004030050446.

Wilkerson, C. L. 1996. Melanocytes and iris color: light microscopic findings. *Arch Ophthalmol* 114:437. doi:10.1001/archopht.1996.01100130433014.

Yamaguchi, K., C. Watanabe, A. Kawaguchi, T. Sato, et al. 2012. Association of melanocortin 1 receptor gene (MC1R) polymorphisms with skin reflectance and freckles in Japanese. *J Hum Gen* 57:700–8. doi:10.1038/jhg.2012.96.

Yang, Z., H. Zhong, J. Chen, X. Zhang, et al. 2016. A genetic mechanism for convergent skin lightening during recent human evolution. *Mol Biol Evol* 33:1177–87. doi:10.1093/molbev/msw003.

3 Cell Processes Underpinning the Evolution of Primate Dental Form and Formula

Cassy M. Appelt, Elsa M. Van Ankum*, Denver F. Marchiori, and Julia C. Boughner*

CONTENTS

3.1 Introduction 55
 3.1.1 Evolutionary Origins of Primate Tooth Morphology and Dental Formula 55
 3.1.2 Primitive Dental Formulae, Evolution, and Tooth Loss in Vertebrates 56
 3.1.3 Evo-Devo Dental Anthropology 59
3.2 A Brief Overview of Early Tooth Development in Mammals 60
3.3 Crown Morphology Appears More Labile among Primate Ante-Molar versus Molar Teeth 64
3.4 Variation in Primate Dental Formula: Comparing Prosimians with Anthropoids 65
 3.4.1 Mechanisms of Evolutionary Tooth Loss in the Primate Dental Formula 68
 3.4.2 Divergent Evolution of Upper and Lower Dental Formulae and Phenotypes, Notably in Prosimians 70
3.5 Evo-Devo of the Primate Dental Lamina 71
3.6 Conclusions 74
3.7 Acknowledgments 74
3.8 References 74

3.1 INTRODUCTION

3.1.1 Evolutionary Origins of Primate Tooth Morphology and Dental Formula

Euprimates appear in the fossil record during the Eocene epoch, ~55 million years ago (mya) (Pozzi et al. 2014). These stem group primates share a number of

* Co-first author.

characteristics with one another, such as grasp-leaping locomotion; (proto)diurnal activity patterns (Ankel-Simons and Rasmussen 2008); and an insectivorous dentition including large canines, sharp premolar and molar cusps, and spatulate incisors (Ni et al. 2013). Arboreal adaptations and insectivorous dental traits, among other characteristics, are shared between primates and their sister groups, *Scandentia* (tree shrews), *Dermoptera* (colugos), and the extinct Plesiadapimorphes (Regan et al. 2001; Sussman et al. 2013). Disagreement persists around which traits distinguish early true primates from closely related groups, particularly the plesiadapiforms (Moffat 2002). Nonetheless, consensus about the emergence of euprimates includes rapid diversification—that coincided and potentially co-evolved with flowering and fruiting flora—following the Paleocene–Eocene thermal maximum (Gingerich 2012; Sussman et al. 2013). While fossils as old as ~57 million years are widely accepted as true primates, molecular studies place primate origins in the early Cretaceous (Pozzi et al. 2014; Herrera and Dávalos 2016), with the divergence of strepsirrhines (lemurs, lorises, galagos) and haplorhines (tarsiers, catarrhines, platyrrhines) at ~80 mya (Martin 1990; Ni et al. 2013). The gap between fossil record and molecular clock estimates equates to ~20–25 million years of primate evolution yet to be clarified (Pozzi et al. 2014). The diversification of extant primate groups occurred partially in the Eocene, when strepsirrhines diverged from the haplorhines. Near the Eocene boundary, within Haplorhini, catarrhines (African and Eurasian anthropoids) and platyrrhines (American monkeys) diverged from one another, while later, in the Oligocene-Miocene, the hominoid lineage (apes, humans) diverged (reviewed in Gingerich 2012).

The dental formulae of extant primates derive from the primitive eutherian mammal heterodont formula of three incisors, one canine, four premolars, and three molars in each quadrant of the upper and lower jaws (i.e., I3/3 C1/1 P4/4 M3/3; upper/lower teeth) (Myers et al. 2020). Primate dental formulae range from, for example, I2/2 C1/1 P3/3 M3/3 in Lemuridae, to I1/1 C0/0 P1/0 M3/3 for Daubentoniidae, to I2/2 C1/1 P2/2 M3/3 in Catarrhini. Dental formulae for all major primate groups are listed in Figure 3.1 and Appendix 3.1. Prosimians appear to show the greatest variation in dental formula. Studies of dental evolution hinge on defining homology among tooth classes, yet homology may be defined in different ways. Classical ways include comparative tooth morphology, function, and position relative to other teeth in the dental arch. More contemporary ways include comparative embryology, timing and process of tooth morphogenesis, and pattern of gene expression (Hautier et al. 2016). Particularly for premolar and molar evolution, theories vary based on definitions of homology, including across different generations of teeth (reviewed in Butler and Clemens 2001; Peterkova et al. 2006).

3.1.2 Primitive Dental Formulae, Evolution, and Tooth Loss in Vertebrates

However homology is defined, within vertebrate dental formulae, tooth loss dominates the fossil record, with wholesale loss of teeth in birds, myriad amphibians, and reptiles and partial edentulism in other groups including, for example, frogs that have lost the mandibular dentition (Davit-Béal et al. 2009). Ultimately, one rule may not fit all (Schwartz 1982): among diversifying groups of early mammals, different dental formulae may have evolved via the loss—and, albeit less often, the gain—of different tooth

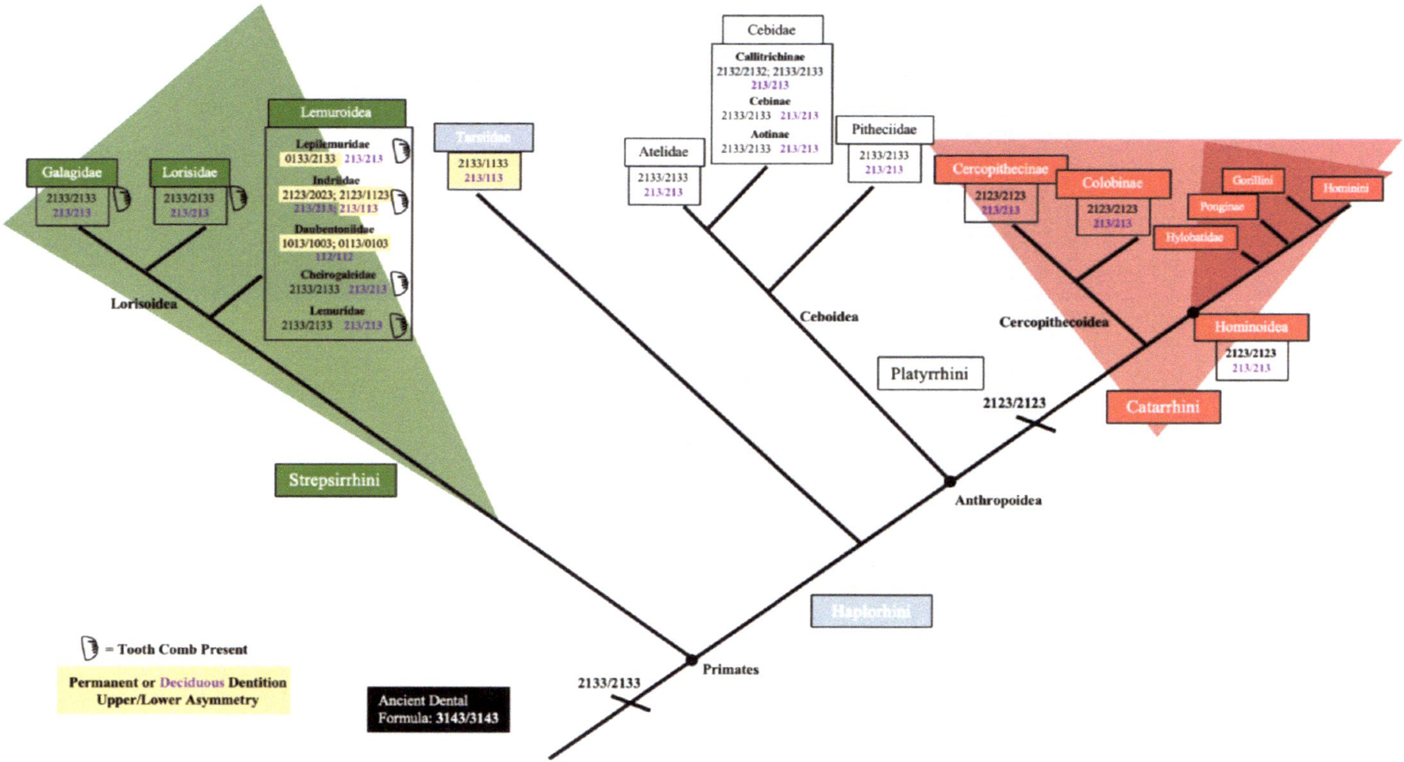

FIGURE 3.1 Phylogeny of living primates, summarizing permanent (black text) and deciduous (purple text) dental formulae, highlighting asymmetry in upper/lower dental formulae (highlighted), and indicating presence or absence of a dental comb ('comb' graphic).

homologues (Luo et al. 2004; Davit-Béal et al. 2009). In this chapter, we use conventional nomenclature and homology whilst appreciating that these definitions are debated, and we return to this debate later in our discussions. We adopt long-held assumptions about tooth generation, class, and identity while recognizing the value of inquiries along the lines of, "If it looks like an incisor, and acts like an incisor, then is it really a canine?"

Primates are vertebrates, and our earliest vertebrate relations are fishes, in whom the evolution of a hinged jaw about 400 million years ago was so immensely advantageous that to this day almost all vertebrate groups are jawed (Brazeau and Friedman 2015). Versus a mouth moved by suction only, a hinged jaw is voluntarily opened and closed. Its defining feature is a lower jaw skeleton, or mandible, attached to the cranial base at a joint—the temporomandibular joint in mammals. The mandible and jaw joint are evolutionary novelties compared to the more ancient upper jaw (Depew et al. 2002, 2005; Kuratani et al. 2016). Exactly when and how teeth appeared in the vertebrate jaw is debated due to gaps in the fossil record (reviewed in Donoghue and Rücklin 2016). Nonetheless, the incorporation of teeth within a kinetic jaw skeleton was strongly selected for by clear functional advantages in eating, hunting, and defense. The jaw and its dentition are tightly integrated via deeply conserved developmental processes but also show clear developmental autonomy that supports the fossil evidence for discreet evolutionary origins.

Teeth evolve according to functional demands of eating, as well as para-functions such as grooming (Butler 2000). The macroevolution of tooth morphology seems particularly conspicuous in primate ante-molar teeth: incisors, canines, and premolars. Primate upper and lower dental formulae typically mirror each other, but the few exceptions to this rule seem to show more tooth loss and greater morphological variation in the lower dentition (Table 3.1, Figure 3.1). The most ancient extant

TABLE 3.1
Summary of Asymmetry (Bold Font) in Primate Dental Formulae between Upper and Lower Dentitions (i.e., Upper/Lower Tooth) and/or between Deciduous and Permanent Dentitions

Family	Dental Formula
Indriidae	Permanent: **I2/1** C1/1 P2/2 M3/3, or I2/2 **C1/0** P2/2 M3/3
	Deciduous: di1/1 dc1/1 dp2/2 or **di1/0** dc1/1 dp2/2
Daubentoniidae (Strepsirrhini)	Permanent: I1/1 C0/0 **P1/0** M3/3, or I0/0 C1/1 **P1/0** M3/3
	Deciduous: di1/1 dc1/1 dp2/2
Tarsiidae	Permanent: **I2/1** C1/1 P3/3 M3/3
	Deciduous: **di2/i** dc1/1 dp2/2
Lemuridae (Strepsirrhini), Callithricidae, Cebidae, Cercopithecoidea, and all apes	No upper/lower asymmetry in either the permanent or deciduous dentitions, with the exception of Lepilemuridae, Permanent: **I0/2** C1/1 P3/3 M3/3

Sources: Information about formulae collected from Ankel-Simons (1996, 2007), Berkovitz and Shellis (2016), Cuozzo and Yamashita (2006), Guthrie and Frost (2011), Hershkovitz (1977), Luckett and Maier (1982), Martin (1990), Swindler (2002), Smith et al. (2015), and Tattersall (1982).

primate group, prosimians, appears to boast the most variation in terms of dental formula if not also tooth form. Together, these observations hint at weaker developmental constraint of the primate lower dentition compared to the upper dentition. Further, these observations imply stronger canalization and developmental stability [i.e., the tendencies of a developmental process to remain unchanged under different, or under same, conditions, respectively (Hallgrímsson et al. 2002)] of dental phenotype in the more recently evolved anthropoid primates.

3.1.3 Evo-Devo Dental Anthropology

The field of evolutionary developmental biology (evo-devo) uses insights about animal development to explain the molecular, cellular, and developmental processes facilitating animal evolution. While evo-devo as a discipline traces its origins to the 1800s, only in recent decades have biological anthropologists applied an evo-devo framework to studies of primate evolutionary biology (reviewed in Boughner and Rolian 2016). Evo-devo work relies on comparative embryology of animal models, including vertebrate groups such as fish, frog, snake, bird, alligator, lizard, and mammal. Mouse is the classic experimental model for mammalian development and disease, and much of what we understand about primate development stems from studies in rodents. As such, most research cited in this chapter is derived from mouse-based studies. However, relative to understanding primate dental development, the mouse model carries some shortcomings that include a rapid-forming monophyodont dentition that lacks premolars and canines and boasts an extensive diastema between incisors and molars (Stembírek et al. 2010). Nonetheless, the developmental-genetic underpinnings of teeth are deeply conserved among vertebrates, even more so within mammals (Jernvall and Thesleff 2000, 2012). As such, it is productive to extrapolate these underpinnings from mouse to primate, including fossil primates.

Among the many mouse (e.g., Anderson et al. 2014) and other vertebrate models (e.g., Lainoff et al. 2015; Hammer et al. 2016) that have helped clarify the evo-devo integration of teeth within jaws, the transgenic mouse mutant for tumor protein 63 (Tp63) (Mills et al. 1999) offers some unique insights. In this *Tp63*-null mouse, tooth development arrests shortly after it begins, and teeth subsequently fail to form, while jaw development proceeds. The upper jaw skeleton suffers some malformations including midfacial clefts (Phen et al. 2018); however, the lower jaw develops virtually unperturbed (Paradis et al. 2013). These findings imply that a Tp63-driven gene regulatory network is required for tooth but not jaw development (Raj and Boughner 2016; Rostampour et al. 2019). These results also suggest that, compared to the upper jaw and teeth, the mandible and its dentition are more labile developmentally, and evolutionarily, since the effects of Tp63 are not pleiotropic: lower dental phenotype can change without obligating change in mandibular phenotype. As biological anthropologists, we remain curious about the extent to which what we have learned from the *Tp63*-null mouse applies to the evo-devo of primate dental formula and tooth form, and so we explore these ideas further here.

In this chapter, we survey primate dental formulae of deciduous and permanent dentitions alongside phenotypic variation in primate tooth morphology. We hypothesize that, compared to the dentition of the upper jaw, the lower dentition is more developmentally flexible and thus more evolvable. Related to this idea, we also hypothesize that, particularly within the lower dentition, ante-molar teeth are more adaptable in terms of their morphological variation across primates. We focus our exploration on extant primates with notes on fossil taxa. We conclude with hypotheses about the evo-devo of the primate dental lamina, specifically the cellular and developmental mechanisms that generate variation in tooth number, size, and shape and that may be targeted by natural selection for dietary and other parafunctional adaptations (summarized in Figure 3.2).

3.2 A BRIEF OVERVIEW OF EARLY TOOTH DEVELOPMENT IN MAMMALS

In dentate vertebrates, teeth develop from a structure called the dental lamina—a thin ribbon of epithelium lining the presumptive tooth-bearing region of the jaw—that becomes seeded with the instructions required to form primary, successional, and additional teeth (reviewed in Peterkova et al. 2014). Mammalian, including primate, teeth develop when oral epithelium is coded to become dental in nature and proceeds to reiteratively signal with underlying neural crest-derived ectomesenchyme (reviewed in Jussila and Thesleff 2012). The reciprocal signaling between dental epithelium and mesenchyme is imperative for the proper morphogenesis and histogenesis of a tooth (Tucker and Sharpe 2004). This reciprocal signaling is triggered by the formation of a dental placode, which initiates odontogenesis and is visible as a thickening of the epithelium. This thickening is the first of four classic stages of tooth morphogenesis that include bud, cap, and bell (early, and late) stages, summarized here with reference to the comprehensive review by (Catón and Tucker 2009). To form the tooth bud, the cells of the dental epithelium proliferate and invaginate the underlying mesenchyme, where cells migrate and condense around the epithelial bud. The tooth bud differentiates into a tooth organ with distinct parts, including the inner and outer enamel epithelium, stratum intermedium, and stellate reticulum that together form the enamel organ that helps shape the tooth crown. Directly beneath the enamel organ is another part of the tooth organ, the dental papilla, which also helps maintain tooth organ shape during morphogenesis. During the cap stage, a structure named the primary enamel knot will form and trigger crown formation. Within this specialized condensation of enamel knot cells, there is no proliferation, only signaling. During the early bell stage, in a tooth organ destined to become a multicuspid tooth type (premolar; molar), secondary enamel knots form, each one at the location of a future cusp. Placodes and enamel knots are transient and disappear via programmed cell death (apoptosis), a process important to sculpting tooth size and shape, organizing the deposition of enamel and dentine, and regulating the tooth-bone interface as a tooth organ grows and, later, erupts (Matalova et al. 2004, 2012). During the late bell stage, histogenesis begins when cells from the ectomesenchyme-derived dental papilla differentiate into odontoblasts, which start secreting dentine. A bit later, cells from the ectodermal epithelium-derived inner enamel organ differentiate into ameloblasts, which start secreting enamel. Variation in rates of cell mitosis and cell differentiation, as well as actomyosin-driven change

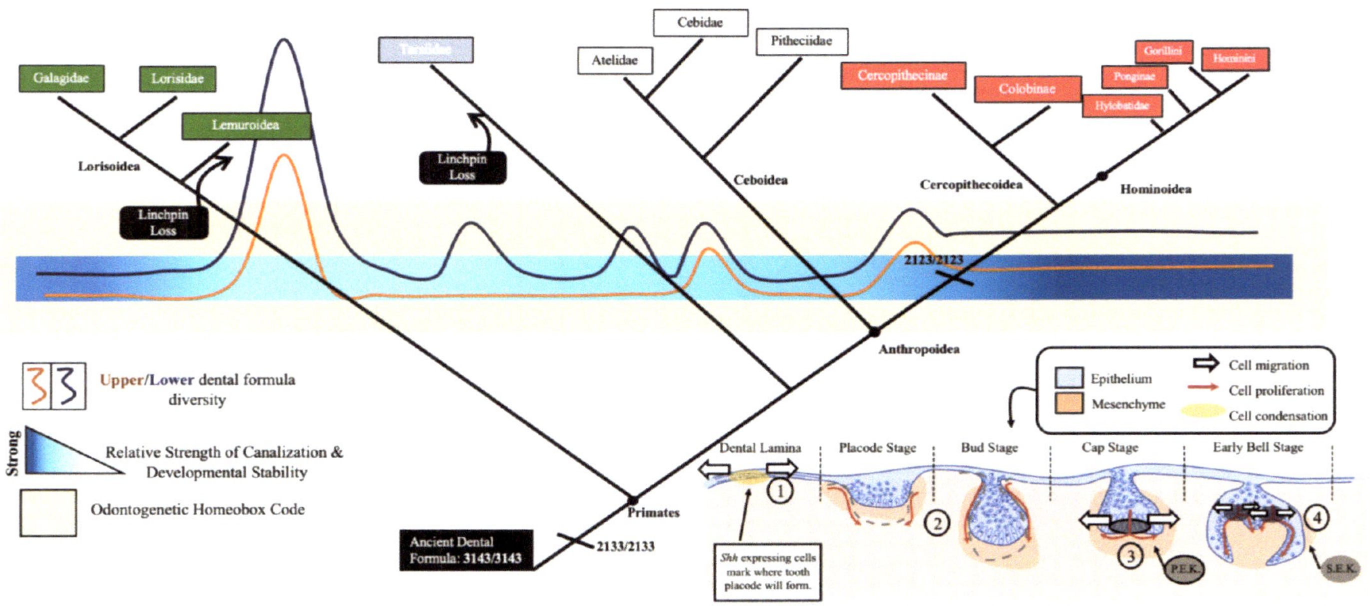

FIGURE 3.2 Phylogeny of living primates summarizing proposed cellular mechanisms acting during primate dental evolution and diversification. Dental formula diversity (blue/lower and orange/upper wavy lines) is high in strepsirrhines (i.e., Lemuroidea, indicated by a large wave amplitude), with fewer instances of tooth loss from dental formulas in haplorhines. Some tooth number losses may be explained by the Linchpin Hypothesis: the loss of a deciduous precursor leads to loss of its permanent successor ("Linchpin Loss"). Cellular mechanisms for dental diversification include weaker developmental constraint (blue gradient bar) in *Strepsirrhini* and stronger canalization and developmental stability within *Haplorhini*. Lability of dental phenotypes such as cusp or crown morphology likely results from strong evolutionary selection to facilitate niche exploitation, dietary changes, and other parafunctions like social grooming. Tinkering (Jernvall and Salazar-Ciudad 2007) of the genetically conserved odontogenic homeobox code (yellow rectangle) likely underpins diversity in tooth morphology by altering: cell migration in the dental lamina (#1, tooth development graphic in lower right); cell proliferation, condensation, and movement that varies the shape and size of dental placodes and bud-stage tooth organs (#2, placode and bud); the position of the primary enamel knot (P.E.K.) during cap stage (#3) and secondary enamel knots (S.E.K.) during bell stage (#4, early and late). Multicuspid teeth develop secondary enamel knots, but unicuspid teeth do not; the stages of odontogenesis represented here depict classic stages of mammalian molar development.

in cell shape (Yamada et al. 2019), cause the folding of the inner enamel epithelium, which now serves as the border between the enamel and dentine, named the enamel-dentine junction, or EDJ. This EDJ contributes to occlusal crown morphology (Guy et al. 2015) and may be sufficiently specific to be used to identify fossil primate taxa such as Neanderthals (Martin et al. 2017). The mesenchymal cells of a tooth organ differentiate into myriad cell types, including cementoblasts, periodontal ligament cells, and alveolar osteoblasts, that contribute to various distinct structures in the complete, erupted tooth (reviewed in Krivanek et al. 2017). In sum, these systematic odontogenic stages and cellular processes are all putative targets of dental macroevolution (Jernvall and Salazar-Ciudad 2007; Jernvall and Thesleff 2000, 2012).

Primates are diphyodont: they have two dentitions (Swindler 2002). The first is the primary, or deciduous ("milk"), dentition composed of incisors, canines, and molars—although some argue that based on developmental homology, these deciduous molars (dms) are actually deciduous premolars (dps) (reviewed in Peterkova et al. 2006). Of course, this argument of dms actually being dps is more difficult to make on the basis of tooth morphology, considering the clear similarities in crown and root forms between dms and permanent molars. We will return to this debate towards the end of this chapter. The deciduous dentition is replaced by the second, or permanent ("adult"), dentition, which is also referred to as "successional" because these incisors, canines, and premolars directly develop from and later "succeed" their deciduous predecessors. The deciduous dentition buds directly off the dental lamina. The permanent successional dentition (i.e., incisors, canines, and premolars) buds off a successional dental lamina that branches from the deciduous dental tissues that produced the deciduous incisors (di), canines (dc) and molars [dm; or perhaps dp (Ooë 1979)]. With regard to the human dental lamina in particular, while Ooë (1979) considered the permanent molars (M1, M2, M3) members of the deciduous dentition, others refer to these three molars as "additional" teeth because they develop *de novo* from the backward-growing tail of the dental lamina (Berkovitz et al. 2009; Juuri et al. 2013; Gaete et al. 2015; reviewed in Peterkova et al. 2014). Regardless of preferences about nomenclature, in primates, the permanent, unreplaced set of teeth is the last to form; work in minipig and other mammals suggests that after the initiation of the successional dental lamina, it degrades and thus cannot give rise to subsequent generations of teeth (Luo et al. 2004; Buchtová et al. 2012).

In humans, mice, and other mammals such as ferret (Line 2003; Jussila et al. 2014) and wallaby (Nasrullah et al. 2018), the dental lamina begins as a thin, elongated line of cells (van Nievelt and Smith 2005), likely derived from at least one founder cell population that migrates anteriorly from its origin near the presumptive joint region of the embryonic jaw (Prochazka et al. 2015). Work in mouse indicates that the mammalian dental lamina has a distinct molecular fingerprint, including markers such as Sox2, Pitx2, and Shh (reviewed in Balic and Thesleff 2015). In contrast to classical descriptions of the dental lamina as a simple, continuous, U-shaped structure, contemporary studies in mouse and human indicate that the dental lamina develops via series of bumps along oral epithelium and that the dental lamina is morphologically variable along its anterior to posterior length (Peterkova et al. 2014). Because of technical challenges to visualizing dental lamina formation, much remains to be learned about exactly how this lamina forms, what it looks like, and how different tooth classes develop from within it. With mouse as the longstanding principal mammalian model system of odontogenesis, only incisor and molar morphogenesis have been characterized in depth. Rudiments of premolars form in the mouse dental lamina (Peterkova et al. 2014), but the earliest embryonic origin of these rudiments is unknown. For tooth classes such as premolars, the dental lamina may form coincident with the placodes that give rise to individual tooth organs (Stock et al. 1997; Jussila et al. 2014). Tooth initiation along the dental lamina is not always unidirectional: for example, in a ferret model, premolars develop in the order P3 P4 P2 (Stock et al. 1997; Jussila et al. 2014). In some mammals, incisors initiate in opposite directions away from the first-forming incisor, which is assumed to be the stem progenitor of that tooth class (Osborn 1973; De Coster et al. 2008).

Sometime before and/or during the development of dental lamina, the precise and complex expression of overlapping sets of genes establishes the identities of tooth classes. Although first characterized in mouse, this odontogenic homeobox code of gene expression appears to be deeply conserved among mammals (Sharpe 2000; McCollum and Sharpe 2001; Wakamatsu et al. 2019) and has been used to model primate dental evolution, including in fossil hominines such as *Paranthropus* (McCollum and Sharpe 2001). The odontogenic code may set gradually, allowing time and opportunity for genetic, cellular, and phenotypic variation as the code crystallizes (Cobourne and Mitsiadis 2006; Townsend et al. 2009). Variation in the odontogenic homeobox code may alter numbers, sizes, and shapes of teeth within a given class (Ahn et al. 2010), while multiple downstream molecular and cellular factors fine-tune primate tooth morphology (Charles et al. 2009; Jernvall and Thesleff 2012). For instance, molar cusp morphology is integral to primate dietary adaptations: insectivore molars have tall, sharp crowns to pierce insect exoskeletons; frugivore molars have shorter shearing crests and small, low crowns for mashing pulp; and folivore molars are large with well-developed leaf-shearing crests (Butler 2000; Swindler 2002). In modern humans, our low-cusped, relatively small molars are well adapted to an omnivorous diet, enabling our ancestors' success in a diverse range of biomes. Variation in the size and location of dental placodes, as well as primary and secondary enamel knots, alters cell proliferation in epithelium and underlying mesenchyme to change the topography of a tooth crown (reviewed in Townsend et al. 2009; Jernvall and Thesleff 2012; Salazar-Ciudad and Marín-Riera 2013), as do cytoskeletal and corresponding cell shape changes (Yamada et al. 2019).

Body size predicts facial length, including jaw length (Cardini and Polly 2013), implying that dental lamina length also tracks facial length and body size: for example, note that the permanent dental formulae (I/C/P/M) of the smallest versus largest living primates, *Microcebus* and *Gorilla*, are almost identical (2/1/3/3 vs. 2/1/2/3, respectively) despite a huge disparity in average body size. Work in mouse suggests that dental lamina length is checked by the length of the oral epithelium from which the lamina derives (Ko et al. 2021 (a full length research article in the J. of Dev. Biol.)). We posit that the dental lamina's intrinsic molecular coding and cellular dynamics must scale up and down, reflecting both strong lability and conservation of this critical precursor to the primate dentition. Classic experimental work supports this idea about scalability: for example, a bisected cultured rabbit molar organ developed into two complete, distinct molars (Glasstone 1962). Natural experiments in humans with cleft palate also show duplicate lateral incisors where a single incisor primordia was bisected (Hovorakova et al. 2018). The dental lamina needs space to elongate into, but a complete heterodont dentition can develop in a tiny length of space, indicating that the odontogenic homeobox code is consistent across primates and that cell number and proliferation rate vary instead to produce teeth and dental arches of different sizes concomitant with different jaw lengths across primate species. Altogether, it is apparent from studies of early tooth morphogenesis across primates and other mammals that the dental lamina is developmentally pliant and variable and thus evolutionarily adaptable.

The development, structure, and function of the dental lamina seems conserved, but not identical, between upper and lower dentitions. Mouse and human show differences in the morphology of the upper versus lower, as well as anterior versus posterior, regions of the dental lamina (Hovorakova et al. 2005, 2007). Among mammals, each side of the upper jaw develops from three embryonic facial primordia, whereas the lower jaw develops from a single primordium. This type of structural variation in the dental lamina has been shown to alter upper dental development (discussed further below). The signaling pathways and gene regulatory networks that drive odontogenesis are comparable between upper and lower dentitions; however, there are key differences in molecular patterning. For example, in mouse, Pitx1 is expressed only in the embryonic mandible, and experimental deletion of Pitx1 expression has jaw-specific effects on downstream targets such as Barx1, resulting in distinct lower versus upper molar phenotypes (reviewed in Ramanathan et al. 2018). Other genes, such as *Dlx5/6*, that are critical for specifying jaw identity (e.g., mandibular fate in the case of *Dlx5/6*) may have been co-opted to also specify lower versus upper tooth identity (Cobourne and Mitsiadis 2006). In this way, the evo-devo origins of the jaws—including the discrete homeobox codes specific to the upper versus lower jaw—help pattern dental phenotype.

3.3 CROWN MORPHOLOGY APPEARS MORE LABILE AMONG PRIMATE ANTE-MOLAR VERSUS MOLAR TEETH

In primate and other mammals, biomechanics related to jaw width, length, gape, and bite force both enable and constrain specialization of tooth form depending on a tooth's location in the dental arch (reviewed in Greaves 1991; Ross and Iriarte-Diaz 2019). The example of a primate sectorial premolar complex is vital not only for eating but also for sexual selection and mate competition, reflecting the multifactorial selective pressures that literally shape the dentition. The highest bite forces along the primate dental arch are closest to the jaw joint, along the molar row, while anterior teeth located furthest from the jaw joint are relieved of the highest bite forces (Helkimo et al. 1977; Spencer 1998; Ferrario et al. 2004). During mastication, M1 has highest bite force, with decreasing force in both directions (Spencer 1998; Ferrario et al. 2004). As a result, ante-molar teeth are afforded greater ranges of motion and (para)function compared to the molars (Brown 1974; Butler 2000).

In the case of a primate dental comb, the lower canines are incisiform, effectively working as extra lateral incisors that contribute to the formation of the procumbent comb. Also noted previously, dental combs are used for grooming, an important social behavior that rids prosimians of pests and builds political alliances, as well as for scraping away tree bark to eat underlying gum and resin (Swindler 2002). Adjacent to the dental comb, the premolars look caniniform. Perhaps the phenocopying of a tooth from one class (e.g., canine) to another (e.g., incisor) is programmed alongside the patterning of tooth classes via combinatorial variation in localized regional gene expression (reviewed in Zhao et al. 2000; Stock 2001). In this way, for example, the canine primordium may be co-opted or influenced by the genetic program for the incisor teeth. Indeed, perhaps the canine is lost and a lateral incisor is gained. Only minor change to the odontogenic homeobox code would be required to effect major phenotypic differences: for example, in opossum and ferret, this code includes Alx3+Msx1 for incisors, Msx1 for canines, Msx1+Barx1 for premolars, and Barx1 for molars (Wakamatsu et al. 2019), and in shrew, a gradient of Barx1 expression levels corresponds with cuspal complexity, being highest in developing molars and absent in unicuspid teeth (Miletich et al. 2011).

In primates with large upper canines, the adjacent lower premolars are often sectorial and so more caniniform in shape. In fossil hominines, most notably paranthropiths, premolar shape shifts in the opposite direction of the dental lamina to look molariform. These subtler examples of phenocopying between tooth classes suggest anterior-posterior shifts in the expression of the odontogenic homeobox code, although such shifts would alter tooth class identity itself. Instead, perhaps tooth class identity remains unchanged and a molariform phenotype is produced by downstream patterning change in the signaling centers that regulate cusp morphology via cell proliferation of the enamel organ, particularly the inner enamel epithelium. Primate incisors, canines, and premolars appear to show the greatest evolutionary lability compared to molars, supported by insights from phenotype-back quantitative genetics studies of African and Eurasian anthropoids showing genetic independence between anterior and posterior teeth and submodules among premolars and molars (Grieco et al. 2013), as well as genetic underpinnings of variation in post-canine tooth proportions (Hlusko et al. 2016). Formally testing this idea about different degrees of lability among tooth classes via, for example, 3D imaging and geometric morphometrics and quantitative genetics would return conclusive results.

3.4 VARIATION IN PRIMATE DENTAL FORMULA: COMPARING PROSIMIANS WITH ANTHROPOIDS

Evolution of the dental formula through loss of teeth is a definitive aspect of mammalian diversification (reviewed in Line 2003) and reiterated throughout the primate lineage. As detailed in Figure 3.1, Table 3.1, and Appendix 3.1, deciduous and permanent dental formulae vary across primates, particularly among prosimians, where in some groups, a tooth class is even lost: for instance, Daubentoniidae lacks all permanent canines (OR all incisors) as well as lower premolars (1013/1003, OR 0113/0103), despite a deciduous dental formula of 112/112. Indriidae presents two permanent dental formulae, 2123/2033 (no lower canine) and 2123/1123 (only one lower incisor), with a single corresponding deciduous dental formula of 213/213. Lepilemuridae shows a permanent dental formula of 0133/2133 and deciduous dental

APPENDIX 3.1
Dental Formulae of Living Primates

Suborder	Infraorder	Superfamily	Family	Subfamily	Dental Formulae (Deciduous)
Prosimii	Lemuriformes	Lemuroidea	Lemuridae[*]		2133/2133 (213/213)
			Lepilemuridae[*]		0133/2133 (213/213)
			Cheirogaleidae[*]		2133/2133 (213/213)
			Indriidae[*]		**2123/2023**; **2123/1123** (213/213; **213/113**)
			Daubentoniidae		**1013/1003**; **0113/0103** (112/112)
		Lorisidae	Lorisidae[*]		2133/2133 (213/213)
			Galagidae[*]		2133/2133 (213/213)
	Tarsiiformes	Tarsoidea	Tarsiidae		**2133/1133** (**212/112**)
Anthropoidea	Platyrrhini	Ceboidea	Cebidae	Callitrichinae	2132/2132, 2133/2133 (213/213)
				Cebinae	2133/2133 (213/213)
				Aotinae	
			Atelidae	Callicebinae	
				Atelinae	
			Pitheciidae	Pitheciinae	
	Catarrhini	Cercopithecoidea	Cercopithecidae	Cercopithecinae	2123/2123 (212/212)
				Colobinae	
		Hominoidea	Hylobatidae		2123/2123 (212/212)
			Pongidae		

Sources: Emboldened dental formulae are asymmetrical between upper and lower dentitions. Upper/lower permanent teeth (upper/lower deciduous teeth). Data from Swindler (2002), Ankel-Simons (1996, 2007), Martin (1990), Tattersall (1982), Cuozzo and Yamashita (2006), Smith et al. (2015), Luckett and Maier (1982), Guthrie and Frost (2011), Berkovitz and Shellis (2016), Hershkovitz (1977).

[*] Dental comb.

formula of 213/213. Tarsiidae has a dental formula of 2133/1122 (212/112). Lemuridae, Cheirogalidae, Lorisidae, and Galagidae all share a permanent dental formula of 2133/2133 and deciduous dental formula of 213/213. Among anthropoids, the deciduous dental formula for all platyrrhines is 213/213, while for catarrhines, this formula is 212/212. Within platyrrhines, Callitrichinae presents two permanent dental formulae, 2132/2132 and 2133/2133, while all other groups share a permanent dental formula of 2133/2133. All catarrhines share a permanent dental formula of 2123/2123.

In summary, for the permanent dental formula: Lepilemuridae lacks an upper incisor, Daubentoniidae lacks either all canines or all incisors as well as all lower premolars, Tarsiidae lacks a lower incisor, and Indriidae has neither lower canine nor lower incisor. For the deciduous dental formula, there is never an instance of upper/lower asymmetry. With the exception of Callitrichinae, which has lost M3 in all genera but one (*Callimico goeldii*), three permanent molars always develop.

In accord with Williston's law that over the course of evolution, serial body parts tend towards specialization and reduction (Williston 1914), there is stronger constraint against the evolutionary gain—versus loss—of teeth across dental formulae. For instance, the mammalian fossil record—including the primate record—shows multiple cases of derived dental formulae defined by loss of incisors, premolars, and molars. That being said, recent evidence suggests that teeth have also been gained in primates, for example, *Callimico* (Scott 2015; Monson et al. 2019). Another example of loss based on modern human tooth development is that, for permanent teeth, hypodontia occurs about twice as frequently (4.4%) as 'extra', or supernumerary, teeth (2.1%), with frequencies of 0.3% agenic versus 0.8% supernumerary deciduous teeth (Brook et al. 2014). Also in modern human populations, various syndromic and isolated (non-syndromic) genetic conditions are associated with tooth agenesis, the extent of which depends on the target genes and their local or systemic disruption (Ramanathan et al. 2018; Williams and Letra 2018).

Within the dental lamina and tooth organs, these target genes, such as *Tp63*, may regulate tissue integrity, signaling, and/or cell cycling, thus potentially varying cell proliferation, differentiation, apoptosis, and altering—or arresting—tooth morphogenesis. In this regard, there are multiple entry points to "tinkering" (Jernvall and Salazar-Ciudad 2007) with tooth number and thus dental formula, as well as with tooth morphology. Tinkering also mitigates constraint against evo-devo gain in tooth number by enabling gain in tooth function; for example, an incisiform canine in a prosimian dental comb or the molariform premolars of *Paranthropus* and some australopithecines, including *A. afarensis*, *A. africanus*, and *A. garhi* (Asfaw et al. 1999; Ward et al. 2010). Function is the key driver of dental macroevolution, in tooth number and form. Each tooth class is typified by its morphology, location, and function. Broadly speaking, incisors catch, canines pierce, premolars crush, and molars grind (Butler 2000). Within each tooth class, individual teeth can evolve specialized forms and functions: for example, a sectorial premolar complex in African and Eurasian monkeys and apes. In this complex, the lower premolars, P_3 and P_4, are heteromorphic: the caniniform P_3 hones C^1 as they occlude, and the molariform P_4 helps crush food (Swindler 2002). Another example is prosimian dental combs: narrowed, elongated lower incisors and canines aligned in a tight row to form one functional unit.

Primate teeth vary in their sizes and shapes, such as sexually dimorphic canines, sectorial premolars, and dental combs. However, in terms of dental formula, prosimians (represented by tarsier, lemurs, and lorises), and anthropoids (represented by monkeys, apes, and humans) each show distinct patterns (Swindler 2002; Evans et al. 2016). Dental formulae differ more widely among prosimians compared to anthropoids, in whom dental formula is similar among platyrrhines and identical among catarrhines (Table 3.1). These observations imply different amounts and directions of lability in the odontogenic homeobox code programming the dental lamina in prosimians versus anthropoids. Downstream of this code, much of the diversity in premolar and molar crowns is likely due to genes including *Sostdc1*, *Fgf4*, *Lef1*, *p21*, and *Shh* (Jernvall et al. 2000; Carter and Worthington 2015) that determine cusp morphology. Traditionally, molar development is the focus of developmental genetic studies, but recent studies of incisor development in mouse are confirming many similarities between developing incisor and molar tooth organs and revealing key differences, some specific to the ever-growing nature of the mouse incisor (Yu and Klein 2020).

3.4.1 Mechanisms of Evolutionary Tooth Loss in the Primate Dental Formula

For reasons that remain unclear and are beyond the scope of this chapter, perhaps the most infamous type of tooth agenesis is that of the third molars (M3s) in modern humans. These last-formed molars are missing in 10% to 41% of the population—a frequency so high that M3s are excluded from the clinical definition of hypodontia, which is the failed formation of one to five teeth (Shimizu and Maeda 2009; Sujon et al. 2016). Human tooth agenesis is a common condition, categorized as either syndromic or non-syndromic, that almost exclusively affects the permanent dentition (Shimizu and Maeda 2009; Gkantidis et al. 2017). The more severe type of tooth agenesis, oligodontia, is defined by the absence of six or more teeth. Oligodontia is typically syndromic and thus linked to a genetic disease, while hypodontia is more often non-syndromic, such as a point mutation in one gene critical to tooth morphogenesis (reviewed in Brook et al. 2014). In about 80% of people with hypodontia, only one to two teeth are missing (Lidral and Reising 2002). As hypodontia is the milder condition that still affords a functional dental phenotype, the cellular mechanisms that underpin loss of five teeth or less are more likely to also be under selection for a reduced dental formula. However, losses of important tooth programming genes are often accompanied by severe developmental abnormalities and complications that are maladaptive phenotypes (i.e., orofacial clefting) (Mills et al. 1999). Genes with non-pleiotropic phenotypes are more likely candidates for the evolutionary loss of teeth in the dental formula.

Other than M3s, the most commonly lost teeth in all primates are, in descending order of frequency: P_4, I^2, I_2, and P^4 (Lavelle and Moore 1973; Miles and Grigson 1990; Shimizu and Maeda 2009; Klein et al. 2013; Gkantidis et al. 2017). The teeth at the end of a particular class (e.g., M3) are more frequently agenic (Juuri and Balic 2017). The evolutionary rationale for tooth agenesis in primate groups continues to be elucidated (e.g., Monson et al. 2019). While the exact developmental-genetic reason(s) that I2s, P4s, and M3s are agenic across primates is unclear, it is likely related to the molecular patterning and outgrowth of the dental lamina (Mostowska et al. 2003; Klein et al. 2013), particularly reduced odontogenic potential, including the capacity of epithelial and mesenchymal cells to properly cycle, proliferate, differentiate, orient, adhere, migrate, and signal (Townsend et al. 2009; Juuri and Balic 2017). Upper and lower dentitions are about equally affected by non-syndromic agenesis (Lavelle and Moore 1973; Miles and Grigson 1990; Klein et al. 2013), although there are exceptions. Based on studies in human and mouse, major targets include MSX1, PAX9, EDA/EDAR, and SPR2/4, and the phenotypes associated with these mutations are asymmetrical or uneven between upper and lower dentitions. For instance, MSX1 is particularly associated with the loss of P^4 and M^3 (Vastardis et al. 1996) as well as with loss of P^3 (Mostowska et al. 2006; Qin et al. 2013; reviewed in Ramanathan et al. 2018); mouse models with mutations in Sprouty homologues yield supernumerary molars anterior to M1 in both upper and lower jaws (Spry4) or in the lower jaw only (Spry2) (Marangoni et al. 2015).

Compared to humans, other primate groups have far lower reported rates of agenesis (Lavelle and Moore 1973; Miles and Grigson 1990). This observation may reflect the paucity of nonhuman primate samples and agenesis data or perhaps the higher dental phenotypic variation (including neutral or non-lethal point mutations) that is invariably generated by a 7-billion-plus breeding population of modern humans. Non-syndromic hypodontia is hereditary (Vastardis et al. 1996; Mostowska et al. 2003). Some of the genes linked to human tooth agenesis are conserved in other primates. For example, PAX9, which is critical to early tooth development, is highly conserved between humans and gorillas and conserved to a lesser extent in American monkeys (Pereira et al. 2006). Haplorhines show more frequent agenesis than strepsirrhines (Miles and Grigson 1990), with the exception of some genetically isolated strepsirrhine populations with unusually high rates of agenesis (Jablonski 1992). These studies establish that tooth agenesis is experienced among living primates, implying that tooth agenesis occurred in fossil primates and that its underlying causes are also mechanisms of evolutionary tooth loss in the primate dental formula.

There are different ways to lose a tooth from a dental formula (reviewed in Juuri and Balic 2017), including among tetrapod groups (Davit-Béal et al. 2009; Tokita et al. 2013). Particular genes appear to specify the identity of a given tooth class (e.g., *Dlx1/2* and maxillary molars) (Thomas et al. 1997) or delete a particular tooth type within a class (e.g., *MSX1*$^{-/-}$ and loss of I2, P4, M3) (Vastardis et al. 1996). The dental lamina may atrophy or lose dental odontogenic potential, or teeth may initiate and then regress, as seen in mouse and human, due to failed expression of genes required to maintain odontogenesis (Juuri and Balic 2017), including the integrity of dental epithelium. Some rudiments may merge either together as they do in the developing mouse incisor or with the organs of other teeth, as happens between the premolar rudiment and the first molar (reviewed in Peterkova et al. 2014). In Tarsiidae, earlier work reported that the di2 and dm2 (dp2) progress as far as histogenesis, then arrest and are resorbed (Luckett and Maier 1982). Shifting expression domains that specify dental epithelium versus dental ectomesenchyme may physically prevent direct contact between these two critical odontogenic tissue types, thus prohibiting the cross-signaling essential for tooth initiation (Harris et al. 2006). Odontogenic gene regulatory networks may mutate, transforming their roles—and cells' fates—in tooth morphogenesis and histogenesis into other functions such as keratinization (Louchart and Viriot 2011). In diphyodont mammals, including primates, the dental lamina also atrophies between deciduous and successional teeth, breaking the connection required for a third tooth generation to form (reviewed in Juuri and Balic 2017). The entire dentition is lost in the *Tp63*-null mouse due to failed formation of the dental lamina and tooth placodes (Laurikkala et al. 2006; Juuri and Balic 2017).

3.4.2 Divergent Evolution of Upper and Lower Dental Formulae and Phenotypes, Notably in Prosimians

In most primate groups, the dental formulae of upper and lower dentitions mirror each other; however, there are interesting cases of upper/lower asymmetry. Incisors, canines, and premolars are typically fewer in the lower dentition (Table 3.1) (Ankel-Simons 1996; Swindler 2002; Smith et al. 2015). In terms of permanent dental formulae and specifically lower teeth: Indriidae have lost C_1 and I_2; Daubentonidae lack P_3; and in Tarsiidae, I_2 is absent. For the deciduous dentition, Indriidae and Lorisidae both lack di_2. In contrast, Lemuridae, and all anthropoids, show no asymmetry between upper and lower dental formulae for either deciduous or permanent teeth. Also among strepsirrhines and platyrrhines, upper and lower incisor and canine morphologies vary considerably (reviewed in Luckett and Maier 1982).

For vertebrates, relative to the upper jaw, the mandible is an evolutionary novelty. This may explain why we see more frequent loss of lower teeth than upper teeth in primate dental formulae (Table 3.1). As such, we might expect to see tooth agenesis occur more often in the mandible than in the upper jaw. In contemporary humans, some reports conclude that agenesis of P4, M1, M2, and M3 is more frequent in the lower dentition (Kirkham et al. 2005; Brook et al. 2006). Other studies of tooth agenesis in human and nonhuman primates report that agenesis occurred at about equal rates in upper and lower jaws (Lavelle and Moore 1973; Klein et al. 2013). The reason for this variation among reports of agenesis may include specific mechanisms of tooth loss, which differ somewhat between upper and lower dentitions due to structural differences in jaw morphogenesis. For example, the upper jaw skeleton and teeth derive from embryonic facial primordia that must fold, merge, and fuse together (Scheuer and Black 2004). If any of these dynamic tissue processes fail, the dental lamina can be physically disrupted and teeth consequently lost from the permanent dentition (Klein et al. 2013). In humans, this problem accounts for the agenesis of I^2 and P^4, two of the most common non-syndromic tooth losses (Klein et al. 2013). Incidentally, this problem of facial primordia outgrowth can also result in a duplicated I^2 in cases of cleft palate where the dental lamina containing the I_2 primordium has split (Hovorakova et al. 2018).

With reference to the evolution of lower versus upper jaws, we posited that the primate mandibular dentition is more evolutionarily labile in terms of dental formula, as well as in terms of tooth size and shape. In addition to tooth loss, we examined tooth gain. In humans, 90% of supernumerary teeth occur in the upper jaw (Klein et al. 2013). In mouse, two premolar rudiments and five rudiments along the diastema appear in the developing upper dentition compared to just one premolar rudiment in the lower dentition, and five to six epithelial primordia converge to form the upper incisor compared to three primordia for the lower incisor (Klein et al. 2013). Among non-human free-lived primates, great apes, baboons, and other anthropoid primates are examples of groups where the presence of an M4 is relatively frequent, particularly in the upper dentition, either unilaterally or bilaterally (Shaw 1927; Jungers and Gingerich 1980; Schwartz 1984). This fourth and supernumerary molar located behind M3 may be enabled by very prognathic jaws that allow abnormal growth of the dental lamina (Shaw 1927) and, we suggest, perhaps in concert with point mutations such as EDAR, enable the invagination of an M4 in the molar row. Again, we see a pattern suggestive of greater retention—or greater inertia against loss—of teeth in the upper dentition, which may reflect jaw-specific patterns of developmental integration of a dentition within the midfacial skeleton versus the dentary bone.

3.5 EVO-DEVO OF THE PRIMATE DENTAL LAMINA

Surveying all reported formulae for deciduous and permanent dentitions across primates, no successional tooth forms without its deciduous precursor. The majority of tooth loss, as well as morphological variation, occurs in the ante-molar teeth and arguably in the lower dentition. Asymmetry between upper and lower dentitions is likely due to differences in odontogenic programs and developmental constraints between lower versus upper jaws. For example, in mouse, experimental deletion of Dlx1 and Dlx2 led to defects of the upper molars only, because lower molar development is likely buffered by Dlx5/6 expression specific to the embryonic mandible (Denaxa et al. 2009).

Regarding the ante-molar permanent dental formula, dropping an incisor, canine, or premolar can occur via different developmental stages and cellular processes discussed elsewhere in this chapter. We suggest that inhibiting the formation of a deciduous tooth is the most robust and definitive mechanism to delete the successional counterpart tooth from the dental formula. Indeed, in humans, deciduous teeth are almost never absent (0.5% to 0.9% prevalence), but when they are, the successional tooth is usually also absent (Gomes et al. 2014; Neville et al. 2015). Others go so far as to consider it a rule that failed primary teeth result in failed secondary teeth (Matalova et al. 2008). Thus, ante-molar deciduous teeth act as the stem progenitors of ante-molar successional teeth. As such, we propose the Linchpin Hypothesis, where a deciduous tooth must develop—or at least initiate even if it later regresses—for a successional tooth to form. We appreciate that some mammals, including murid rodents, are monophyodont (van Nievelt and Smith 2005), where the primary dentition serves as the permanent dentition. Our hypothesis concerns diphyodont mammals and primates specifically. That being said, it is salient to note that at least shrews, a phenotypically monophyodont mammalian group, are developmentally diphyodont: the developing deciduous teeth regress around cap stage, and lingual extensions of the deciduous dental lamina give rise to the permanent teeth (Jarvinen et al. 2008). In the context of this chapter, the Linchpin Hypothesis aligns with what we understand about the dental lamina producing the ante-molar primary (deciduous) dentition and then budding as a successional dental lamina to produce secondary (permanent) ante-molar teeth. Even human patients with agenesis of permanent teeth—either syndromic or non-syndromic—appear to develop normal deciduous dentitions. In cases of hypodontia of deciduous teeth, 75% of these patients showed agenesis of permanent teeth in the same region of the dental arch (Brook et al. 2014). In minipig and human, the eruption of the deciduous canine releases mechanical compression on the successional dental lamina, triggering initiation of the permanent canine via reduced RUNX2 activity in dental (lamina) mesenchyme, resulting in higher Wnt signaling in dental epithelium (Wu et al. 2020). Perhaps in shrew, the cap-stage deciduous tooth organ creates via cell rearrangement and condensation (Jarvinen et al. 2008) some biomechanical trigger for the onset of the replacement tooth. Also, without the deciduous dental lamina, the successional lamina does not form. There would seem to be strong selection for deciduous teeth to develop for three reasons: (1) for teeth to fit within the smaller jaw of a young primate, (2) who must also chew solid food, and (3) to enable the initiation as well as the proper eruption of successional permanent teeth.

Instead of successional teeth, the permanent molars may be considered additional teeth because they lack deciduous precursors. M3 is the most frequently agenic tooth. We suggest that rather than variation in absence or presence of deciduous precursors—and whether a successional lamina branches lingually from the deciduous lamina—initiation of M3, and possibly M2, is acutely influenced by the *de novo* elongation of the dental lamina via cell proliferation and subsequent changes in gene expression and signaling that activate molar onset (Kavanagh et al. 2007; Carter and Worthington 2015). This process relies on the lengthening of the dental lamina in the antero-posterior axis, perpendicular to the labial-lingual plane of the successional lamina's extension. The posterior-ward growth of the molar lamina and the initiation of each molar seems to be influenced by multiple factors (Marchiori et al. 2019). We suggest that the *de novo* mode of molar lamina extension somehow translates to variation in cell proliferation, differentiation, adhesion, and migration that raises the frequency of M3 agenesis compared to agenesis of successional ante-molar teeth.

While the Linchpin loss may seem an obvious idea—that is, no successors can form without predecessors—we hope that it stimulates deeper thought around key mechanisms of tooth development, replacement, and evolution. These mechanisms include the cellular and tissue dynamics of the dental lamina, which is somehow deciduous when extending along an antero-posterior axis and somehow successive when projecting lingually from a predecessor tooth. Related to this process, it is perplexing that deciduous molars do not give rise to larger versions of themselves but rather to the morphologically distinct premolars. Does the programming of the dental lamina shift between tooth generations? Or do the premolars develop from a differently patterned region of interdental lamina persisting between deciduous molars or between deciduous molars and canines? Clearly more work needs to be done to explain.

Premolars are named from anterior to posterior: P1 P2 P3 P4. However, developmentally, premolars tend to form in the opposite direction, from posterior to anterior: P4 P3 P2 P1 (reviewed in Stock et al. 1997). However, among primates, P4, P3, and, when present, P2 often initiate either simultaneously or with little time between them (Hillson

2014; Smith et al. 2015). To further complicate things, premolars are typically lost via agenesis from posterior to anterior (i.e., P4 is absent more often than is P3) (Kirkham et al. 2005), but premolars are lost via evolution from anterior to posterior (i.e., P1 and in many cases P2 are dropped from the primate dental formula) (Stock et al. 1997). Strong canalization likely hampers the loss of the first initiating member—the stem progenitor—within each tooth class: for example, in mouse, despite the loss of the premolar tooth class, transient premolar rudiments still form (Peterkova et al. 2006, 2014). It would seem particularly important to study cell migration and differentiation in the premolar region of the dental lamina in order to resolve the developmental nuances and variation on which selection may act for this particular ante-molar tooth class.

In most instances, in a primate dental formula of three deciduous molars—or, as others have argued, three deciduous premolars—three permanent premolars typically develop (Table 3.1, Appendix 3.1). Three permanent molars also develop. However, in a few cases, three dms (or three dps?) are associated with only two (Indriidae), one, or no Ps (Daubentoniidae), and with two Ms (Callitrichinae, albeit with three Ms in *Callimico goeldii*). These observations are consistent with our Linchpin Hypothesis that, at least among primates, the number of successional teeth can be no more than the number of deciduous teeth. Also, in the exclusively catarrhine primate groups that develop only two dms (or two dps), two Ps form, as do three Ms. In which case, three molars developing from two dms would violate our hypothesis. If taken at face value, then these observations imply that at least in catarrhines, the deciduous 'molars' are indeed developmentally homologous with premolars and thus are best identified as dps and not as dms.

The mouse incisor forms from several fused primordia. In the dental lamina of prosimians with dental combs, perhaps the canine is actually lost and a third incisor gained, especially considering the ancestral mammalian dental formula (I3-C1-P4-M3) includes three incisors per quadrant, and the interdental lamina retains dental capability (Jussila et al. 2014). Evidence for this idea comes from observations of Daubentoniidae: two dental formulae have been suggested for the permanent dentition because it is unclear whether the continuously growing maxillary anterior tooth is an incisor or a canine (Tattersall and Schwartz 1974; Cuozzo and Yamashita 2006). For several reasons, we are inclined to consider the aye-aye's upper anterior tooth an incisor rather than a canine. First, based on developmental biology data about the specialization of the ever-growing (hypselodont) rodent incisor and its stem cell niche located within the labial cervical loop that enables continuous ameloblast activity (Yu and Klein 2020), the aye-aye incisor closely resembles the self-sharpening rodent incisor, with ongoing enamel production on the labial facet only (Berkovitz and Shellis 2018). Also, a hypselodont incisor tooth type occurs at higher frequency in recent mammals (compared to much rarer hypselodont canines) (Renvoisé and Michon 2014). Further, the evolutionary origins and rise of hypselodonty within the rodent lineage during epochs are coincident with primate origins and ancient rodent-primate shared ancestry (Renvoisé and Michon 2014). If our inclination holds true, then in aye-aye, this anterior region of the dental lamina would be coded for an incisor tooth class identity instead of for canine (Figure 3.2). Lemur adaptive radiation on the island of Madagascar was likely linked to rapid dietary specialization (Godfrey et al. 2004; Herrera 2017), which may explain the high degree of dental phenotype diversity in this group, whose origin lies among the stem members of all modern primates (Herrera and Dávalos 2016) (Figure 3.2). A unique feature of this group includes the high rate of continued diversification (Herrera 2017), and modern genetic studies continue to support that genetic diversity in lemur clades is high (Pastorini et al. 2003; Aleixo-Pais et al. 2019; Sgarlata et al. 2019). Dental diversity exhibited by Lemuroidea is telling of a long evolutionary history laden with strong selective pressure for dietary specialization (Herrera 2017).

3.6 CONCLUSIONS

This chapter explored patterns of primate dental macroevolution with the aim of gleaning deeper insight into the cell processes underpinning the evo-devo of primate dental formula and form. Broadly speaking, the lower dentition, and the lower ante-molar teeth in particular, appears to be most labile in number and morphology, and this lability is most notable in prosimians compared to anthropoids. Referring back to the Tp63 mouse model of craniodental development, results in primates generally support the hypothesis that stronger developmental and functional integration among midfacial structures, including the upper jaw and teeth, constrain dental macroevolution in the upper dentition compared to the lower dentition. We posit that the primate dental formula for permanent dentitions evolves via molecular and/or spatial changes in the odontogenic homeobox code that differentially patterns the nascent deciduous dental lamina. We posit that subtler macroevolution in tooth form occurs downstream of the dental code. Evolutionary losses in the dental formula may occur via more than one type of change in the cellular dynamics of odontogenesis. More studies of diphyodont, heterodont animal models are needed to clarify the cell processes evolving in all four tooth classes under strong selection for diet and other parafunctions in primates, including humans and fossil relations.

3.7 ACKNOWLEDGMENTS

We thank the editors of this volume and its corresponding book series for the invitation to contribute this chapter and for helpful guidance alongside the thoughtful critical feedback of two external reviewers. JCB is supported by Discovery Grant #2016–05177 awarded by the Natural Sciences and Engineering Research Council of Canada (NSERC) that also supports DFM, while EMV and CMA are supported by MSc and PhD NSERC Canada Graduate Scholarships, respectively.

3.8 REFERENCES

Ahn, Y., B. W. Sanderson, O. D. Klein, and R. Krumlauf. 2010. Inhibition of Wnt signaling by wise (Sostdcl) and negative feedback from Shh controls tooth number and patterning. *Development* 137:3221–31.

Aleixo-Pais, I., J. Salmona, G. M. Sgarlata, A. Rakotonanahary, et al. 2019. The genetic structure of a mouse lemur living in a fragmented habitat in Northern Madagascar. *Conserv Gen* 20:229–43. doi:10.1007/s10592-018-1126-z.

Anderson, P. S. L., S. Renaud, and E. J. Rayfield. 2014. Adaptive plasticity in the mouse mandible. *BMC Evol Biol* 14:1–9.

Ankel-Simons, F. 1996. Deciduous dentition of the aye aye, *Daubentonia madagascariensis*. *Am J Primatol* 39:87–97.

Ankel-Simons, F. 2007. *Primate Anatomy: An Introduction*. Burlington, MA: Elsevier.

Ankel-Simons, F., and D. T. Rasmussen. 2008. Diurnality, nocturnality, and the evolution of primate visual systems. *Am J Phys Anthropol* 137:100–17.

Asfaw, B., T. White, O. Lovejoy, B. Latimer, et al. 1999. *Australopithecus garhi*: a new species of early hominid from Ethiopia. *Science* 284:629–35.

Balic, A., and I. Thesleff. 2015. Tissue interactions regulating tooth development and renewal. *Curr Top Dev Biol* 115:157–86.

Berkovitz, B. K., G. R. Holland, and B. J. Moxham. 2009. *Oral Anatomy, Histology and Embryology*. E-Book: Mosby.

Berkovitz, B., and R. P. Shellis. 2016. *The Teeth of Non-Mammalian Vertebrates*. San Diego, CA: Academic Press.

Berkovitz, B., and R. P. Shellis. 2018. *The Teeth of Mammalian Vertebrates*. San Diego, CA: Academic Press.

Boughner, J. C., and C. Rolian. 2016. *Developmental Approaches to Human Evolution*. Hoboken, NJ: John Wiley & Sons.

Brazeau, M. D., and M. Friedman. 2015. The origin and early phylogenetic history of jawed vertebrates. *Nature* 520:490–7.

Brook, A. H., J. Jernvall, R. N. Smith, T. E. Hughes, et al. 2014. The dentition: the outcomes of morphogenesis leading to variations of tooth number, size and shape. *Austral Dent J* 59:131–42.

Brook, A. H., C. Underhill, L. K. Foo, and M. Hector. 2006. Approximal attrition and permanent tooth crown size in a Romano-British population. *Dent Anthropol J* 19:23–8.

Brown, T. 1974. Mandibular movements. *Monographs in Oral Science* 4:126–50.

Buchtová, M., J. Stembírek, K. Glocová, E. Matalová, et al. 2012. Early regression of the dental lamina underlies the development of diphyodont dentitions. *J Dent Res* 91:491–8.

Butler, P. M. 2000. The evolution of tooth shape and tooth function in primates. In *Development, Function and Evolution of Teeth*, ed. M. F. Teaford, M. M. Smith, and M. W. J. Ferguson, 201–11. Cambridge: Cambridge University Press.

Butler, P. M., and W. A. Clemens. 2001. Dental morphology of the Jurassic Holotherian mammal *Amphitherium*, with a discussion of the evolution of mammalian post-canine dental formulae. *Palaeontology* 44:1–20. doi:10.1111/1475-4983.00166.

Cardini, A., and P. D. Polly. 2013. Larger mammals have longer faces because of size-related constraints on skull form. *Nat Comm* 4:1–7. doi:10.1038/ncomms3458.

Carter, K., and S. Worthington. 2015. Morphologic and demographic predictors of third molar agenesis: a systematic review and meta-analysis. *J Dent Res* 94:886–94.

Catón, J., and A. S. Tucker. 2009. Current knowledge of tooth development: patterning and mineralization of the murine dentition. *J Anat* 214:502–15.

Charles, C., V. Lazzari, P. Tafforeau, T. Schimmang, et al. 2009. Modulation of Fgf3 dosage in mouse and men mirrors evolution of mammalian dentition. *PNAS* 106:22364–8.

Cobourne, M. T., and T. Mitsiadis. 2006. Neural crest cells and patterning of the mammalian dentition. *J Exp Zool B* 306:251–60.

DeCoster, L. A. Marks, L. C. Martens, and A. Huysseune. 2008. Dental agenesis: genetic and clinical perspectives. *J Oral Pathol & Med* 38:1–17. doi:10.1111/j.1600-0714.2008.00699.x.

Cuozzo, F. P., and N. Yamashita. 2006. Impact of ecology on the teeth of extant lemurs: a review of dental adaptations, function, and life history. In *Lemurs: Ecology and Adaptation*, ed. L. Gould and M. L. Sauther, 67–96. New York: Springer.

Davit-Béal, T., A. S. Tucker, and J.-Y. Sire. 2009. Loss of teeth and enamel in tetrapods: fossil record, genetic data and morphological adaptations. *J Anat* 14:477–501. doi:10.1111/j.1469-7580.2009.01060.x.

Denaxa, M., P. T. Sharpe, and V. Pachnis. 2009. The LIM homeodomain transcription factors Lhx6 and Lhx7 are key regulators of mammalian dentition. *Dev Biol* 333:324–36.

Depew, M. J., T. Lufkin, and J. L. R. Rubenstein. 2002. Specification of jaw subdivisions by Dlx genes. *Science* 298:381–5.

Depew, M. J., C. A. Simpson, M. Morasso, and J. L. R. Rubenstein. 2005. Reassessing the Dlx code: the genetic regulation of branchial arch skeletal pattern and development. *J Anat* 207:501–61.

Donoghue, P. C. J., and M. Rücklin. 2016. The ins and outs of the evolutionary origin of teeth. *Evol Dev* 18:19–30.

Evans, A. R., E. S. Daly, K. K. Catlett, K. S. Paul, et al. 2016. A simple rule governs the evolution and development of hominin tooth size. *Nature* 530:477–80.

Ferrario, V. F., C. Sforza, G. Serrao, C. Dellavia, et al. 2004. Single tooth bite forces in healthy young adults. *J Oral Rehabil* 31:18–22.

Gaete, M., J. M. Fons, E. M. Popa, L. Chatzeli, et al. 2015. Epithelial topography for repetitive tooth formation. *Biology Open* 4:1625–34.

Gingerich, P. D. 2012. Primates in the Eocene. *Palaeobiodiver Palaeoenviron* 92:649–63. doi:10.1007/s12549-012-0093-5.

Gkantidis, N., H. Katib, E. Oeschger, M. Karamolegkou, et al. 2017. Patterns of non-syndromic permanent tooth agenesis in a large orthodontic population. *Arch Oral Biol* 79:42–7.

Glasstone, S. 1962. Tissue culture and tooth development. *Arch Oral Biol* 7:125–31. doi:10.1016/0003-9969(62)90092-4.

Godfrey, L. R., G. M. Semprebon, W. L. Jungers, M. R. Sutherland, et al. 2004. Dental use wear in extinct lemurs: evidence of diet and niche differentiation. *J Hum Evol* 47:145–69.

Gomes, R. R., J. A. C. Fonseca, L. M. Paula, A. C. Acevedo, et al. 2014. Dental anomalies in primary dentition and their corresponding permanent teeth. *Clin Oral Investig* 18:1361–7.

Greaves, W. S. 1991. A relationship between premolar loss and jaw elongation in selenodont artiodactyls. *Zool J Linn Soc* 101:121–9. doi:10.1111/j.1096-3642.1991.tb00889.x.

Grieco, T. M., O. T. Rizk, and L. J. Hlusko. 2013. A modular framework characterizes micro- and macroevolution of Old World monkey dentitions. *Evolution* 67:241–59.

Guthrie, E. H., and S. R. Frost. 2011. Pattern and pace of dental eruption in tarsius. *Am J Phys Anthropol* 145:446–51.

Guy, F., V. Lazzari, E. Gilissen, and G. Thiery. 2015. To what extent is primate second molar enamel occlusal morphology shaped by the enamel-dentine junction? *PLoS One* 10:e0138802.

Hallgrímsson, B., K. Willmore, and B. K. Hall. 2002. Canalization, developmental stability, and morphological integration in primate limbs. *Am J Phys Anthropol* Suppl 35:131–58.

Hammer, C. L., A. D. S. Atukorala, and T. A. Franz-Odendaal. 2016. What shapes the oral jaws? Accommodation of complex dentition correlates with premaxillary but not mandibular shape. *Mech Dev* 141:100–8. doi:10.1016/j.mod.2016.04.001.

Harris, M. P., S. M. Hasso, M. W. J. Ferguson, and J. F. Fallon. 2006. The development of archosaurian first-generation teeth in a chicken mutant. *Curr Biol* 16:371–7 doi:10.1016/j.cub.2005.12.047.

Hautier, L., H. G. Rodrigues, G. Billet, and R. J. Asher. 2016. The hidden teeth of sloths: evolutionary vestiges and the development of a simplified dentition. *Sci Rep* 6:27763.

Helkimo, E., G. E. Carlsson, and M. Helkimo. 1977. Bite force and state of dentition. *Acta Odontol Scand* 35:297–303. doi:10.3109/00016357709064128.

Herrera, J. P. 2017. Testing the adaptive radiation hypothesis for the lemurs of Madagascar. *R Soc Open Sci* 4:161014.

Herrera, J. P., and L. M. Dávalos. 2016. Phylogeny and divergence times of lemurs inferred with recent and ancient fossils in the tree. *Syst Biol* 65:772–91.

Hershkovitz, P. 1977. *Living New World Monkeys (Platyrrhini), Volume 1: With an Introduction to Primates*. Chicago, IL: University of Chicago Press.

Hillson, S. 2014. *Tooth Development in Human Evolution and Bioarchaeology*. Cambridge: Cambridge University Press. doi:10.1017/cbo9780511894916.

Hlusko, L. J., C. A. Schmitt, T. A. Monson, M. F. Brasil, et al. 2016. The integration of quantitative genetics, paleontology, and neontology reveals genetic underpinnings of primate dental evolution. *PNAS* 113:9262–7. doi:10.1073/pnas.1605901113.

Hovorakova, M., H. Lesot, M. Peterka, and R. Peterkova. 2005. The developmental relationship between the deciduous dentition and the oral vestibule in human embryos. *Anat Embryol.* 209:303–13.

Hovorakova, M., H. Lesot, M. Peterka, and R. Peterkova. 2018. Early development of the human dentition revisited. *J Anat* 233:135–45.

Hovorakova, M., H. Lesot, J.-L. Vonesch, M. Peterka, et al. 2007. Early development of the lower deciduous dentition and oral vestibule in human embryos. *Eur J Oral Sci* 115:280–7.

Jablonski, N. G. 1992. Dental agenesis as evidence of possible genetic isolation in the colobine monkey. *Rhinopithecus Roxellana. Primates.* 33:371–6. doi:10.1007/bf02381198.

Jarvinen, E., K. Vlimki, M. Pummila, I. Thesleff, et al. 2008. The taming of the shrew milk teeth. *Evol Dev* 10:477–86. doi:10.1111/j.1525-142x.2008.00258.x.

Jernvall, J., S. V. Keränen, and I. Thesleff. 2000. Evolutionary modification of development in mammalian teeth: quantifying gene expression patterns and topography. *PNAS* 97:14444–8.

Jernvall, J., and I. Salazar-Ciudad. 2007. The economy of tinkering mammalian teeth. In *Tinkering: The Microevolution of Development*, ed. G. Bock and J. Goode. Novartis Foundation Symposia. Chichester: John Wiley & Sons, Inc. doi:10.1002/9780470319390.ch14.

Jernvall, J., and I. Thesleff. 2000. Reiterative signaling and patterning during mammalian tooth morphogenesis. *Mech Dev* 92:19–29.

Jernvall, J., and I. Thesleff. 2012. Tooth shape formation and tooth renewal: evolving with the same signals. *Development* 139:3487–97.

Jungers, W. L., and P. D. Gingerich. 1980. Supernumerary molars in Anthropoidea, Adapidae, and Archaeolemur: implications for primate dental homologies. *Am J Phys Anthropol* 52:1–5.

Jussila, M., and I. Thesleff. 2012. Signaling networks regulating tooth organogenesis and regeneration, and the specification of dental mesenchymal and epithelial cell lineages. *Cold Spring Harb Perspect Biol* 4:a008425.

Jussila, M., X. C. Yanez, and I. Thesleff. 2014. Initiation of teeth from the dental lamina in the ferret. *Differentiation* 87:32–43.

Juuri, E., and A. Balic. 2017. The biology underlying abnormalities of tooth number in humans. *J Dent Res* 96:1248–56.

Juuri, E., M. Jussila, K. Seidel, S. Holmes, P. Wu, et al. 2013. Sox2 marks epithelial competence to generate teeth in mammals and reptiles. *Development* 140:1424–32.

Kavanagh, K. D., A. R. Evans, and J. Jernvall. 2007. Predicting evolutionary patterns of mammalian teeth from development. *Nature* 449:427–32.

Kirkham, J., R. Kaur, E. C. Stillman, P. G. Blackwell, et al. 2005. The patterning of hypodontia in a group of young adults in Sheffield, UK. *J Hum Evol* 50:287–91. doi:10.1016/j.archoralbio.2004.11.015.

Klein, O. D., S. Oberoi, A. Huysseune, M. Hovorakova, et al. 2013. Developmental disorders of the dentition: an update. *Am J Med Genet C* 163:318–32.

Ko, D., T. Kelly, L. Thompson, J. K. Uppal, et al. 2021. Timing of mouse molar formation is independent of jaw length including retromolar space. *J Dev Biol* 9(1):8. doi:10.3390/jdb9010008.

Krivanek, J., I. Adameyko, and K. Fried. 2017. Heterogeneity and developmental connections between cell types inhabiting teeth. *Front Physiol* 8:376.

Kuratani, S., Y. Oisi, and K. G. Ota. 2016. Evolution of the vertebrate cranium: viewed from hagfish developmental studies. *Zool Sci* 33:229–38.

Lainoff, A. J., J. E. Moustakas-Verho, D. Hu, A. Kallonen, et al. 2015. A comparative examination of odontogenic gene expression in both toothed and toothless amniotes. *J Exp Zool Part B* 324:255–69.

Laurikkala, J., M. L. Mikkola, M. James, M. Tummers, et al. 2006. P63 regulates multiple signalling pathways required for ectodermal organogenesis and differentiation. *Development* 133:1553–63.

Lavelle, C. L., and W. J. Moore. 1973. The incidence of agenesis and polygenesis in the primate dentition. *Am J Phys Anthropol* 38:671–9.

Lidral, A. C., and B. C. Reising. 2002. The role of MSX1 in human tooth agenesis. *J Dent Res* 81:274–8.

Line, S. R. P. 2003. Variation of tooth number in mammalian dentition: connecting genetics, development, and evolution. *Evol Dev* 5:295–304.

Louchart, A., and L. Viriot. 2011. From snout to beak: the loss of teeth in birds. *Trends Ecol Evol* 26:663–73.

Luckett, W. P., and W. Maier. 1982. Development of deciduous and permanent dentition in *Tarsius* and its phylogenetic significance. *Folia Primatol* 37:1–36.

Luo, Z.-X., Z. Kielan-Jaworowska, and R. L. Cifelli. 2004. Evolution of dental replacement in mammals. *Bull Carnegie Mus Nat Hist* 36:159–75. doi:10.2992/0145-9058(2004)36[159:eodrim]2.0.co;2.

Marangoni, P., C. Charles, P. Tafforeau, V. Laugel-Haushalter, et al. 2015. Phenotypic and evolutionary implications of modulating the ERK-MAPK cascade using the dentition as a model. *Sci Rep* 5:11658.

Marchiori, D. F., G. V. Packota, and J. C. Boughner. 2019. Initial third molar development is delayed in jaws with short distal space: an early impaction sign? *Arch Oral Biol* 106:104475.

Martin, R. D. 1990. *Primate Origins and Evolution: A Phylogenetic Reconstruction.* Princeton, NJ: Princeton University Press.

Martin, R. M. G., J.-J. Hublin, P. Gunz, and M. M. Skinner. 2017. The morphology of the enamel-dentine junction in Neanderthal molars: gross morphology, non-metric traits, and temporal trends. *J Hum Evol* 103:20–44.

Matalova, E., J. Fleischmannova, P. T. Sharpe, and A. S. Tucker. 2008. Tooth agenesis: from molecular genetics to molecular dentistry. *J Dent Res* 87:617–23. doi:10.1177/154405910808700715.

Matalova, E., E. Svandova, and A. S. Tucker. 2012. Apoptotic signaling in mouse odontogenesis. *OMICS* 16:60–70.

Matalova, E., A. S. Tucker, and P. T. Sharpe. 2004. Death in the life of a tooth. *J Dent Res* 83:11–16. doi:10.1177/154405910408300103.

McCollum, M., and P. T. Sharpe. 2001. Evolution and development of teeth. *J Anat* 199:153–9.

Miles, A. E. W., and C. Grigson. 1990. *Colyer's Variations and Diseases of the Teeth of Animals*. Cambridge: Cambridge University Press.

Miletich, I., W.-Y. Yu, R. Zhang, K. Yang, et al. 2011. Developmental stalling and organ-autonomous regulation of morphogenesis. *PNAS* 108:19270–5.

Mills, A. A., B. Zheng, X. J. Wang, H. Vogel, et al. 1999. P63 is a p53 homologue required for limb and epidermal morphogenesis. *Nature* 398:708–13.

Moffat, A. S. 2002. Primate origins meeting: new fossils and a glimpse of evolution. *Science* 295:613–15. doi:10.1126/science.295.5555.613.

Monson, T. A., J. L. Coleman, and L. J. Hlusko. 2019. Craniodental allometry, prenatal growth rates, and the evolutionary loss of the third molars in New World Monkeys. *Anat Rec* 302:1419–33.

Mostowska, A., B. Biedziak, and W. H. Trzeciak. 2006. A novel c.581C>T transition localized in a highly conserved homeobox sequence of MSX1: is it responsible for oligodontia? *J Appl Genet* 47:159–64. doi:10.1007/bf03194616.

Mostowska, A., A. Kobielak, and W. H. Trzeciak. 2003. Molecular basis of non-syndromic tooth agenesis: mutations of MSX1 and PAX9 reflect their role in patterning human dentition. *Eur J Oral Sci* 111:365–70.

Myers, P., R. Espinosa, C. S. Parr, T. Jones, et al. 2020. *The Animal Diversity Web*. https://animaldiversity.org.

Nasrullah, Q., M. B. Renfree, and A. R. Evans. 2018. Three-dimensional mammalian tooth development using diceCT. *Arch Oral Biol* 85:183–91.

Neville, B. W., D. D. Damm, C. Allen, and A. C. Chi, eds. 2015. Abnormalities of teeth. In *Oral and Maxillofacial Pathology*, 54–99. St. Louis: Elsevier Health Sciences.

Ni, X., D. L. Gebo, M. Dagosto, J. Meng, et al. 2013. The oldest known primate skeleton and early haplorhine evolution. *Nature* 498:60–4.

Ooë, T. 1979. Development of human first and second permanent molar, with special reference to the distal portion of the dental lamina. *Anat Embryol* 155:221–40.

Osborn, J. W. 1973. The evolution of dentitions: the study of evolution suggests how the development of mammalian dentitions may be controlled. *Am Sci* 61:548–59.

Paradis, M. R., M. T. Raj, and J. C. Boughner. 2013. Jaw growth in the absence of teeth: the developmental morphology of edentulous mandibles using the p63 mouse mutant. *Evol Dev* 15:268–79.

Pastorini, J., U. Thalmann, and R. D. Martin. 2003. A molecular approach to comparative phylogeography of extant Malagasy lemurs. *PNAS* 100:5879–84.

Pereira, T. V., F. M. Salzano, A. Mostowska, W. H. Trzeciak, et al. 2006. Natural selection and molecular evolution in primate PAX9 gene, a major determinant of tooth development. *PNAS* 103:5676–81.

Peterkova, R., M. Hovorakova, M. Peterka, and H. Lesot. 2014. Three-dimensional analysis of the early development of the dentition. *Aust Dent J* 59:S55–S80.

Peterkova, R., H. Lesot, and M. Peterka. 2006. Phylogenetic memory of developing mammalian dentition. *J Exp Biol Part B* 306:234–50.

Phen, A., J. Greer, J. Uppal, J. Der, et al. 2018. Upper jaw development in the absence of teeth: new insights for craniodental evo-devo integration. *Evol Dev* 20:146–59.

Pozzi, L., J. A. Hodgson, A. S. Burrell, K. N. Sterner, et al. 2014. Primate phylogenetic relationships and divergence dates inferred from complete mitochondrial genomes. *Mol Phylogenet Evol* 75:165–83.

Prochazka, J., M. Prochazkova, W. Du, F. Spoutil, et al. 2015. Migration of founder epithelial cells drives Proper molar tooth positioning and morphogenesis. *Dev Cell* 35:713–24.

Qin, H., H.-Z. Xu, and K. Xuan. 2013. Clinical and genetic evaluation of a Chinese family with isolated oligodontia. *Arch Oral Biol* 58:1180–6.
Raj, M. T., and J. C. Boughner. 2016. Detangling the evolutionary developmental integration of dentate jaws: evidence that a p63 gene network regulates odontogenesis exclusive of mandible morphogenesis. *Evol Dev* 18:317–23.
Ramanathan, A., T. C. Srijaya, P. Sukumaran, R. B. Zain, et al. 2018. Homeobox genes and tooth development: understanding the biological pathways and applications in regenerative dental science. *Arch Oral Biol* 85:23–39.
Regan, B. C., C. Julliot, B. Simmen, F. Viénot, et al. 2001. Fruits, foliage and the evolution of primate colour vision. *Philos Trans R Soc Lond B Biol Sci* 356:229–83.
Renvoisé, E., and F. Michon. 2014. An evo-devo perspective on ever-growing teeth in mammals and dental stem cell maintenance. *Front Physiol* 5:324 doi:10.3389/fphys.2014.00324.
Ross, C. F., and J. Iriarte-Diaz. 2019. Evolution, constraint, and optimality in primate feeding systems. In *Feeding in Vertebrates*, ed. V. Bels and I. Q. Whishaw, 787–829. Heidelberg: Springer. doi:10.1007/978-3-030-13739-7_20.
Rostampour, N., C. M. Appelt, A. Abid, and J. C. Boughner. 2019. Expression of new genes in vertebrate tooth development and p63 signaling. *Dev Dyn* 248:744–55.
Salazar-Ciudad, I., and M. Marín-Riera. 2013. Adaptive dynamics under development-based genotype—phenotype maps. *Nature* 487:361–4. doi:10.1038/nature12142.
Scheuer, L., and S. Black. 2004. *The Juvenile Skeleton*. London: Academic Press, Elsevier.
Schwartz, J. H. 1982. Morphological approach to heterodonty and homology. In *Teeth: Form, Function, and Evolution*, ed B. Kurten, 123–44. New York: Columbia University Press.
Schwartz, J. H. 1984. Supernumerary teeth in anthropoid primates and models of tooth development. *Arch Oral Biol* 29:833–42.
Scott, J. E. 2015. Lost and found: the third molars of *Callimico goeldii* and the evolution of the callitrichine postcanine dentition. *J Hum Evol* 83:65–73. doi:10.1016/j.jhevol.2015.03.006.
Sgarlata, G. M., J. Salmona, B. Le Pors, E. Rasolondraibe, et al. 2019. Genetic and morphological diversity of mouse lemurs (*Microcebus Spp.*) in Northern Madagascar: the discovery of a putative new species? *Am J Primatol* 81:e23070.
Sharpe, P. T. 2000. Homeobox genes in initiation and shape of teeth during development in mammalian embryos. In *Development, Function and Evolution of Teeth*, ed. M. F. Teaford, M. M. Smith, and M. W. J. Ferguson, 3–12. Cambridge: Cambridge University Press.
Shaw, J. C. 1927. Four cases of fourth molar teeth in South African baboons. *J Anat* 62:79–85.
Shimizu, T., and T. Maeda. 2009. Prevalence and genetic basis of tooth agenesis. *Jpn Dent Sci Rev* 45:52–8. doi:10.1016/j.jdsr.2008.12.001.
Smith, T. D., M. N. Muchlinski, K. D. Jankord, A. J. Progar, et al. 2015. Dental maturation, eruption, and gingival emergence in the upper jaw of newborn primates. *Anat Rec* 298:2098–131. doi:10.1002/ar.23273.
Spencer, M. A. 1998. Force production in the primate masticatory system: electromyographic tests of biomechanical hypotheses. *J Hum Evol* 34:25–54.
Stembírek, J., M. Buchtová, T. Král, E. Matalová, et al. 2010. Early morphogenesis of heterodont dentition in minipigs. *Eur J Oral Sci* 118:547–58.
Stock, D. W. 2001. The genetic basis of modularity in the development and evolution of the vertebrate dentition. *Philos Trans R Soc Lond B Biol Sci* 356:1633–53.
Stock, D. W., K. M. Weiss, and Z. Zhao. 1997. Patterning of the mammalian dentition in development and evolution. *BioEssays* 19:481–90.
Sujon, M. K., M. K. Alam, and S. A. Rahman. 2016. Prevalence of third molar agenesis: associated dental anomalies in non-syndromic 5923 patients. *PLoS One* 11:e0162070.

Sussman, R. W., D. T. Rasmussen, and P. H. Raven. 2013. Rethinking primate origins again. *Am J Primatol* 75:95–106.

Swindler, D. R. 2002. *Primate Dentition: An Introduction to the Teeth of Non-Human Primates*. Cambridge: Cambridge University Press.

Tattersall, I. 1982. *The Primates of Madagascar*. New York: Columbia University Press.

Tattersall, I., and J. H. Schwartz. 1974. Craniodental morphology and the systematics of the Malagasy lemurs (*Primates, Prosimii*; Pt. 3). *Anthropol Pap Am Mus Nat Hist* 52:141–90.

Thomas, B. L., A. S. Tucker, M. Qui, C. A. Ferguson, et al. 1997. Role of Dlx-1 and Dlx-2 genes in patterning of the murine dentition. *Development* 124:4811–18.

Tokita, M., W. Chaeychomsri, and J. Siruntawineti. 2013. Developmental basis of toothlessness in turtles: insight into convergent evolution of vertebrate morphology. *Evolution* 6:260–73.

Townsend, G., E. F. Harris, H. Lesot, F. Clauss, et al. 2009. Morphogenetic fields within the human dentition: a new, clinically relevant synthesis of an old concept. *Arch Oral Biol* 54:S34–S44.

Tucker, A., and P. T. Sharpe. 2004. The cutting-edge of mammalian development: how the embryo makes teeth. *Nat Rev Gen* 5:499–508.

van Nievelt, A. F. H., and K. K. Smith. 2005. To replace or not to replace: the significance of reduced functional tooth replacement in marsupial and placental mammals. *Paleobiology* 31:324–46. doi:10.1666/0094-8373(2005)031[0324:trontr]2.0.co;2.

Vastardis, H., N. Karimbux, S. W. Guthua, J. G. Seidman, et al. 1996. A human MSX1 homeodomain missense mutation causes selective tooth agenesis. *Nat Genet* 13:417–21.

Wakamatsu, Y., S. Egawa, Y. Terashita, H. Kawasaki, et al. 2019. Homeobox code model of heterodont tooth in mammals revised. *Sci Rep* 9:12865.

Ward, C. V., J. M. Plavcan, and F. K. Manthi. 2010. Anterior dental evolution in the *Australopithecus anamensis–afarensis* lineage. *Philos Trans R Soc Lond B Biol Sci* 365:3333–44. doi:10.1098/rstb.2010.0039.

Williams, M. A., and A. Letra. 2018. The changing landscape in the genetic etiology of human tooth agenesis. *Genes* 9:255 doi:10.3390/genes9050255.

Williston, S. W. 1914. *Water Reptiles of the Past and Present*. Chicago, IL: University of Chicago Press.

Wu, X., J. Hu, G. Li, Yan Li, et al. 2020. Biomechanical stress regulates mammalian tooth replacement via the integrin β1-RUNX2-Wnt pathway. *EMBO J* 39:e102374.

Yamada, S., R. Lav, J. Li, A. S. Tucker, et al. 2019. Molar bud-to-cap transition is proliferation independent. *J Dent Res* 98:1253–61. doi:10.1177/0022034519869307.

Yu, T., and O. D. Klein. 2020. Molecular and cellular mechanisms of tooth development, homeostasis and repair. *Development* 147:dev184754. doi:10.1242/dev.184754.

Zhao, Z., K. M. Weiss, and D. W. Stock. 2000. Development and evolution of dentition patterns and their genetic basis. In *Development, Function and Evolution of Teeth*, ed. M. F. Teaford, M. M. Smith, and M. W. J. Ferguson, 152–72. Cambridge: Cambridge University Press. doi:10.1017/cbo9780511542626.011.

4 Gene Regulatory Processes in the Development and Evolution of Primate Skeletal Traits

Genevieve Housman

CONTENTS

4.1 Introduction 83
4.2 Cells in the Skeleton 85
4.3 Obtaining Molecular Data from Human and Nonhuman Primate Skeletal Tissues 86
4.4 Inferences Made from Comparative Skeletal Tissue Studies in Human and Nonhuman Primates 89
4.5 Associations between Gene Regulation and Trait Variation in Primate Skeletons 91
4.5.1 Primate Postcranial Skeletal Morphologies 92
4.5.2 Primate Postcranial Skeletal Pathologies 93
4.6 Future Directions for Examining Cell Processes Contributing to Skeletal Traits in Primates 95
4.7 Conclusion 97
4.8 References 97

4.1 INTRODUCTION

The skeletal system of vertebrates serves a variety of functions, including supporting the body, facilitating movement, protecting internal organs, storing and releasing minerals and fat, and producing blood cells. Because of its important roles in the body and ability to fossilize in archaeological and paleontological records, the skeleton is essential in anthropological and evolutionary research (Percival and Richtsmeier 2017).

The skeleton holds up surrounding soft tissues and anchors skeletal muscles for mobility through complex morphologies in each skeletal element. For example, the rounded depression of the sphenoid bone at the sella turcica helps to encase and protect the pituitary gland of the brain, and the robust and extended portion of the proximal ulna bone at the olecranon process helps to secure the triceps brachii and enable elbow extension. Because the sizes, positioning, and shapes of such bony features vary across species, they are often used to distinguish primate groups from one another (Fleagle 1999; Ankel-Simons 2007). Comparative skeletal anatomy provides insight into the timing and types of evolutionary forces that resulted in species-specific differences. One important set of skeletal traits that differentiate species are morphologies adapted to specific forms of locomotion in particular ecological niches. These morphologies can be examined in extant species, and because the extracellular matrices produced by skeletal cells to form bones can fossilize and preserve in geological strata, such morphologies can also be studied in the fossilized remains of extinct species. In combination, these studies of extant and extinct primate skeletons have been critical for informing primate evolutionary theories.

In addition to evolutionary changes in skeletal traits, the skeleton can also develop and acquire features during an individual's life that reflect underlying genetic variation or that arise in response to external conditions. Such attributes on skeletal elements can be studied to glean aspects of human demography (e.g., sex and age) and life experiences (e.g., injury, disease burden, nutritional changes, and other stressors) (Larsen 2018). Macro- and microscopic changes in the skeleton, such as epiphyseal fusion in response to aging, dental hypoplasia in response to malnutrition, or bone lesions as a result of prolonged tuberculosis infection, are a primary focus in bioarchaeology and forensic research (Temple 2018; Dirkmaat et al. 2018).

The development and evolution of healthy and pathological skeletal traits are influenced by genetic (Valdes and Spector 2011; Reynard and Loughlin 2013; Joganic et al. 2017; Ritzman et al. 2017; Zengini et al. 2018) and environmental (Cooper et al. 2020; Blagojevic et al. 2010; Macrini et al. 2013; Lewton 2017; Lewton et al. 2019) factors. Gene regulation, or the timing and degree to which different genes are expressed in cells, can also impact phenotypes. Genes can be regulated through several mechanisms, including epigenetic mechanisms, which are heritable molecular modifications that can change gene expression without altering the underlying genetic code. Recent research has found that these regulatory and epigenetic mechanisms can also impact skeletal traits (Cotney et al. 2013; Delgado-Calle et al. 2013; Gokhman et al. 2014, 2020; Rushton et al. 2014; Guo et al. 2017; Simon and Jeffries 2017; Peffers et al. 2018; Housman et al. 2019, 2020a; Richard et al. 2020).

Thus, skeletal traits evolve and develop via numerous complex processes. Determining how skeletal variability arises both inter- and intra-specifically requires examining how genetic, environmental, regulatory, and epigenetic factors, both individually and in combination, contribute to the expression of different phenotypes. Moreover, because skeletal tissue is composed of several types of cells (Hall 2015), it is important to examine molecular processes and environmental effects in the cells that make up the skeleton. How genetic variation and environmental disturbances may influence the development of skeletal traits and how the molecular regulation between genetic and environmental effects contributes to these processes must be examined at the cellular level in order to provide deeper insight into skeletal trait evolution in primates.

This chapter explores recent scientific endeavors that evaluate the contribution of gene regulation in skeletal cells to primate evolution. It begins with an overview of the cell types present in skeletal tissues and how molecular data can be obtained from both modern and ancient skeletal tissues. It next discusses how comparative studies using such data have informed our understanding of primate skeletal trait evolution with respect to vocal and facial anatomies. It further explores how studies in extant primates are identifying tighter associations between gene regulation and morphological and pathological phenotypes of the postcranial skeleton. Last, it concludes with a proposal of future in vitro models that will enable more focused evaluations of gene-by-environment interactions contributing to complex skeletal traits in human and nonhuman primates.

4.2 CELLS IN THE SKELETON

The skeleton is composed of several cell types. While some cells directly form and maintain the hard tissues that make up skeletal elements, other cells contribute to different functions in the body. In particular, skeletal elements contain bone marrow, which consists of fat stores (yellow marrow) and sites of hematopoiesis (red marrow) at which red blood cells, white blood cells, and platelets are produced. Bone marrow cell types are essential for metabolic, circulatory, and immune functions but do not produce skeletal extracellular matrices. Thus, they are not discussed further in this chapter.

Broadly speaking, there are four types of vertebrate skeletal tissues—bone, cartilage, dentine, and enamel—as well as several intermediate tissues between these broad classes (e.g., chondroid, chondroid bone, cementum, and enameloid) (Hall 2015). However, dentine matrix, which is synthesized by odontoblasts, and enamel matrix, which is synthesized by ameloblasts, are often classified as dental tissues given their unique role in teeth, so these tissues and cell types are not covered here. Instead, this chapter focuses on bone and cartilage, which are the primary tissue types in primate skeletons.

Bone is a hard, dense connective tissue that is formed by osteoblasts and osteocytes. In addition to its mechanical functions (support the body, facilitate movement, protect internal organs), the bony matrix serves as a reservoir for calcium and phosphorus, which can be released into the bloodstream to support physiological processes when needed. On the other hand, cartilage is a semi-rigid connective tissue that is formed by chondroblasts and chondrocytes, and cartilaginous matrix enables low-friction articulation of skeletal elements and facilitates the transmission of loads to underlying bony matrices.

Bone and cartilage cells are located in different portions of skeletal elements and serve different roles during skeletal development and maintenance. Specifically, bone-forming cells are located in the compact cortical and spongy trabecular bone of skeletal elements, while cartilage-forming cells are primarily located at the articulating ends of skeletal elements. In general, long bones (e.g., femur) develop via endochondral ossification in which skeletal elements begin as cartilaginous tissues and then ossify. Thus, chondrogenic cells and cartilage are the preliminary cell types and tissues that establish much of the prenatal skeleton structure before being replaced by osteogenic cells and bone, which then make up the majority of the adult skeleton. Conversely, flat and irregular bones (e.g., fontal bone) develop via intramembranous ossification in which skeletal elements begin as concentrations of osteoblasts that produce osteoid. Several cranial bones also receive developmental contributions from cells known as neural crest cells, which emerge from the central nervous system early in gestation. In addition to migrating to and adding to aspects of facial skeletal tissues, neural crest cells also differentiate into a variety of other nonskeletal cell types (Jheon and Schneider 2009; Cordero et al. 2011; Bronner and LeDouarin 2012). In both developmental processes, basic skeletal shape is determined *in utero* before ossification occurs, and the initial patterning of matrix condensations that are replaced by osteoblasts and hard bony tissue can influence the final morphologies that skeletal elements form. After skeletal elements are established and fully grown, bone continues remodeling itself throughout an organism's life—with bone-forming cells (osteoblasts) producing bone and their hematopoietic-derived counterparts (osteoclasts) breaking down bone. Bone turnover continually acts to either maintain the initial skeletal morphology set during development or to reshape morphology so that skeletal elements can better support an organism experiencing particular environmental constraints. Conversely, cartilage at articulating ends of skeletal elements does not remodel or repair itself.

Given the similarities and differences between bone- and cartilage-forming cells, it is important to consider the roles of each cell type when evaluating skeletal trait variation within and among primate species. Specifically, understanding how genetic variation, environmental disturbances, and regulatory processes shape skeletal cell function is crucial for informing our interpretations of skeletal elements.

4.3 OBTAINING MOLECULAR DATA FROM HUMAN AND NONHUMAN PRIMATE SKELETAL TISSUES

Obtaining information about factors that influence skeletal trait development and evolution can be challenging. An organisms' nutrition, body weight, and locomotion, among many other traits, can often be reconstructed through observation and measurements of skeletal elements. Conversely, assessing acute, cellular responses to specific external stimuli requires more controlled experimentation. Similarly, obtaining molecular data from skeletal tissues has its own set of ethical and logistic considerations.

From an ethical perspective, skeletal tissues cannot be deliberately removed from living human and nonhuman primates. Bone marrow biopsies are possible, but these procedures can be painful (Lidén et al. 2012) and only produce a limited number of cells, very few of which are bone- or cartilage-forming cells (Baccin et al. 2020). Alternatively, fresh skeletal samples can be obtained during surgeries—osteosarcoma removal, joint replacements due to osteoarthritis, bone fractures requiring surgical implants—but in these cases, molecular patterns may be confounded with disease-state or injury-induction. Further, acquiring consent and ethical board approvals, as well as organizing the logistics of sample collections during surgical events, is difficult. On the other hand, skeletal tissue can be procured post-mortem (Kuliwaba et al. 2005). The length of time between death and sample collection can impact the preservation of molecular patterns (Bahar et al. 2007; Garneau et al. 2007). For many genetic techniques, samples should be assayed immediately following collection or should be flash frozen in liquid nitrogen to sufficiently preserve molecular patterns of interest (Wolf 2013; Klemm et al. 2019). However, more often than not, skeletal samples are not collected or stored appropriately for molecular assays, making many current skeletal collections of limited use or unusable for such molecular studies. Further, because skeletal tissues can be heterogeneous, isolating molecular patterns from specific cell types is also an important consideration.

Given the inherent challenges in obtaining skeletal molecular data from human and nonhuman primates, little has yet been learned about gene regulation in skeletal tissues from these species. For instance, while the Genotype-Tissue Expression project (The GTEx Consortium 2015) documents regulatory information from dozens of human tissues, it contains no data on gene expression in human skeletal tissues. Similarly, the Roadmap Epigenomics Mapping Consortium (Bernstein et al. 2010), which catalogs human epigenomic data across tissues and cell types, only lists regulatory data from primary human osteoblasts and articular chondrocytes from a couple cell lines. Instead, most molecular data from human skeletal tissues come from studies of ancient DNA (aDNA) or skeletal diseases. Although much more data on gene regulation of bone- and cartilage-forming cells in nonprimate animal models exist, knowing about the regulation of cells making mineralized tissues in human and nonhuman primates will better inform skeletal studies in the primate order.

Several methods exist to collect and examine molecular information from skeletal tissues. Extracting genetic information from cells in bony tissues involves pulverizing the hard, extracellular matrix surrounding cells, lysing the cell membranes, and isolating the DNA molecules from other organic matter (Cabana et al. 2013). Conversely, in cartilaginous and fetal skeletal tissues, cells are often isolated through tissue homogenization and enzymatic collagenase treatments that further break down extracellular matrices (Cotney et al. 2013; Guo et al. 2017; Richard et al. 2020). For fresh or well-preserved samples, standard phenol-chloroform protocols can be used to extract DNA (Barnett and Larson 2012). Conversely, for poorly preserved samples, such as archaeological and forensic remains, silica-based extraction methods have proven to be more efficient at extracting low-quality and fragmented aDNA (Nieves-Colón and Stone 2018). Once DNA molecules are isolated, they can be amplified and sequenced to obtain genetic information.

Patterns of gene regulation can also be assessed from extracted DNA. As previously mentioned, variation in gene regulation impacts the timing and degree to which different genes are expressed in cells, and this in turn can affect cell function and tissue phenotypes. Epigenetic mechanisms, which include DNA methylation and histone modifications, are a set of molecular processes that can modify gene expression levels without altering the underlying genetic code. In mammals, DNA methylation involves the binding of methyl groups to cytosine nucleotides. Methylation in gene promoter regions can silence gene expression (Suzuki and Bird 2008), while methylation in gene bodies is correlated with the activation of gene expression (Singer et al. 2015). Alternatively, histone modifications affect chromatin conformation and the accessibility of DNA to transcription and expression. The addition of acetyl groups to histones modifies histone protein conformation such that DNA is unwound, making it less compact and more accessible for transcription. Conversely, when histone proteins are deacetylated, DNA becomes compact and not accessible for transcription. Additionally, the binding or removal of methyl groups to histones can either activate (e.g., H3K4me1, H3K4me3) or silence gene expression (e.g., H3K27me3) (Bannister and Kouzarides 2011; Ernst et al. 2011). Because variation in epigenetic patterns has been associated with clinical and evolutionary traits in human and nonhuman primates (Lettice et al. 2003; Wittkopp and Kalay 2012; Hernando-Herraez et al. 2015; Lappalainen and Greally 2017), epigenetic markers are useful to consider with respect to skeletal traits.

When skeletal samples are well preserved, DNA methylation can be evaluated using a bisulfite treatment which converts unmethylated cytosines to uracils and keeps methylated cytosines as cytosines. Bisulfite-treated DNA can then be evaluated at a genome-wide scale using a variety of methods—microarrays (Housman et al. 2019, 2020a), reduced representation bisulfite sequencing (Vilgalys et al. 2019), or whole-genome bisulfite sequencing (Blake et al. 2020). Although directly assessing methylation patterns in aDNA is sometimes possible (Smith et al. 2015), the degradation of these epigenetic markers over time usually makes conventional methods impractical. Nevertheless, patterns of degradation present in aDNA are reflective of prior methylation patterns—with unmethylated cytosines decaying to uracils and methylated cytosines decaying to thymines (Briggs et al. 2010). Given this, researchers have developed methods for reconstructing DNA methylation maps of ancient genomes using patterns of deamination present in aDNA from skeletal tissues (Gokhman et al. 2014).

Evaluating patterns of chromatin accessibility in skeletal tissues also depends on sample preservation, as most methods require fresh tissue samples. ATAC-seq is a common method for isolating, amplifying, and sequencing regions of the genome that are completely accessible (i.e., not bound by proteins) and involves initially treating chromatin with hyperactive Tn5 transposase (Guo et al. 2017; Richard et al. 2020). Alternatively, isolating, amplifying, and sequencing regions of the genome that are bound by specific proteins, including histones with epigenetic modifications, is accomplished using ChIP-seq methods, in which DNA and bound proteins are cross-linked with formaldehyde before proteins of interest and their attached DNA molecules are immunoprecipitated (Cotney et al. 2012). Some markers of chromatin conformation, such as nucleosome occupancy, are stable over time, and the tendency of aDNA to fragment in linker regions between nucleosomes has enabled the reconstruction of nucleosome occupancy in ancient samples (Pedersen et al. 2014).

Last, studying the effects of regulatory processes on gene expression by directly measuring RNA expression levels is methodologically challenging. RNA rapidly degrades (Garneau et al. 2007), which makes it difficult to examine in extant primate skeletal tissues and unfeasible to examine in extinct primate skeletal tissues. Nevertheless, novel in vitro models may enable further exploration of these and other molecular processes.

Using the genetic and epigenetic methods described here, researchers have started uncovering how molecular patterns in skeletal tissues vary within and among primates and contribute to complex skeletal traits. The next two sections describe how comparative studies using molecular data from skeletal tissues have informed our understanding of human and nonhuman primate skeletal trait evolution with respect to vocal and facial anatomies and how molecular studies in extant primates are identifying tighter associations between gene regulation and morphological and pathological phenotypes of the postcranial skeleton.

4.4 INFERENCES MADE FROM COMPARATIVE SKELETAL TISSUE STUDIES IN HUMAN AND NONHUMAN PRIMATES

Recent efforts have examined gene regulation in skeletal tissues from extant (Housman et al. 2020a) and extinct (Gokhman et al. 2014, 2020) primates using a comparative perspective to identify species-specific changes that correspond with complex skeletal trait differences. These endeavors have primarily focused on DNA methylation patterns, as the epigenetic marker most accessible to assay in skeletal tissues.

Soon after the discovery of the Denisovan population, researchers initiated comparative skeletal epigenetics studies by reconstructing ancient methylation maps for this archaic hominin and Neandertals, comparing these patterns to those present in modern humans. Levels of methylation across the genome were highly similar among hominin species (99% identical) (Gokhman et al. 2014), suggesting that gene regulation and expression in skeletal tissues were also comparable among species. Of the regions found to be differentially methylated, several displayed lineage-specific patterns unique to Neandertals ($n = 307$), to the Denisovans ($n = 295$), or to modern humans ($n = 891$) (Gokhman et al. 2014). One region showing high levels of interspecific methylation differences was the *HOXD10* gene. This gene is part of the HOXD cluster which is involved in limb development (Zakany and Duboule 2007). While the gene body of *HOXD10* is hypomethylated in modern human skeletal tissues, it has intermediate levels of methylation in Neandertals and is hypermethylated in Denisovans (Gokhman et al. 2014). Methylation variation in skeletal tissues along *HOXD10* and other HOXD cluster genes implies differential regulation and expression of genes in this genomic region, which may have played a role in the evolution of features in limb skeletal elements that differentiate more gracile modern humans from more robust archaic hominins (Weaver 2003; Mittra et al. 2007; De Groote 2011).

Another study examining DNA methylation patterns in skeletal tissues from nonhuman primates (baboons, macaques, vervets, chimpanzees, and marmosets) identified intermediate levels of methylation across the upstream region and body of *HOXD10* in all species (Housman et al. 2020a). While the results of this study cannot lead to speculation about how substantially the methylation changes in *HOXD10* impact limb morphology across nonhuman primates, they better inform the evolutionary picture of methylation across the primate lineage. Specifically, they indicate that intermediate methylation of *HOXD10* is an ancestral molecular feature and human-specific hypomethylation and Denisovan-specific hypermethylation are derived molecular features. Although neither of the studies described here (Gokhman et al. 2014; Housman et al. 2020a) explicitly tests whether methylation changes in primate skeletal tissues result in skeletal trait differences, interspecific DNA methylation differences do appear to be related to complex skeletal trait differences. Indeed, nonhuman primate species-specific methylation patterns are associated with genes involved in anatomical development and skeletal development (Housman et al. 2020a), suggesting that interspecific DNA methylation differences may have functional consequences that contribute to complex trait differences between species.

Because nonhuman primate skeletal epigenetics data can better inform ancient skeletal epigenetic findings, a more recent study utilized this design by combining skeletal DNA methylation data from Neandertals (n = 2), a Denisovan (n = 1), ancient humans (n = 5), modern humans (n = 54), and chimpanzees (n = 6) (Gokhman et al. 2020). Using these data, differentially methylated regions were identified across species, and chimpanzees were used as an outgroup to determine whether differential methylation along each hominin lineage is ancestral or derived. The study identified hundreds of regions with methylation patterns specific to each hominin lineage, and specifically 873 human-derived regions that are linked with 588 genes (Gokhman et al. 2020).

Using a tool called Gene ORGANizer (Gokhman et al. 2017), organs and regions of the body were tested to identify those phenotypically affected by the human-derived genes. The tool statistically assesses which regions of the body are significantly affected by a set of genes (e.g., which regions are enriched or depleted in human-derived genes). Three regions of the body were found to be particularly affected by human-derived genes: the pelvis, face, and larynx (Gokhman et al. 2020). Human-specific gene regulation changes that impact the pelvis and the face are of interest as the morphologies of these features are highly divergent between modern and archaic hominins (Aiello and Christopher 2002). Further, human-specific gene regulation changes that impact the larynx are of interest because this feature has an important role in speech production, which is necessary for the uniquely human trait of oral language.

The larynx, which is primarily made of cartilage tissues, is the organ by which humans produce sound and vocalize language. In humans, the larynx is positioned much more inferiorly relative to the oral cavity than in other mammals, and its horizontal dimension equals its vertical height (Lieberman 2007). The position and shape of the larynx allow humans to produce a wide range of sounds that are optimal for speech. Conversely, the more superior position of the larynx in nonhuman primates, while still allowing vocalization, impedes the precise articulation capabilities that enable complex oral speech (Lieberman 2007, 2017). It is not clear how or when the lowering of the larynx evolved within the hominin lineage, but laryngeal decent is thought to be driven or at least accompanied by the flattening of the face (Lieberman 2011; Nishimura et al. 2006). While Neandertals and other hominins have flatter faces than nonhuman primates, laryngeal reconstructions are heavily debated (Fitch 2000; Steele et al. 2013), because laryngeal cartilage does not preserve in the fossil record.

Gokhman and colleagues strove to further inform human laryngeal evolution by analyzing skeletal epigenetic data (Gokhman et al. 2020). Specifically, the group focused on one gene containing a large number of human-derived differentially methylated regions—*NFIX*, which is a key regulator of facial and laryngeal development (Shaw et al. 2010; Malan et al. 2010; van Balkom et al. 2011). Across this gene, both modern and ancient humans are hypermethylated, while Neandertals, the Denisovans, and chimpanzees are hypomethylated (Gokhman et al. 2020). The most parsimonious explanation for this is that *NFIX* became methylated along the human lineage after the split from archaic hominins. Because the expression of *NFIX* is negatively correlated with methylation (Maunakea et al. 2010; Carrió et al. 2015; Ford et al. 2017), researchers conjecture that *NFIX* became less active in modern humans, a human-specific change in gene activity that may be associated with the evolution of modern human facial and vocal tract anatomies (Gokhman et al. 2020).

The hypothesis proposed by Gokhman and colleagues is partially supported by a case of *NFIX* expression variation in humans—the genetic disorder Marshall-Smith syndrome in which mutations in *NFIX* deplete the gene's expression. Among other phenotypes, individuals with Marshall-Smith syndrome have extremely flat faces, with a receding midface and prominent forehead (Malan et al. 2010). Facial flatness in otherwise healthy humans has also been documented to vary with NFIX expression; individuals with lower NFIX expression show flatter faces than those with higher NFIX expression. Along this same trajectory, Neandertals, who likely had high *NFIX* expression due to low methylation in this gene, had projecting midfaces and sloping foreheads. Thus, there is evidence of a relationship between *NFIX* activity and facial shape. Additionally, *NFIX* mutations in patients with Marshall-Smith syndrome and other disorders are also associated with laryngeal malformations (Cullen et al. 1997; Shaw et al. 2010). Because changes in laryngeal anatomy co-occur with variation in facial morphology among modern humans (Nishimura et al. 2006), *NFIX* downregulation along the modern human lineage may have contributed to the evolution of the laryngeal and facial morphologies of modern humans. Since the fossil record of archaic laryngeal apparatuses is sparse, gene knockout mouse models are currently being examined to fully evaluate the effect of *NFIX* expression changes on laryngeal anatomy (Gokhman et al., unpublished data). Using a nonprimate animal model will enable a more complete experimental test of these research findings. In conclusion, by integrating extant nonhuman primate and ancient hominin skeletal methylation data, researchers have been able to isolate human-specific changes that may have contributed to the evolution of human-specific vocal and facial anatomies.

4.5 ASSOCIATIONS BETWEEN GENE REGULATION AND TRAIT VARIATION IN PRIMATE SKELETONS

The comparative skeletal research described previously hints that DNA methylation, and gene regulation more broadly, can impact phenotypic variation. To more explicitly examine the connection between gene regulation and complex skeletal traits, researchers have begun examining two broad categories of phenotypes—nonpathological skeletal morphologies of nonhuman primate femora (Housman et al. 2020a) and pathological traits comprising the presence or absence of the skeletal disorder osteoarthritis in nonhuman primates (Housman et al. 2019, 2020b). While laboratory mice represent a critical model for understanding the connections between gene regulation and skeletal traits, mice are phylogenetically distant from primates, limiting the applications of mouse study findings to primates. To address the drawbacks of traditional mouse models, new research has begun evaluating how skeletal DNA methylation variation in individual primates is associated with the phenotypic variation present in those same individuals.

4.5.1 Primate Postcranial Skeletal Morphologies

Different primate species are characterized by divergent skeletal traits that contribute to anatomical variation and enable differential niche occupations and forms of locomotion. Postcranial skeletal elements have been shown to reflect locomotor adaptations particularly well (Knüsel and Sparacello 2018; McGraw 2018). Although the range of mechanisms that enable the development and maintenance of postcranial skeletal features is not entirely understood, as described, genetic and environmental factors, as well as regulatory and epigenetic mechanisms are all involved. Epigenetic variation contributing to primate phenotypic variation was first considered by King and Wilson (1975), and since then a substantial amount of work has been done to characterize regulatory changes in humans (Petronis et al. 2003; Shelnutt et al. 2004; Fraga et al. 2005; Flanagan et al. 2006; Heyn et al. 2013; Slieker et al. 2013) and non-human primates (Martin et al. 2011; Molaro et al. 2011; Pai et al. 2011; Kei Fukuda et al. 2013; Hernando-Herraez et al. 2013; Lindskog et al. 2014; Lea et al. 2016; Gao et al. 2017; Vilgalys et al. 2019). However, previous studies have predominantly focused on blood cells and soft tissues, and direct tests of primate methylation-phenotype relationships have been limited to brain tissues (Enard et al. 2004; Farcas et al. 2009; Provencal et al. 2012; Zeng et al. 2012; Mendizabal et al. 2016; Madrid et al. 2018).

To address the knowledge gaps that exist with respect to epigenetic variation and primate skeletal complexity, researchers have begun characterizing gene regulation patterns in primate appendicular skeletal tissues, which are important in enabling and facilitating locomotion. Similar to the studies described previously (Gokhman et al. 2014, 2020; Housman et al. 2020a), early work in primate postcranial gene regulation focused on identifying molecular variation between species. In one study, researchers examined chromatin accessibility in developmental limb bud tissues from humans, macaques, and mice in order to identify human-specific regulatory patterns that may contribute to human-specific limb morphology (Cotney et al. 2013). By examining histone acetylation patterns in tissues from the initial limb bud stage through the onset of digit emergence and separation, Cotney et al. (2013) were able to isolate several gene promoter and enhancer regions that display human-specific gains of activity. Human-specific regulatory changes are associated genes enriched for functions related to bone morphogenesis and connective tissue, suggesting that they may play molecular roles in the evolution of human-specific limb traits. While this research presents regulatory regions that may be relevant for clarifying primate postcranial skeletal morphology evolution, the explicit relationship between regulatory variation and phenotypic consequences was not tested.

To assess whether epigenetic variation is associated with primate postcranial skeletal trait variation, other researchers evaluated genome-wide DNA methylation patterns in femoral trabecular bone cores from baboons (n = 28), macaques (n = 10), vervets (n = 10), chimpanzees (n = 4), and marmosets (n = 6) to test the hypothesis that specific features of femur morphology are correlated with specific variations in methylation (Housman et al. 2020a). The species included in the study represent a broad phylogenetic sampling of the primate order and display anatomical and locomotor differences—terrestrial and arboreal quadrupedal locomotion in the cercopithecoids (baboons, macaques, vervets), knuckle-walking quadrupedal locomotion along with vertical climbing and occasional bipedal locomotion in the apes (chimpanzees), and vertical clinging and leaping locomotion in the platyrrhines (marmosets) (Cawthon Lang 2005, 2006). Overall, this sample provided a unique opportunity to examine skeletal epigenetic differences in relation to skeletal morphology differences in the primate order. Linear measurements of each femora ($n = 29$) chosen to quantify overall femur shape were also collected to assess how intraspecific methylation patterns directly relate to intraspecific variation in femur morphology.

Previous interspecific studies found connections between methylation variation and phenotypic variation (Cotney et al. 2013; Gokhman et al. 2014, 2020; Housman et al. 2020a), so similar associations were predicted for an intraspecific study. However, the analysis revealed few associations between differential methylation and morphological variation within species. Specifically, differential methylation was only identified in association with baboon femur length, macaque proximal femur width, macaque medial condyle width, vervet femur shaft width, vervet anatomical neck height, and chimpanzee anatomical neck length (Housman et al. 2020a).

Although the study is underpowered due to relatively small sample sizes (Graw et al. 2019), the morphology-associated differentially methylated sites identified likely have weak functional effects and thus no biological role in the development or maintenance of morphological variation in the femur. The sites identified represent individual site-specific methylation changes, known to have limited impact on gene expression changes (Bork et al. 2010; Chen et al. 2011; Koch et al. 2011) rather than accumulations of several methylation changes across gene regions, which are more likely to alter gene expression (Suzuki and Bird 2008; Singer et al. 2015). Further, these sites generally show small changes in methylation which are thought to have little biological relevance (Hernando-Herraez et al. 2013). Overall, the findings of this study suggest that while some nonhuman primate DNA methylation patterns are associated with nonpathological morphological variation in the femur, the data do not yet support a causal or functional link between the two (Housman et al. 2020a).

4.5.2 Primate Postcranial Skeletal Pathologies

While nonpathological skeletal traits do not appear to be strongly associated with skeletal gene regulation changes in primates, other research suggests that skeletal DNA methylation does influence pathological traits (Delgado-Calle et al. 2013; den Hollander et al. 2014; Moazedi-Fuerst et al. 2014; Rushton et al. 2014; Jeffries et al. 2016; Morris et al. 2017; Simon and Jeffries 2017; Ostanek et al. 2018). One skeletal disease that is particularly relevant for primates is osteoarthritis (OA), a complex degenerative joint disease characterized by the degradation of cartilage and the underlying bone in joints. In humans, OA is a leading cause of disability world-wide (Cross et al. 2014), and other nonhuman primates vary in their susceptibilities to this disorder. For example, the prevalence of OA in humans and baboons is very similar, 66% and 59% of older individuals, respectively (O'Connor 2006; Cox et al. 2013), while the prevalence of OA in wild (Rothschild and Rühli 2005) and captive (Videan et al. 2011; Magden et al. 2013) chimpanzees is very low.

Some researchers have conjectured that OA, specifically at the knee joint, may be more prevalent in humans than in other animals due to changes in selective pressures that have acted on the knee joint during human evolution so that it can withstand the biomechanical loads experienced during bipedal locomotion (Richard et al. 2020). To examine this, Richard et al. (2020) profiled chromatin accessibility in human cartilaginous joint regions of developing long bones to examine the relationship between genetic variants selected during the evolution of human-specific knee morphology and genetic variants associated with OA. By comparing chromatin accessibility across different limb joints, they were able to isolate open chromatin regions that are unique to the knee joint. These knee-specific regulatory elements experienced past positive selection and recent negative selection, constraint, and drift in humans, and they overlap with several loci associated with OA pathogenesis (Richard et al. 2020).

While focusing on the potential mechanisms underlying human OA is important, it is also essential to compare human findings with regulatory patterns and disease susceptibilities in other nonhuman primates in order to clarify the conservation or divergence of these genotype-phenotype relationships in the primate order. Because of their similar prevalence rate and pathological progression of OA to humans, cercopithecoids like baboons are particularly good models of OA (Cox et al. 2013; Macrini et al. 2013) that can be studied from an evolutionary medicine perspective. Thus, researchers have examined baboons to test whether the strong association between gene regulation and OA pathogenesis that has been identified in humans (Delgado-Calle et al. 2013; Fernández-Tajes et al. 2014; Moazedi-Fuerst et al. 2014; Rushton et al. 2014; Aref-Eshghi et al. 2015; Alvarez-Garcia et al. 2016; Jeffries et al. 2016) is evolutionarily conserved across other primates.

An initial study of skeletal gene regulation underlying baboon OA examined a cohort of age-matched adult baboons that were skeletally healthy (n = 5) or had severe knee OA (n = 5) (Housman et al. 2019). Genome-wide DNA methylation patterns were assessed in articular cartilage and underlying trabecular bone sampled from the right medial femoral condyle. This methodology was selected because both articular cartilage and trabecular bone are affected during OA progression (Mahjoub et al. 2012) and because the medial femoral condyle is most severely and consistently degraded in OA (Peters et al. 2018). Although the study had limited statistical power to detect differential methylation associated with OA, some cartilaginous changes were identified between healthy and OA individuals (Housman et al. 2019). In particular, *RUNX1* was found to be hypomethylated in OA cartilage as compared to healthy cartilage, matching what has been previously identified in studies of human OA cartilage (Fernández-Tajes et al. 2014). The remaining differentially methylated genes identified in the study are novel (Housman et al. 2019), having not been previously associated with OA in humans (Delgado-Calle et al. 2013; Fernández-Tajes et al. 2014; Moazedi-Fuerst et al. 2014; Rushton et al. 2014; Aref-Eshghi et al. 2015; Alvarez-Garcia et al. 2016; Jeffries et al. 2016).

Expanding upon the preliminary findings in Housman et al. (2019), the team expanded their sample to better evaluate the contributions of epigenetics to OA development in baboons. Using data from 28 skeletally healthy adult baboons and 28 adult baboons with knee OA, a larger number of differentially methylated positions that differentiate bone and cartilage tissues, as well as healthy and OA disease states within each tissue, have been identified. Approximately 25% of the almost 200,000 sites examined were found to be differentially methylated between cartilage samples and bone samples, and this number persists when considering either healthy baboon tissues or OA baboon tissues (Housman et al. 2020b). The findings suggest that as OA progresses, bone and cartilage tissues maintain distinct regulatory functions and are not de-differentiating or becoming more like one another (Charlier et al. 2019). Additionally, when comparing healthy baboons and baboons with OA, about 2% of sites examined show differential methylation in cartilage, while many fewer sites are differentially methylated in bone (Housman et al. 2020b). The difference in disease-associated methylation patterns across tissues may have some etiological implications—specifically, that cartilage epigenetics may have a more influential role in the development of OA than bone epigenetics. While some OA-related methylation changes observed in baboons appear evolutionarily conserved, others do not (Iliopoulos et al. 2008; Saito et al. 2010; Goldring and Marcu 2012; Delgado-Calle et al. 2013; García-Ibarbia et al. 2013; Fernández-Tajes et al. 2014; Moazedi-Fuerst et al. 2014; Ramos et al. 2014; Reynard et al. 2014; Rushton et al. 2014; Aref-Eshghi et al. 2015; Alvarez-Garcia et al. 2016; Jeffries et al. 2016). Overall, an important finding to take away from this study is that intraspecific skeletal methylation variation can be directly associated with intraspecific skeletal phenotypic variation. However, this relationship is more strongly apparent in disease phenotypes than in nonpathological phenotypes.

4.6 FUTURE DIRECTIONS FOR EXAMINING CELL PROCESSES CONTRIBUTING TO SKELETAL TRAITS IN PRIMATES

The comparative and intraspecific work described previously forms an initial foundation for future explorations of gene regulation and skeletal trait evolution in primates. Some insight into the processes regulating the development and evolution of primate skeletons can be obtained from studies associating general genetic and environmental factors with skeletal trait variation, but analyses of molecular regulation in hard tissues that contain skeletal cells, which can be influenced by both genetic and environmental factors, has greatly advanced this field of research (Housman et al. 2019, 2020a, 2020b; Gokhman et al. 2014, 2020; Cotney et al. 2013; Richard et al. 2020). Nevertheless, these endeavors only reveal molecular patterns averaged across the heterogenous populations of cells within skeletal tissues rather than probing molecular patterns in individual skeletal cell types. Further, they are unable to disentangle which genetic and environmental factors are contributing to skeletal cell processes or the mechanisms by which these factors act on molecular patterns to influence skeletal traits. Therefore, it is essential that in vitro models be developed that enable focused evaluations of gene-by-environmental interactions directly in cell types that contribute to complex skeletal traits.

Given the ethical and logistic difficulties of obtaining primate skeletal tissue samples, alternative in vitro methods of studying skeletal cells, such as cell culture systems, would be highly beneficial. Cell culture systems allow individual cell types to be isolated, subjected to controlled experimentation, and examined for changes in molecular processes. Several researchers use cell culture methods to study bone- and cartilage-producing cells at a higher resolution (e.g., Oldershaw et al. 2010; Hynes et al. 2014; Kanke et al. 2014; Nejadnik et al. 2015). Additionally, relevant environmental factors, such as biomechanical stress, that are known to impact skeletal growth, development, and health can be used as treatments on cell cultures to assess molecular responses. Bone- and cartilage-producing cells are sensitive to external loads (Murray et al. 2001; Henrotin et al. 2016), and in vitro systems have been used to measure how skeletal cells respond to different mechanical stresses, including hydrostatic pressure (Takano-Yamamoto et al. 1991; Parkkinen et al. 1993, 1995; Takahashi et al. 1997, 1998), shearing stress (Mohtai et al. 1996), compression (Bougault et al. 2008, 2009, 2012; Sanchez et al. 2009, 2012), and tension (De Witt et al. 1984; Banes et al. 1990; Holmvall et al. 1995; Fukuda et al. 1997; Fujisawa et al. 1999; Honda et al. 2000; Lin et al. 2010; Gao et al. 2016; Wang et al. 2016; Xu et al. 2016). Researchers have also developed in vitro models of OA which involve specific types and amounts of biomechanical stress (Fujisawa et al. 1999; Honda et al. 2000; Lin et al. 2010; Johnson et al. 2016). Studying environmental treatments, such as biomechanical strain, in an in vitro cell culture model is advantageous, as such systems enable highly controlled experimentation that can be used to assess acute, cellular, and molecular responses to different types, magnitudes, and durations of external stimuli—features that are difficult and in some cases impossible to evaluate in animal models.

Despite the wealth of in vitro methods available for studying skeletal cell functions and gene-by-environment effects, current research is limited to only using a small set of cell lines from humans or other model organisms. As research in this area expands, studying cell types derived from larger samples of humans and from additional nonhuman primate species will improve primate skeletal research at both population and evolutionary levels.

While not specifically focused on the genetic regulation of skeletal traits, several recent projects have begun to examine a variety of cell types (Advani et al. 2016; Pizzollo et al. 2018) and organoid models (Pollen et al. 2019) in primate species (Edsall et al. 2019; Zhang et al. 2019) to study primate functional genomics more broadly (Grogan and Perry 2020). While primary skeletal cells can be cultured from primates, obtaining these has the same complications as obtaining primary primate skeletal tissues. Thus, the availability of a comparative panel of matched human and chimpanzee induced pluripotent stem cell (iPSC) lines (Gallego Romero et al. 2015) may be a more appropriate alternative. iPSCs are renewable sample resources reprogrammed from differentiated cells (Shi et al. 2017). Their potential to differentiate into any cell type makes them ideal for studying difficult to obtain cell types, such as bone- and cartilage-forming cells. Additionally, the specific panel of human and chimpanzee iPSCs described by Gallego Romero et al. (2015) is particularly valuable given that sample collections from chimpanzees are no longer allowed in the United States. Several levels of molecular regulation have been examined in these iPSCs (Gallego Romero et al. 2018; Eres et al. 2019; Ward et al. 2018) and iPSC-derived cell types (Pavlovic et al. 2018; Blake et al. 2018), as well as dynamic gene expression responses to environmental perturbations (Ward and Gilad 2019). Additionally, researchers have evaluated epigenomic and transcriptomic divergences in skeletally relevant precursor cells, such as iPSC-derived neural crest cells in two human and two chimpanzee cell lines (Prescott et al. 2015). Similarly, ongoing research is examining regulatory patterns in bone- and cartilage-producing cell types from human and chimpanzee iPSCs and how molecular processes change in response to biomechanical treatments (Housman et al. unpublished data).

Pursuing such in vitro methods for examining primate skeletal cell functions should reveal more insight into the factors and mechanisms contributing to skeletal trait variation within and among primates. Simultaneously, it is important to recognize that in vitro cell culture systems are models. While some models are useful, all models are imperfect. It is critical to consider the benefits and limitations of in vitro models before using them to answer research questions and when interpreting results. This is true for cell culture systems, and similar considerations must also be made when using nonprimate animal models, such as mice, to inform our understanding of human and nonhuman primate biology. Overall, in order to most effectively study primate skeletal evolution, a combination of in vitro and in vivo methods is necessary to successfully interrogate functional mechanisms and pathways and frame these processes in the molecular, cellular, and biological variation present in primates.

4.7 CONCLUSION

The development and evolution of primate skeletal anatomy is a fascinating area of exploration, but until recently, research in the field has been unable to fully incorporate data regarding primate skeletal cell functions and processes into its understanding of skeletal variation. However, some recent comparative studies have examined gene regulation pattern in skeletal tissues from extant and extinct primates, revealing new information about primate skeletal trait evolution related to vocal and facial anatomies. Additionally, some investigations have tested associations between gene regulation and morphological and pathological phenotypes of primate skeletal tissues. Nevertheless, current studies only hint at underlying cellular processes that contribute to primate skeletons. More focused evaluations using in vitro model systems are viable avenues for understanding how genetic variation, environmental disturbances, and regulatory processes shape skeletal cells and the resulting skeletal traits they produce in primates.

4.8 REFERENCES

Advani, A. S., A. Y. Chen, and C. C. Babbitt. 2016. Human fibroblasts display a differential focal adhesion phenotype relative to chimpanzee. *Evol Med Public Health* 2016:110.

Aiello, L., and D. Christopher. 2002. *An Introduction to Human Evolutionary Anatomy*. Boston: Elsevier. doi:10.1016/C2009-0-02515-X.

Alvarez-Garcia, O., K. M. Fisch, N. E. Wineinger, R. Akagi, et al. 2016. Increased DNA methylation and reduced expression of transcription factors in human osteoarthritis cartilage: differential DNA methylation in OA cartilage. *Arthritis Rheumatol* 68:1876–86. doi:10.1002/art.39643.

Ankel-Simons, F. 2007. *Primate Anatomy*, 3rd ed. New York: Academic Press.

Aref-Eshghi, E., Y. Zhang, M. Liu, P. E. Harper, et al. 2015. Genome-wide DNA methylation study of hip and knee cartilage reveals embryonic organ and skeletal system morphogenesis as major pathways involved in osteoarthritis. *BMC Musculoskel Dis* 16:1–10. doi:10.1186/s12891-015-0745-5.

Baccin, C., J. Al-Sabah, L. Velten, P. M. Helbling, et al. 2020. Combined single-cell and spatial transcriptomics reveal the molecular, cellular and spatial bone marrow niche organization. *Nat Cell Biol* 22:38–48. doi:10.1038/s41556-019-0439-6.

Bahar, B., F. J. Monahan, A. P. Moloney, O. Schmidt, et al. 2007. Long-term stability of RNA in post-mortem bovine skeletal muscle, liver and subcutaneous adipose tissues. *BMC Mol Biol* 8:108. doi:10.1186/1471-2199-8-108.

Balkom, I. D. C. van, A. Shaw, P. J. Vuijk, M. Franssens, et al. 2011. Development and behaviour in Marshall-Smith syndrome: an exploratory study of cognition, phenotype and autism. *J Intell Disabil Res* 55:973–87. doi:10.1111/j.1365-2788.2011.01451.x.

Banes, A. J., G. W. Link, J. W. Gilbert, R. Tran Son Tay, et al. 1990. Culturing cells in a mechanically active environment. *Am Biotechnol Lab* 8:12–22.

Bannister, A. J., and T. Kouzarides. 2011. Regulation of chromatin by histone modifications. *Cell Res* 21:381–95. doi:10.1038/cr.2011.22.

Barnett, R., and G. Larson. 2012. A phenol-chloroform protocol for extracting DNA from ancient samples. In *Ancient DNA*, ed. B. Shapiro and M. Hofreiter, 13–19. Methods in Molecular Biology 840. New York: Humana Press. doi:10.1007/978-1-61779-516-9_2.

Bernstein, B. E., J. A. Stamatoyannopoulos, J. F. Costello, B. Ren, et al. 2010. The NIH roadmap epigenomics mapping consortium. *Nat Biotechnol* 28:1045–8. doi:10.1038/nbt1010-1045.

Blagojevic, M., C. Jinks, A. Jeffery, and K. P. Jordan. 2010. Risk factors for onset of osteoarthritis of the knee in older adults: a systematic review and meta-analysis. *Osteoarthr Cartilage* 18:24–33. doi:10.1016/j.joca.2009.08.010.

Blake, L. E., J. Roux, I. Hernando-Herraez, N. E. Banovich, et al. 2020. A comparison of gene expression and DNA methylation patterns across tissues and species. *Genome Res* 30:250–62. doi:10.1101/gr.254904.119.

Blake, L. E., S. M. Thomas, J. D. Blischak, C. J. Hsiao, et al. 2018. A comparative study of endoderm differentiation in humans and chimpanzees. *Genome Biol* 19:162. doi:10.1186/s13059-018-1490-5.

Bork, S., S. Pfister, H. Witt, P. Horn, et al. 2010. DNA methylation pattern changes upon long-term culture and aging of human mesenchymal stromal cells. *Aging Cell* 9:54–63. doi:10.1111/j.1474-9726.2009.00535.x.

Bougault, C., E. Aubert-Foucher, A. Paumier, E. Perrier-Groult, et al. 2012. Dynamic compression of chondrocyte-agarose constructs reveals new candidate mechanosensitive genes. *PLoS One* 7:e36964. doi:10.1371/journal.pone.0036964.

Bougault, C., A. Paumier, E. Aubert-Foucher, and F. Mallein-Gerin. 2008. Molecular analysis of chondrocytes cultured in agarose in response to dynamic compression. *BMC Biotechnol* 8:71. doi:10.1186/1472-6750-8-71.

Bougault, C., A. Paumier, E. Aubert-Foucher, and F. Mallein-Gerin. 2009. Investigating conversion of mechanical force into biochemical signaling in three-dimensional chondrocyte cultures. *Nature Protoc* 4:928–38. doi:10.1038/nprot.2009.63.

Briggs, A. W., U. Stenzel, M. Meyer, J. Krause, et al. 2010. Removal of deaminated cytosines and detection of in vivo methylation in ancient DNA. *Nucleic Acids Res* 38:e87. doi:10.1093/nar/gkp1163.

Bronner, M. E., and N. M. LeDouarin. 2012. Development and evolution of the neural crest: an overview. *Dev Biol* 366:2–9. doi:10.1016/j.ydbio.2011.12.042.

Cabana, G. S., B. I. Hulsey, and F. L. Pack. 2013. Molecular methods. In *Research Methods in Human Skeletal Biology*, 449–80. Waltham: Elsevier. doi:10.1016/B978-0-12-385189-5.00016-9.

Carrió, E., A. Díez–Villanueva, S. Lois, I. Mallona, et al. 2015. Deconstruction of DNA methylation patterns during myogenesis reveals specific epigenetic events in the establishment of the skeletal muscle lineage. *Stem Cells* 33:2025–36. doi:10.1002/stem.1998.

Cawthon Lang, K. A. 2005. *Primate Factsheets: Common Marmoset (Callithrix jacchus) Taxonomy, Morphology, & Ecology*. http://pin.primate.wisc.edu/factsheets/entry/common_marmoset.

Cawthon Lang, K. A. 2006. *Primate Factsheets: Chimpanzee (Pan troglodytes) Taxonomy, Morphology, & Ecology*. http://pin.primate.wisc.edu/factsheets/entry/chimpanzee.

Charlier, E., C. Deroyer, F. Ciregia, O. Malaise, et al. 2019. Chondrocyte dedifferentiation and osteoarthritis (OA). *Biochem Pharmacol* 165:49–65. doi:10.1016/j.bcp.2019.02.036.

Chen, Y., S. Choufani, J. C. Ferreira, D. Grafodatskaya, et al. 2011. Sequence overlap between autosomal and sex-linked probes on the Illumina HumanMethylation27 microarray. *Genomics* 97:214–22. doi:10.1016/j.ygeno.2010.12.004.

Cooper, C., S. Snow, T. E. McAlindon, S. Kellingray, et al. 2000. Risk factors for the incidence and progression of radiographic knee osteoarthritis. *Arthrit Rheum* 43:995–1000. doi:10.1002/1529-0131(200005)43:5<995::AID-ANR6>3.0.CO;2-1.

Cordero, D. R., S. Brugmann, Y. Chu, R. Bajpai, et al. 2011. Cranial neural crest cells on the move: their roles in craniofacial development. *Am J Med Genet A* 155:270–9. doi:10.1002/ajmg.a.33702.

Cotney, J., J. Leng, S. Oh, L. E. DeMare, et al. 2012. Chromatin state signatures associated with tissue-specific gene expression and enhancer activity in the embryonic limb. *Genome Res* 22:1069–80. doi:10.1101/gr.129817.111.

Cotney, J., J. Leng, J. Yin, S. K. Reilly, et al. 2013. The evolution of lineage-specific regulatory activities in the human embryonic limb. *Cell* 154:185–96. doi:10.1016/j.cell.2013.05.056.

Cox, L. A., A. G. Comuzzie, L. M. Havill, G. M. Karere, et al. 2013. Baboons as a model to study genetics and epigenetics of human disease. *ILAR J* 54:106–21. doi:10.1093/ilar/ilt038.

Cross, M., E. Smith, D. Hoy, S. Nolte, et al. 2014. The global burden of hip and knee osteoarthritis: estimates from the Global Burden of Disease 2010 study. *Ann Rheum Dis* 73:1323–30. doi:10.1136/annrheumdis-2013-204763.

Cullen, A., T. A. Clarke, and T. P. O'Dwyer. 1997. The Marshall-Smith syndrome: a review of the laryngeal complications. *Eur J Pediatr* 156:463–4. doi:10.1007/s004310050640.

De Groote, I. 2011. The Neanderthal lower arm. *J Hum Evol* 61:396–410. doi:10.1016/j.jhevol.2011.05.007.

Delgado-Calle, J., A. F. Fernández, J. Sainz, M. T. Zarrabeitia, et al. 2013. Genome-wide profiling of bone reveals differentially methylated regions in osteoporosis and osteoarthritis. *Arthrit Rheum* 65:197–205. doi:10.1002/art.37753.

De Witt, M. T., C. J. Handley, B. W. Oakes, and D. A. Lowther. 1984. In vitro response of chondrocytes to mechanical loading: the effect of short term mechanical tension. *Connect Tissue Res* 12:97–109.

Dirkmaat, D., H. Garvin, and L. L. Cabo, 2018. Forensic anthropology. In *The International Encyclopedia of Biological Anthropology*, 1–17. Hoboken, NJ: Wiley Blackwell. doi:10.1002/9781118584538.ieba0183.

Edsall, L. E., A. Berrio, W. H. Majoros, D. Swain-Lenz, et al. 2019. Evaluating chromatin accessibility differences across multiple primate species using a joint modeling approach. *Genome Biol Evol* 11:3035–53. doi:10.1093/gbe/evz218.

Enard, W., A. Fassbender, F. Model, P. Adorján, et al. 2004. Differences in DNA methylation patterns between humans and chimpanzees. *Curr Biol* 14:R148–9.

Eres, I. E., K. Luo, C. J. Hsiao, L. E. Blake, et al. 2019. Reorganization of 3D genome structure may contribute to gene regulatory evolution in primates. *PLoS Genet* 15:e1008278. doi:10.1371/journal.pgen.1008278.

Ernst, J., P. Kheradpour, T. S. Mikkelsen, N. Shoresh, et al. 2011. Systematic analysis of chromatin state dynamics in nine human cell types. *Nature* 473:43–9. doi:10.1038/nature09906.

Farcas, R., E. Schneider, K. Frauenknecht, I. Kondova, et al. 2009. Differences in DNA methylation patterns and expression of the CCRK gene in human and nonhuman primate cortices. *Mol Biol Evol* 26:1379–89. doi:10.1093/molbev/msp046.

Fernández-Tajes, J., A. Soto-Hermida, M. E. Vázquez-Mosquera, E. Cortés-Pereira, et al. 2014. Genome-wide DNA methylation analysis of articular chondrocytes reveals a cluster of osteoarthritic patients. *Ann Rheum Dis* 73:668–77. doi:10.1136/annrheumdis-2012-202783.

Fitch, W. T. 2000. The evolution of speech: a comparative review. *Trends Cog Sci* 4:258–67. doi:10.1016/S1364-6613(00)01494-7.

Flanagan, J. M., V. Popendikyte, N. Pozdniakovaite, M. Sobolev, et al. 2006. Intra- and inter-individual epigenetic variation in human germ cells. *Am J Hum Genet* 79:67–84.

Fleagle, J. G. 1999. *Primate Adaptation and Evolution*. New York: Academic Press.

Ford, E., M. R. Grimmer, S. Stolzenburg, O. Bogdanovic, et al. 2017. Frequent lack of repressive capacity of promoter DNA methylation identified through genome-wide epigenomic manipulation. *BioRxiv*:170506. doi:10.1101/170506.

Fraga, M. F., E. Ballestar, M. F. Paz, S. Ropero, et al. 2005. Epigenetic differences arise during the lifetime of monozygotic twins. *P Natil Acad Sci* 102:10604–9. doi:10.1073/pnas.0500398102.

Fujisawa, T., T. Hattori, K. Takahashi, T. Kuboki, et al. 1999. Cyclic mechanical stress induces extracellular matrix degradation in cultured chondrocytes via gene expression of matrix metalloproteinases and interleukin-1. *J Biochem* 125:966–75.

Fukuda, K., S. Asada, F. Kumano, M. Saitoh, et al. 1997. Cyclic tensile stretch on bovine articular chondrocytes inhibits protein kinase C activity. *J Lab Clin Med* 130:209–15. doi:10.1016/S0022-2143(97)90098-6.

Fukuda, K., K. Ichiyanagi, Y. Yamada, Y. Go, et al. 2013. Regional DNA methylation differences between humans and chimpanzees are associated with genetic changes, transcriptional divergence and disease genes. *J Hum Genet* 58:446–54. doi:10.1038/jhg.2013.55.

Gallego Romero, I., S. Gopalakrishnan, and Y. Gilad. 2018. Widespread conservation of chromatin accessibility patterns and transcription factor binding in human and chimpanzee induced pluripotent stem cells. *BioRxiv*:466631. doi:10.1101/466631.

Gallego Romero, I., B. J. Pavlovic, I. Hernando-Herraez, X. Zhou, et al. 2015. A panel of induced pluripotent stem cells from chimpanzees: a resource for comparative functional genomics. *eLife* 4:e07103. doi:10.7554/eLife.07103.

Gao, F., Y. Niu, Y. E. Sun, H. Lu, et al. 2017. De novo DNA methylation during monkey pre-implantation embryogenesis. *Cell Res* 2017:1–14. doi:10.1038/cr.2017.25.

Gao, J., S. Fu, Z. Zeng, F. Li, et al. 2016. Cyclic stretch promotes osteogenesis-related gene expression in osteoblast-like cells through a cofilin-associated mechanism. *Mol Med Rep* 14:218–24. doi:10.3892/mmr.2016.5239.

García-Ibarbia, C., J. Delgado-Calle, I. Casafont, J. Velasco, et al. 2013. Contribution of genetic and epigenetic mechanisms to WNT pathway activity in prevalent skeletal disorders. *Gene* 532:165–72. doi:10.1016/j.gene.2013.09.080.

Garneau, N. L., J. Wilusz, and C. J. Wilusz. 2007. The highways and byways of mRNA decay. *Nature Reviews. Mol Cell Biol* 8:113–26. doi:10.1038/nrm2104.

Gokhman, D., G. Kelman, A. Amartely, G. Gershon, et al. 2017. Gene ORGANizer: linking genes to the organs they affect. *Nucleic Acids Res* 45:W138–45. doi:10.1093/nar/gkx302.

Gokhman, D., E. Lavi, K. Prüfer, M. F. Fraga, et al. 2014. Reconstructing the DNA methylation maps of the Neandertal and the Denisovan. *Science* 344:523–7. doi:10.1126/science.1250368.

Gokhman, D., M. Nissim-Rafinia, L. Agranat-Tamir, G. Housman, et al. 2020. Differential DNA methylation of vocal and facial anatomy genes in modern humans. *Nat Commun* 11:1–21. doi:10.1038/s41467-020-15020-6.

Goldring, M. B., and K. B. Marcu. 2012. Epigenomic and microRNA-mediated regulation in cartilage development, homeostasis, and osteoarthritis. *Trends Mol Med* 18:109–18. doi:10.1016/j.molmed.2011.11.005.

Graw, S., R. Henn, J. A. Thompson, and D. C. Koestler. 2019. PwrEWAS: a user-friendly tool for comprehensive power estimation for epigenome wide association studies (EWAS). *BMC Bioinformatics* 20:218. doi:10.1186/s12859-019-2804-7.

Grogan, K. E., and G. H. Perry. 2020. Studying human and nonhuman primate evolutionary biology with powerful in vitro and in vivo functional genomics tools. *Evo Anthropol* 29:143–58. doi:10.1002/evan.21825.

The GTEx Consortium. 2015. The Genotype-Tissue Expression (GTEx) pilot analysis: multitissue gene regulation in humans. *Science* 348:648–60. doi:10.1126/science.1262110.

Guo, M., Z. Liu, J. Willen, C. P. Shaw, et al. 2017. Epigenetic profiling of growth plate chondrocytes sheds insight into regulatory genetic variation influencing height. *eLife* 6:e29329. doi:10.7554/eLife.29329.

Hall, B. K. 2015. *Bones and Cartilage: Developmental and Evolutionary Skeletal Biology*, 2nd ed. Amsterdam: Elsevier/AP.

Henrotin, Y., C. Sanchez, A. C. Bay-Jensen, and A. Mobasheri. 2016. Osteoarthritis biomarkers derived from cartilage extracellular matrix: current status and future perspectives. *Ann Phys Rehabil Med* 59:145–8. doi:10.1016/j.rehab.2016.03.004.

Hernando-Herraez, I., R. Garcia-Perez, A. J. Sharp, and T. Marques-Bonet. 2015. DNA methylation: insights into human evolution. *PLoS Genet* 11:e1005661. doi:10.1371/journal.pgen.1005661.

Hernando-Herraez, I., J. Prado-Martinez, P. Garg, M. Fernandez-Callejo, et al. 2013. Dynamics of DNA methylation in recent human and great ape evolution. *PLoS Genet* 9:e1003763. doi:10.1371/journal.pgen.1003763.

Heyn, H., S. Moran, I. Hernando-Herraez, S. Sayols, D. Monk, et al. 2013. DNA methylation contributes to natural human variation. *Genome Res* 23:1363–72. doi:10.1101/gr.154187.112.

den Hollander, W., Y. F. M. Ramos, S. D. Bos, N. Bomer, et al. 2014. Knee and hip articular cartilage have distinct epigenomic landscapes: implications for future cartilage regeneration approaches. *Ann Rheum Dis* 73:2208–12. doi:10.1136/annrheumdis-2014-205980.

Holmvall, K., L. Camper, S. Johansson, J. H. Kimura, et al. 1995. Chondrocyte and chondrosarcoma cell integrins with affinity for collagen type II and their response to mechanical stress. *Exp Cell Res* 221:496–503. doi:10.1006/excr.1995.1401.

Honda, K., S. Ohno, K. Tanimoto, C. Ijuin, et al. 2000. The effects of high magnitude cyclic tensile load on cartilage matrix metabolism in cultured chondrocytes. *Eur J Cell Biol* 79:601–9. doi:10.1078/0171-9335-00089.

Housman, G., L. M. Havill, E. E. Quillen, A. G. Comuzzie, et al. 2019. Assessment of DNA methylation patterns in the bone and cartilage of a nonhuman primate model of osteoarthritis. *Cartilage* 10:335–45. doi:10.1177/1947603518759173.

Housman, G., E. E. Quillen, and A. C. Stone. 2020a. Intraspecific and interspecific investigations of skeletal DNA methylation and femur morphology in primates. *American J Phy Anthropol* 173:34–49. doi:10.1002/ajpa.24041.

Housman, G., E. E. Quillen, and A. C. Stone. 2020b. An evolutionary perspective of DNA methylation patterns in skeletal tissues using a baboon model of osteoarthritis. *J Orthop Res* 2020:1–10. doi:10.1002/jor.24957.

Hynes, K., D. Menicanin, S. Gronthos, and M. P. Bartold. 2014. Differentiation of IPSC to mesenchymal stem-like cells and their characterization. In *Induced Pluripotent Stem (IPS) Cells*, 353–74. Methods in Molecular Biology. New York: Humana Press. doi:10.1007/7651_2014_142.

Iliopoulos, D., K. N. Malizos, P. Oikonomou, and A. Tsezou. 2008. Integrative microRNA and proteomic approaches identify novel osteoarthritis genes and their collaborative metabolic and inflammatory networks. *PLoS One* 3:e3740. doi:10.1371/journal.pone.0003740.

Jeffries, M. A., M. Donica, L. W. Baker, M. E. Stevenson, et al. 2016. Genome-wide DNA methylation study identifies significant epigenomic changes in osteoarthritic subchondral bone and similarity to overlying cartilage. *Arthritis Rheumatol* 68:1403–14. doi:10.1002/art.39555.

Jheon, A. H., and R. A. Schneider. 2009. The cells that fill the bill: neural crest and the evolution of craniofacial development. *J Dent Res* 88:12–21. doi:10.1177/0022034508327757.

Joganic, J. L., K. E. Willmore, J. T. Richtsmeier, K. M. Weiss, et al. 2017. Additive genetic variation in the craniofacial skeleton of baboons (genus Papio) and its relationship to body and cranial size. *Am J Phys Anthropol* 165:269–85. doi:10.1002/ajpa.23349.

Johnson, C. I., D. J. Argyle, and D. N. Clements. 2016. In vitro models for the study of osteoarthritis. *Vet J* 209:40–9. doi:10.1016/j.tvjl.2015.07.011.

Kanke, K., H. Masaki, T. Saito, Y. Komiyama, et al. 2014. Stepwise differentiation of pluripotent stem cells into osteoblasts using four small molecules under serum-free and feeder-free conditions. *Stem Cell Rep* 2:751–60. doi:10.1016/j.stemcr.2014.04.016.

King, M. C., and A. C. Wilson. 1975. Evolution at two levels in humans and chimpanzees. *Science* 188:107–16. doi:10.1126/science.1090005.

Klemm, S. L., Z. Shipony, and W. J. Greenleaf. 2019. Chromatin accessibility and the regulatory epigenome. *Nat Rev Genet* 20:207–20. doi:10.1038/s41576-018-0089-8.

Knüsel, C. J., and V. Sparacello. 2018. Functional morphology, postcranial, human. In *The International Encyclopedia of Biological Anthropology*, 1–8. Hoboken, NJ: Wiley Blackwell. doi:10.1002/9781118584538.ieba0187.

Koch, C. M., C. V. Suschek, Q. Lin, S. Bork, et al. 2011. Specific age-associated DNA methylation changes in human dermal fibroblasts. *PLoS One* 6:e16679. doi:10.1371/journal.pone.0016679.

Kuliwaba, J. S., N. L. Fazzalari, and D. M. Findlay. 2005. Stability of RNA isolated from human trabecular bone at post-mortem and surgery. *BBA-Mol Basis Dis* 1740:1–11. doi:10.1016/j.bbadis.2005.03.005.

Lappalainen, T., and J. M. Greally. 2017. Associating cellular epigenetic models with human phenotypes. *Nat Rev Genet* 18:441–51. doi:10.1038/nrg.2017.32.

Larsen, C. S. 2018. Bioarchaeology in perspective: from classifications of the dead to conditions of the living. *Am J Phys Anthropol* 165:865–78. doi:10.1002/ajpa.23322.

Lea, A. J., J. Altmann, S. C. Alberts, and J. Tung. 2016. Resource base influences genome-wide DNA methylation levels in wild baboons (*Papio cynocephalus*). *Mol Ecol* 25:1681–96. doi:10.1111/mec.13436.

Lettice, L. A., S. J. H. Heaney, L. A. Purdie, L. Li, et al. 2003. A long-range Shh enhancer regulates expression in the developing limb and fin and is associated with preaxial polydactyly. *Hum Mol Genet* 12:1725–35. doi:10.1093/hmg/ddg180.

Lewton, K. L. 2017. The effects of captive versus wild rearing environments on long bone articular surfaces in common chimpanzees (*Pan troglodytes*). *PeerJ* 5:e3668. doi:10.7717/peerj.3668.

Lewton, K. L., T. Ritzman, L. E. Copes, T. Garland, et al. 2019. Exercise-induced loading increases ilium cortical area in a selectively bred mouse model. *Am J Phys Anthropol* 168:543–51. doi:10.1002/ajpa.23770.

Lidén, Y., N. Olofsson, O. Landgren, and E. Johansson. 2012. Pain and anxiety during bone marrow aspiration/biopsy: comparison of ratings among patients versus health-care professionals. *Eur J Oncol Nurs* 16:323–9. doi:10.1016/j.ejon.2011.07.009.

Lieberman, D. 2011. *The Evolution of the Human Head*. Cambridge, MA: Harvard University Press. www.hup.harvard.edu/catalog.php?isbn=9780674046368.

Lieberman, P. 2007. The evolution of human speech: its anatomical and neural bases. *Curr Anthropol* 48:39–66. doi:10.1086/509092.

Lieberman, P. 2017. Comment on 'monkey vocal tracts are speech-ready'. *Sci Adv* 3:e1700442. doi:10.1126/sciadv.1700442.

Lin, Y.-Y., N. Tanaka, S. Ohkuma, Y. Iwabuchi, et al. 2010. Applying an excessive mechanical stress alters the effect of subchondral osteoblasts on chondrocytes in a co-culture system. *Eur J Oral Sci* 118:151–8. doi:10.1111/j.1600-0722.2010.00710.x.

Lindskog, C., M. Kuhlwilm, A. Davierwala, N. Fu, et al. 2014. Analysis of candidate genes for lineage-specific expression changes in humans and primates. *J Proteome Re* 13:3596–606.

Macrini, T. E., H. B. Coan, S. M. Levine, T. Lerma, et al. 2013. Reproductive status and sex show strong effects on knee OA in a baboon model. *Osteoarthr Cartilage* 21:839–48. doi:10.1016/j.joca.2013.03.003.

Madrid, A., P. Chopra, and R. S. Alisch. 2018. Species-specific 5 MC and 5 HmC genomic landscapes indicate epigenetic contribution to human brain evolution. *Front Mol Neurosci* 11:39. doi:10.3389/fnmol.2018.00039.

Magden, E. R., R. L. Haller, E. J. Thiele, S. J. Buchl, et al. 2013. Acupuncture as an adjunct therapy for osteoarthritis in chimpanzees (*Pan troglodytes*). *J Am Assoc Lab Anim Sci* 52:475–80.

Mahjoub, M., F. Berenbaum, and X. Houard. 2012. Why subchondral bone in osteoarthritis? The importance of the cartilage bone interface in osteoarthritis. *Osteoporosis Int* 23:841–6. doi:10.1007/s00198-012-2161-0.

Malan, V., D. Rajan, S. Thomas, A. C. Shaw, et al. 2010. Distinct effects of allelic NFIX mutations on nonsense-mediated mRNA decay engender either a Sotos-like or a Marshall-Smith syndrome. *Am J Hum Genet* 87:189–98. doi:10.1016/j.ajhg.2010.07.001.

Martin, D. I. K., M. Singer, J. Dhahbi, G. Mao, et al. 2011. Phyloepigenomic comparison of great apes reveals a correlation between somatic and germline methylation states. *Genome Res* 21:2049–57. doi:10.1101/gr.122721.111.

Maunakea, A. K., R. P. Nagarajan, M. Bilenky, T. J. Ballinger, et al. 2010. Conserved role of intragenic DNA methylation in regulating alternative promoters. *Nature* 466:253–7. doi:10.1038/nature09165.

McGraw, W. S. 2018. Postcranial morphology (primates). In *The International Encyclopedia of Biological Anthropology*, 1–3. Hoboken, NJ: Wiley Blackwell. doi:10.1002/9781118584538.ieba0394.

Mendizabal, I., L. Shi, T. E. Keller, G. Konopka, et al. 2016. Comparative methylome analyses identify epigenetic regulatory loci of human brain evolution. *Mol Biol Evol* 33:2947–59. doi:10.1093/molbev/msw176.

Mittra, E. S., H. F. Smith, P. Lemelin, and W. L. Jungers. 2007. Comparative morphometrics of the primate apical tuft. *Am J Phys Anthropol* 134:449–59. doi:10.1002/ajpa.20687.

Moazedi-Fuerst, F. C., M. Hofner, G. Gruber, A. Weinhaeusel, et al. 2014. Epigenetic differences in human cartilage between mild and severe OA. *J Orthop Res* 32:1636–45. doi:10.1002/jor.22722.

Mohtai, M., M. K. Gupta, B. Donlon, B. Ellison, et al. 1996. Expression of interleukin-6 in osteoarthritic chondrocytes and effects of fluid-induced shear on this expression in normal human chondrocytes in vitro. *J Orthop Res* 14:67–73. doi:10.1002/jor.1100140112.

Molaro, A., E. Hodges, F. Fang, Q. Song, et al. 2011. Sperm methylation profiles reveal features of epigenetic inheritance and evolution in primates. *Cell* 146:1029–41. doi:10.1016/j.cell.2011.08.016.

Morris, J. A., P.-C. Tsai, R. Joehanes, J. Zheng, et al. 2017. Epigenome-wide association of DNA methylation in whole blood with bone mineral density. *J Bone Miner Res* 32:1644–50. doi:10.1002/jbmr.3148.

Murray, R. C., S. Vedi, H. L. Birch, K. H. Lakhani, et al. 2001. Subchondral bone thickness, hardness and remodelling are influenced by short-term exercise in a site-specific manner. *J Orthop Res* 19:1035–42. doi:10.1016/S0736-0266(01)00027-4.

Nejadnik, H., S. Diecke, O. D. Lenkov, F. Chapelin, et al. 2015. Improved approach for chondrogenic differentiation of human induced pluripotent stem cells. *Stem Cell Rev* 11:242–53. doi:10.1007/s12015-014-9581-5.

Nieves-Colón, M. A., and A. C. Stone. 2018. Ancient DNA analysis of archaeological remains. In *Biological Anthropology of the Human Skeleton*, 515–44. Hoboken, NJ: John Wiley & Sons, Ltd. doi:10.1002/9781119151647.ch16.

Nishimura, T., A. Mikami, J. Suzuki, and T. Matsuzawa. 2006. Descent of the hyoid in chimpanzees: evolution of face flattening and speech. *J Hum Evol* 51:244–54. doi:10.1016/j.jhevol.2006.03.005.

O'Connor, M. I. 2006. Osteoarthritis of the hip and knee: sex and gender differences. *Orthop Clin North Am* 37:559–68. doi:10.1016/j.ocl.2006.09.004.

Oldershaw, R. A., M. A. Baxter, E. T. Lowe, N. Bates, et al. 2010. Directed differentiation of human embryonic stem cells toward chondrocytes. *Nat Biotechnol* 28:1187–94. doi:10.1038/nbt.1683.

Ostanek, B., T. Kranjc, N. Lovšin, J. Zupan, et al. 2018. Epigenetic mechanisms in osteoporosis. In *Epigenetics of Aging and Longevity*, ed. A. Moskalev and A. M. Vaiserman, 365–88. Translational Epigenetics. Boston: Academic Press. doi:10.1016/B978-0-12-811060-7.00018-8.

Pai, A. A., J. T. Bell, J. C. Marioni, J. K. Pritchard, et al. 2011. A genome-wide study of DNA methylation patterns and gene expression levels in multiple human and chimpanzee tissues. *PLoS Genet* 7:e1001316. doi:10.1371/journal.pgen.1001316.

Parkkinen, J. J., J. Ikonen, M. J. Lammi, J. Laakkonen, et al. 1993. Effects of cyclic hydrostatic pressure on proteoglycan synthesis in cultured chondrocytes and articular cartilage explants. *Arch Biochem Biophys* 300:458–65. doi:10.1006/abbi.1993.1062.

Parkkinen, J. J., M. J. Lammi, R. Inkinen, M. Jortikka, et al. 1995. Influence of short-term hydrostatic pressure on organization of stress fibers in cultured chondrocytes. *J Orthop Res* 13:495–502. doi:10.1002/jor.1100130404.

Pavlovic, B. J., L. E. Blake, J. Roux, C. Chavarria, et al. 2018. A comparative assessment of human and chimpanzee IPSC-derived cardiomyocytes with primary heart tissues. *Sci Rep* 8:1–14. doi:10.1038/s41598-018-33478-9.

Pedersen, J. S., E. Valen, A. M. V. Velazquez, B. J. Parker, et al. 2014. Genome-wide nucleosome map and cytosine methylation levels of an ancient human genome. *Genome Res* 24:454–66. doi:10.1101/gr.163592.113.

Peffers, M. J., P. Balaskas, and A. Smagul. 2018. Osteoarthritis year in review 2017: genetics and epigenetics. *Osteoarthr Cartilage* 26:304–11. doi:10.1016/j.joca.2017.09.009.

Percival, C. J., and J. T. Richtsmeier, eds. 2017. *Building Bones: Bone Formation and Development in Anthropology*. Cambridge: Cambridge University Press. doi:10.1017/978 1316388907.

Peters, A. E., R. Akhtar, E. J. Comerford, and K. T. Bates. 2018. The effect of ageing and osteoarthritis on the mechanical properties of cartilage and bone in the human knee joint. *Sci Rep* 8:5931. doi:10.1038/s41598-018-24258-6.

Petronis, A., I. I. Gottesman, P. Kan, J. L. Kennedy, et al. 2003. Monozygotic twins exhibit numerous epigenetic differences: clues to twin discordance? *Schizophrenia Bull* 29:169–78.

Pizzollo, J., W. J. Nielsen, Y. Shibata, A. Safi, et al. 2018. Comparative serum challenges show divergent patterns of gene expression and open chromatin in human and chimpanzee. *Genome Biol Evol* 10:826–39. doi:10.1093/gbe/evy041.

Pollen, A. A., A. Bhaduri, M. G. Andrews, T. J. Nowakowski, et al. 2019. Establishing cerebral organoids as models of human-specific brain evolution. *Cell* 176:743–56.e17. doi:10.1016/j.cell.2019.01.017.

Prescott, S. L., R. Srinivasan, M. C. Marchetto, I. Grishina, et al. 2015. Enhancer divergence and cis-regulatory evolution in the human and chimp neural crest. *Cell* 163:68–83. doi:10.1016/j.cell.2015.08.036.

Provencal, N., M. J. Suderman, C. Guillemin, R. Massart, et al. 2012. The signature of maternal rearing in the methylome in rhesus macaque prefrontal cortex and T cells. *J Neurosci* 32:15626–42. doi:10.1523/JNEUROSCI.1470-12.2012.

Ramos, Y. F. M., W. den Hollander, J. V. M. G. Bovée, N. Bomer, et al. 2014. Genes involved in the osteoarthritis process identified through genome wide expression analysis in articular cartilage; the RAAK study. *PLoS One* 9:e103056. doi:10.1371/journal.pone.0103056.

Reynard, L. N., C. Bui, C. M. Syddall, and J. Loughlin. 2014. CpG methylation regulates allelic expression of GDF5 by modulating binding of SP1 and SP3 repressor proteins to the osteoarthritis susceptibility SNP Rs143383. *Hum Genet* 133:1059–73. doi:10.1007/s00439-014-1447-z.

Reynard, L. N., and J. Loughlin. 2013. Insights from human genetic studies into the pathways involved in osteoarthritis. *Nat Rev Rheumatol* 9:573–83. doi:10.1038/nrrheum.2013.121.

Richard, D., Z. Liu, J. Cao, A. M. Kiapour, et al. 2020. Evolutionary selection and constraint on human knee chondrocyte regulation impacts osteoarthritis risk. *Cell* 181:362–81.e28. doi:10.1016/j.cell.2020.02.057.

Ritzman, T. B., N. Banovich, K. P. Buss, J. Guida, et al. 2017. Facing the facts: the Runx2 gene is associated with variation in facial morphology in primates. *J Hum Evol* 111:139–51. doi:10.1016/j.jhevol.2017.06.014.

Rothschild, B. M., and F. J. Rühli. 2005. Etiology of reactive arthritis in *Pan paniscus, P. troglodytes troglodytes*, and *P. troglodytes schweinfurthii*. *Am J Primatol* 66:219–31. doi:10.1002/ajp.20140.

Rushton, M. D., L. N. Reynard, M. J. Barter, R. Refaie, et al. 2014. Characterization of the cartilage DNA methylome in knee and hip osteoarthritis. *Arthritis Rheumatol* 66:2450–60. doi:10.1002/art.38713.

Saito, T., A. Fukai, A. Mabuchi, T. Ikeda, et al. 2010. Transcriptional regulation of endochondral ossification by HIF-2α during skeletal growth and osteoarthritis development. *Nat Med* 16:678–86. doi:10.1038/nm.2146.

Sanchez, C., O. Gabay, C. Salvat, Y. E. Henrotin, et al. 2009. Mechanical loading highly increases IL-6 production and decreases OPG expression by osteoblasts. *Osteoarthr Cartilage* 17:473–81. doi:10.1016/j.joca.2008.09.007.

Sanchez, C., L. Pesesse, O. Gabay, J.-P. Delcour, et al. 2012. Regulation of subchondral bone osteoblast metabolism by cyclic compression. *Arthrit Rheum* 64:1193–203. doi:10.1002/art.33445.

Shaw, A. C., I. D. C. van Balkom, M. Bauer, T. R. P. Cole, et al. 2010. Phenotype and natural history in Marshall-Smith syndrome. *Am J Med Genet A* 152A:2714–26. doi:10.1002/ajmg.a.33709.

Shelnutt, K. P., G. P. A. Kauwell, J. F. Gregory III, D. R. Maneval, et al. 2004. Methylenetetrahydrofolate reductase 677C→T polymorphism affects DNA methylation in response to controlled folate intake in young women. *J Nutr Biochem* 15:554–60. doi:10.1016/j.jnutbio.2004.04.003.

Shi, Y., H. Inoue, J. C. Wu, and S. Yamanaka. 2017. Induced pluripotent stem cell technology: a decade of progress. *Nat Rev Drug Discov* 16:115–30. doi:10.1038/nrd.2016.245.

Simon, T. C., and M. A. Jeffries. 2017. The epigenomic landscape in osteoarthritis. *Curr Rheumatol Rep* 19:30. doi:10.1007/s11926-017-0661-9.

Singer, M., I. Kosti, L. Pachter, and Y. Mandel-Gutfreund. 2015. A diverse epigenetic landscape at human exons with implication for expression. *Nucleic Acids Res* 43:3498–508. doi:10.1093/nar/gkv153.

Slieker, R. C., S. D. Bos, J. J. Goeman, J. V. M. G. Bovée, et al. 2013. Identification and systematic annotation of tissue-specific differentially methylated regions using the Illumina 450k array. *Epigenet Chromatin* 6:26. doi:10.1186/1756-8935-6-26.

Smith, R. W. A., C. Monroe, and D. A. Bolnick. 2015. Detection of cytosine methylation in ancient DNA from five Native American populations using bisulfite sequencing. *PLoS One* 10:e0125344. doi:10.1371/journal.pone.0125344.

Steele, J., M. Clegg, and S. Martelli. 2013. Comparative morphology of the hominin and African ape hyoid bone, a possible marker of the evolution of speech. *Hum Biol* 85:639–72. doi:10.3378/027.085.0501.

Suzuki, M. M., and A. Bird. 2008. DNA methylation landscapes: provocative insights from epigenomics. *Nat Rev Genet* 9:465–76. doi:10.1038/nrg2341.

Takahashi, K., T. Kubo, Y. Arai, I. Kitajima, et al. 1998. Hydrostatic pressure induces expression of interleukin 6 and tumour necrosis factor α mRNAs in a chondrocyte-like cell line. *Ann Rheum Dis* 57:231–6. doi:10.1136/ard.57.4.231.

Takahashi, K., T. Kubo, K. Kobayashi, J. Imanishi, et al. 1997. Hydrostatic pressure influences mRNA expression of transforming growth factor-B1 and heat shock protein 70 in chondrocyte-like cell line. *J Orthop Res* 15:150–8. doi:10.1002/jor.1100150122.

Takano-Yamamoto, T., S. Soma, K. Nakagawa, Y. Kobayashi, et al. 1991. Comparison of the effects of hydrostatic compressive force on glycosaminoglycan synthesis and proliferation in rabbit chondrocytes from mandibular condylar cartilage, nasal septum, and spheno-occipital synchondrosis in vitro. *Am J Orthod Dentofac Orthop* 99:448–55. doi:10.1016/S0889-5406(05)81578-1.

Temple, D. H. 2018. Bioarchaeology, evolving paradigms. In *The International Encyclopedia of Biological Anthropology*, 1–8. Hoboken, NJ: Wiley Blackwell. doi:10.1002/9781118584538.ieba0169.

Valdes, A. M., and T. D. Spector. 2011. Genetic epidemiology of hip and knee osteoarthritis. *Nat Rev Rheumatol* 7:23–32. doi:10.1038/nrrheum.2010.191.

Videan, E. N., M. L. Lammey, and D. R. Lee. 2011. Diagnosis and treatment of degenerative joint disease in a captive male chimpanzee (*Pan troglodytes*). *J Am Assoc Lab Anim Sci* 50:263–6.

Vilgalys, T. P., J. Rogers, C. J. Jolly, Baboon Genome Analysis, et al. 2019. Evolution of DNA methylation in Papio baboons. *Mol Biol Evol* 36:527–40. doi:10.1093/molbev/msy227.

Wang, J., C. D. Wang, N. Zhang, W. X. Tong, et al. 2016. Mechanical stimulation orchestrates the osteogenic differentiation of human bone marrow stromal cells by regulating HDAC1. *Cell Death Dis* 7:e2221. doi:10.1038/cddis.2016.112.

Ward, M. C., and Y. Gilad. 2019. A generally conserved response to hypoxia in IPSC-derived cardiomyocytes from humans and chimpanzees. *eLife* 8:e42374. doi:10.7554/eLife.42374.

Ward, M. C., S. Zhao, K. Luo, B. J. Pavlovic, et al. 2018. Silencing of transposable elements may not be a major driver of regulatory evolution in primate IPSCs. *eLife* 7:e33084. doi:10.7554/eLife.33084.

Weaver, T. D. 2003. The shape of the Neandertal femur is primarily the consequence of a hyperpolar body form. *P Nat Acad Sci* 100:6926–9. doi:10.1073/pnas.1232340100.

Wittkopp, P. J., and G. Kalay. 2012. Cis-regulatory elements: molecular mechanisms and evolutionary processes underlying divergence. *Nat Rev Genet* 13:59–69. doi:10.1038/nrg3095.

Wolf, J. B. W. 2013. Principles of transcriptome analysis and gene expression quantification: an RNA-seq tutorial. *Mol Ecol Resour* 13:559–72. doi:10.1111/1755-0998.12109.

Xu, H.-G., Q. Zheng, J.-X. Song, J. Li, H. Wang, et al. 2016. Intermittent cyclic mechanical tension promotes endplate cartilage degeneration via canonical WNT signaling pathway and e-cadherin/β-catenin complex cross-talk. *Osteoarthr Cartilage* 24:158–68. doi:10.1016/j.joca.2015.07.019.

Zakany, J., and D. Duboule. 2007. The role of Hox genes during vertebrate limb development. *Curr Opinion Genet Develop* 17:359–66. doi:10.1016/j.gde.2007.05.011.

Zeng, J., G. Konopka, B. G. Hunt, T. M. Preuss, et al. 2012. Divergent whole-genome methylation maps of human and chimpanzee brains reveal epigenetic basis of human regulatory evolution. *Am J Hum Genet* 91:455–65. doi:10.1016/j.ajhg.2012.07.024.

Zengini, E., K. Hatzikotoulas, I. Tachmazidou, J. Steinberg, et al. 2018. Genome-wide analyses using UK Biobank data provide insights into the genetic architecture of osteoarthritis. *Nat Genet* 50:549–58. doi:10.1038/s41588-018-0079-y.

Zhang, Y., T. Li, S. Preissl, M. L. Amaral, et al. 2019. Transcriptionally active HERV-H retrotransposons demarcate topologically associating domains in human pluripotent stem cells. *Nat Genet* 51:1380–8. doi:10.1038/s41588-019-0479-7.

5 Processes That Generate Modularity in the Mammalian Skull

Implications for Primate Evolution

Nandini Singh

CONTENTS

5.1 Introduction 109
 5.1.1 Developmental/Functional Modularity and Integration in the Skull 111
 5.1.2 Craniofacial Development and Sonic Hedgehog Signaling 112
5.2 Modularity Generating Processes in the Skull: A Case Study of SHH Signaling 114
 5.2.1 Data and Analyses 114
 5.2.2 Results and Discussion 116
  5.2.2.1 SHH and Craniofacial Modularity and Integration 116
  5.2.2.2 SHH and Cranio-Mandibular Modularity 120
5.3 SHH and Craniofacial Evolution in Primates 123
5.4 Conclusions 124
5.5 Acknowledgments 125
5.6 References 125

5.1 INTRODUCTION

Morphological integration and modularity are active subjects of investigation in evolutionary biology and physical anthropology. This chapter discusses the role of Sonic Hedgehog (SHH) signaling as a covariation generating mechanism in the craniofacial skeleton and its implications for primate, particularly, modern human skull evolution. Morphological integration refers to patterns of coordinated variation or covariation among phenotypic structures (Klingenberg 2008). The conceptual framework of morphological integration is credited to the work of George Cuvier (Mayr 1982). Olson and Miller (1999) further developed this concept by providing quantitative models to analyze patterns and magnitudes of integration in multivariate datasets. Modularity is a counterpart of morphological integration in that it explicitly refers to the degree of relatedness or integration, usually quantified as correlation, within phenotypic elements or modules. Modules are units that are strongly integrated or correlated within themselves due to shared biological properties, making them semi-independent from other units (Bolker 2000). Modular elements of a phenotype can evolve independently to some extent while still retaining the fundamental organization and structural identity of an organism as a whole (Klingenberg 2005). Phenotypic modularity and integration can offer important insights into how variation is structured within and between modules, providing a framework to study the direction and pattern of morphological evolution in species.

A number of studies have highlighted the role of integration and modularity in driving phenotypic evolution, development, and diversity in the primate skull form (Cheverud 1982; Bookstein et al. 2003; Ackermann and Cheverud 2004; Ackermann 2005; Bastir and Rosas 2005; Cardini and Elton 2008; Mitteroecker and Bookstein 2008; Ackermann 2009; Singh et al. 2012; Neaux et al. 2018, 2019). However, in the case of primates, due to lack of experimental data, information about underlying genetic interactions is largely inferred from morphological traits used to represent patterns of covariation among structures. Powerful techniques such as geometric morphometrics have offered a number of ways to analyze covariation and correlation patterns in morphological datasets (Rohlf and Slice 1990; Cheverud 1995; Adams et al. 2004; Mitteroecker and Bookstein 2007; Slice 2007; Klingenberg 2009; Pavlicev et al. 2009; Goswami and Polly 2010; Klingenberg and Navarro 2012). The design of these studies has by and large centered on the structural relationships among *a priori* defined modules based on developmental information gathered from the molecular-genetics and function of model systems such as the laboratory mouse (Hallgrímsson and Lieberman 2008). Model organisms, particularly mouse models, have allowed a shift from simply quantifying patterns and magnitudes of integration to investigating the genetic, developmental and functional variables that produce covariance structures in biological forms (Hallgrimsson et al. 2007; Martínez-Abadías et al. 2011). For example, studies on mouse mandibles have revealed that the mammalian mandible comprises two distinct developmental modules, the anterior and posterior components (Atchley 1993). Based on embryological and functional considerations, the cranium can also be divided into semi-autonomous developmentally and/or functionally related parts, all relatively connected to each other within the framework of integration (Moss and Young 1960; Cheverud 1982; Moss-Salentijn 1997; Jiang et al. 2002; McBratney-Owen et al. 2008).

As widely demonstrated in murine and avian experimental models, Sonic Hedgehog signaling, a highly conserved developmental pathway across diverse organismal systems, is critical for anterio-posterior outgrowth of the face, midline width variation, palatal fusion, and mandibular embryogenesis (Hu and Helms 1999; Young et al. 2010; Billmyre and Klingensmith 2015). SHH has also been implicated as a contributor to the evolution of human-specific traits in the nervous and skeletal systems (Dorus et al. 2006). In nonhuman primates compared to other mammals, particularly among hominins, midfacial skeletal elements—in part regulated by SHH signaling—experienced tremendous evolutionary transformations (Lacruz et al. 2019). The hominin lineage represents the extreme trends of craniofacial shape changes among primates, comprising a combination of traits marked by a short-orthognathic face and globular neurocranium, as seen in modern humans.

Here, I focus on SHH's contribution in delimiting the facial skeleton as a developmental module in the skull. To that end, the objectives of this chapter are as follows: (1) to explore the effects of a pharmacological upregulation of SHH signaling on the modular organization of craniofacial structures and (2) to determine potential differential responses of elevated SHH signaling in the cranium versus the mandible. Addressing these objectives will provide important insight into the evolutionary role of SHH signaling on primate skull evolution and development, particularly in the context of modularity—a concept used to explain the semi-independence, but also interdependence, among traits or structures within an organism.

5.1.1 Developmental/Functional Modularity and Integration in the Skull

Modularity describes varying levels of concentrated interactions within modules produced by shared developmental-genetic and functional processes that generate those traits. Examples of processes that produce developmental and genetic modules in organisms include common embryonic growth patterns, embryonic tissue origin, gene regulatory networks, epigenetic interactions (e.g., gene activity), and pleiotropic effects (Bolker 2000; Klingenberg 2008). To provide context for understanding the craniofacial subdivisions used in this study, in this section, I discuss the various ways in which the mammalian skull is divided into modules based on developmental and/or functional considerations.

Various studies on the integration of the mammalian skull have focused on the association between *a priori* defined developmental/functional modules (Hallgrímsson et al. 2004; Mitteroecker and Bookstein 2008; Klingenberg and Navarro 2012; Singh et al. 2012). A common subdivision of cranial components follows the functional matrix hypothesis proposed by Melvin Moss (1968). According to this hypothesis, bone growth is influenced by the function of soft tissues and spaces in which skeletal units develop. Moss' original hypothesis outlined two major functional matrices: periosteal and capsular (Moss and Young 1960). In the cranium, those functional units included: (1) pharyngeal capsules and orofacial bones, including the orbits, nasal bones, and masticatory and oral traits, and (2) neurocranial components such as the frontal, parietal, and occipital bones. Following the principle of modularity, traits that are part of the same functional matrix are more tightly integrated than elements from a different functional matrix. Several studies on human and nonhuman primates, pioneered by the work of Jim Cheverud, have shown that developmentally and functionally related traits in the skull are tightly correlated across different primates (Cheverud 1982; Mitteroecker and Bookstein 2008; Ackermann 2009; Singh et al. 2012; Neaux et al. 2019).

Some other ways to partition the cranium are based on embryonic tissue origin (Jiang et al. 2002; McBratney-Owen et al. 2008). Accordingly, cranial components that originate from cranial neural crest cells (CNCCs) make up a module, and tissues derived from paraxial mesoderm form the other major cranial module. Structures derived from CNCC include elements of the facial skeleton and frontal bone. Mesoderm-derived regions consist of the more posteriorly placed bones of the cranial base and cranial vault. Cranial bones can also be grouped according to their mode of ossification as either intramembranously or endochondrally forming bones (Richtsmeier and Flaherty 2013). The case study presented in this chapter uses Jiang et al.'s (2002) tissue origin hypothesis to partition the cranium into neural crest- and mesoderm-derived modules.

Though the focus of this chapter is on cranial modularity and integration, the mandible is also analyzed as a distinct developmental module in the skull. Like the cranium, the mandible, too, has received considerable attention as a model for investigating the developmental and genetic underpinnings that cause covariation among complex phenotypes (Klingenberg et al. 2003). The mandible can be partitioned into embryonically distinct units such as the tooth-bearing alveolar region and the muscle-bearing ramus (Atchley 1993; Klingenberg et al. 2003). Given the diversity in dento-mandibular traits from strepsirrhine to haplorhine primates, exploring how alterations in SHH signaling impact variation in the mandible can provide further insights into cell processes that contribute to species-specific trajectories of morphological change.

5.1.2 Craniofacial Development and Sonic Hedgehog Signaling

The skull is a complex arrangement of bones that are largely distinct embryonically and evolutionarily and display hierarchical levels of integration. Knowledge of craniofacial embryogenesis provides insight into the dynamic interactions among the skeletal components and the mechanisms that generate species-specific variation. This section outlines the critical role of SHH signaling in craniofacial embryogenesis.

The developing facial structures are formed around an ectodermal depression, the stomodeum (Osumi-Yamashita et al. 2002). As embryogenesis progresses, the first pharyngeal arch (PA1) separates into maxillary and mandibular prominences (Chai and Maxson 2006). The tissues of the frontonasal prominence (two medial nasal and two lateral nasal processes) are formed by mesenchyme from CNCCs from the posterior forebrain (Osumi-Yamashita et al. 2002). By the end of the 4th week of gestation in humans, bilateral oval-shaped ectodermal "thickenings" form on the lower lateral part of the frontonasal prominence, the nasal placodes (Baker and Bronner-Fraser 2001; Szabo-Rogers et al. 2010). In the 5th week, the mesenchymal growth in PA1 results in the maxillary and the mandibular processes on either side of the face. As growth occurs, each maxillary prominence merges medially with the medial nasal processes toward the midline to form the upper lip (Senders et al. 2003; Jiang et al. 2006).

The fusion of the medial nasal processes and maxillary prominences form the primary palate in the midline (Bush and Jiang 2012). The maxillary prominence forms large components of the upper lip, maxilla, zygomatic bone, and secondary palate. The nasal passages are demarcated into left and right sides initially by the septal cartilage and eventually by the perpendicular ethmoid plate and the vomer. The lateral walls and base of the nasopharynx comprise the pterygoid plates of the sphenoid, the palatine bones, maxillae, and premaxillae. The mandibular prominence fuses and gives rise to the chin, lower lip, and mandible. CNCCs differentiate into prechondrocytes to form Meckel's cartilage, which is followed by proliferation and expansion of chondrocytes within Meckel's cartilage and subsequent formation of the mandible (Amano et al. 2010). While there are several pathways such as fibroblast growth factor (FGF), Wingless (WNT), and bone morphogenetic protein (BMP) that contribute to skull embryogenesis, animal models have demonstrated that SHH is crucial for regulating epithelial-mesenchymal interactions that control the proximodistal growth and dorsoventral patterning of the vertebrate mid and upper facial skeleton (Hu et al. 2003; Hu and Marcucio 2009a). In the earliest stages of the developing mouse embryo, *Shh* is expressed in the ventral-most part of the neural tube, the floor plate, the notochord, and the ventral forebrain (Helms 2005; Hu and Marcucio 2009a). *Shh* expression is particularly important for the survival of CNCC during the early stages of embryonic development (~E8.5) and for promoting cell proliferation in the later stages (E10.5) to mediate the size of the facial primordia (Ahlgren and Bronner-Fraser 1999; Helms 2005). There are three SHH signaling domains in the embryonic head between E10.5–E11.5: the neuroectoderm of the ventral forebrain, ectoderm of the facial midline, and pharyngeal endoderm (Xavier et al. 2016). Coordinated SHH activity from these domains have a direct effect on the facial processes during the organization of the craniofacial region, particularly the midline structures of the mid-upper facial skeleton.

Expression studies using avian and mouse models have provided detailed knowledge of the signaling boundary between *Shh* and *Fgf8*-expressing cells in the ectoderm of the frontonasal process called the frontonasal ectodermal zone (FEZ), which directly influences the formation and patterning of upper and mid-facial structures (Hu et al. 2003). The murine FEZ that represents the mammalian condition is not a single signaling center as it is in the avian FEZ (Hu and Marcucio 2009b). In the mouse, *Shh* is expressed bilaterally in the ectoderm of the median nasal processes (MNP) extending into the base of the MNP and nasal pit creating a left and right FEZ (Hu and Marcucio 2009b). NCCs are present in the lateral regions but almost absent in the midline. Similar bilateral expression of *SHH* is observed in the maxilla of human embryos (Roessler et al. 1996; Odent et al. 1999).

SHH signaling also plays an integrative role between the developing midline structures of the face and forebrain, made evident by the etiology of holoprosencephaly (HPE) (Chiang et al. 1996; Xavier et al. 2016). HPE is a heterogeneous developmental abnormality characterized by a failure of the forebrain vesicle to separate into distinct left and right hemispheres (Geng and Oliver 2009). This brain malformation has significant consequences for subsequent development of the midline elements of the face. *Shh* expression initiates in the forebrain prior to outgrowth of the facial prominences (Marcucio et al. 2005). During CNCC migration to the midface, *Shh* is activated in the epithelium. Young et al. (2010) have further demonstrated that blocking SHH signaling in the forebrain reduces *Shh* expression in the FEZ, leading to an overall narrow face. Conversely, elevating SHH signaling in the forebrain expands *Shh* expression pattern in the FEZ resulting in a mediolaterally expanded face (Young et al. 2010). These findings mimic midline malformations in human diseases such as hypotelorism, cyclopia, and midline clefts that result from altered SHH signaling (Muenke and Cohen 2000; Balk and Biesecker 2008). Together, these findings provide compelling evidence of an intimate connection between the magnitude and spatial distribution of SHH signaling and morphogenesis of the face during embryonic development.

SHH has also been implicated as a regulator of mandibular growth and development (Kronmiller et al. 1995; Billmyre and Klingensmith 2015; Xu et al. 2019). Mouse models have revealed that inactivation of *Shh* in the pharyngeal endoderm causes apoptosis of CNCC in PA1 resulting in micrognathia (Billmyre and Klingensmith 2015). Jeong et al. (2004) additionally demonstrated that SHH is specifically required in postmigratory CNCC in the facial primordia rather than in early events of CNCC-generation and migration. Their experiments showed that along with severe abnormalities in CNCC-derived regions of the skull, the development of Meckel's cartilage was truncated resulting in mandibular hypoplasia.

Model organisms have facilitated the study of variation from simply quantifying patterns and magnitudes of integration to gaining a deeper understanding of the underlying processes that contribute to these patterns and to their variation across taxa and across organismal structures. This chapter focuses on SHH signaling during postnatal development and its role in maintaining the arrangement of cranial modules and pattern of modularity in the skull form.

5.2 MODULARITY GENERATING PROCESSES IN THE SKULL: A CASE STUDY OF SHH SIGNALING

5.2.1 Data and Analyses

This study utilizes an experimental dataset that was designed to investigate the effects of SAG on cerebellar development in a mouse model for Down Syndrome (DS), Ts65Dn (Das et al. 2013; Singh et al. 2015). See Das et al. (2013) for detailed information on the initial study design and mouse models. The sample includes only euploid—no trisomic—specimens and provides a model dataset to examine the correspondence between targeted stimulation of SHH signaling postnatally and consequential phenotypic outcomes.

Three-dimensional (3D) shape analysis is used to examine the effects of pharmacological stimulation of the SHH pathway via a SHH-based agonist, SAG, on the skull morphology of adult mice. The groups in our sample comprise adult euploid mice that were injected with either a 20 µg/g dose or a 40 µg/g dose of the SAG agonist or with vehicle (saline solution) on the day of birth (P0) (Table 5.1). SAG is a Smoothened (SMO) agonist, a chlorobenzothiophene-based compound, which is an activator of the G protein-coupled receptor SMO (Chen et al. 2002). SMO is a key component of the Hedgehog (HH) signaling pathway. SAG elevates the canonical SHH pathway, replicating many SHH-responsive activities in vivo.

Data comprise micro-computed tomography (µCT) images of the mice in our sample. µCT was done at the Johns Hopkins Medical Institutions through the Small Animal Resource Imaging Program at the Research Building Imaging Center using

TABLE 5.1
Specimen List with Ploidy and Treatment Information

Ploidy	Treatment	Sample Size
Euploid	20 µg/g SAG injection	30
Euploid	40 µg/g SAG injection	10
Euploid	Vehicle injection	34
Total		**74**

a Gamma Medica X-SPECT/CT scanner (Northridge, CA, USA) with 0.05-mm resolution along *x*, *y*, and *z* axes for each mouse. Isosurfaces to represent all cranial and mandibular bones were reconstructed from 8-bit µCT images using the software package Avizo 6.3 (Visualization Sciences Group, VSG). Additional technical information regarding the same data is available in Singh et al. (2015). Twenty-six 3D landmarks were measured on the cranium and mandible, respectively, of each specimen in Avizo (Figure 5.1a, b, c). The specimen landmark configurations for the cranial and mandibular landmarks were superimposed separately using generalized Procrustes analysis (GPA). This method extracts shape coordinates from the original landmark data by translating, scaling, and rotating the data, subsequently yielding Procrustes shape coordinates. A commonly used measure for size is centroid size (CS), which is the square root of the sum of squared Euclidean distances from a set of landmarks to their centroid ("center of gravity—the average of the *x* and *y* coordinates of all landmarks") (Rohlf and Slice 1990; Slice 2007; Klingenberg 2016). All the specimens are scaled according to CS.

A two-block partial least squares (2B-PLS) analysis was performed to examine general patterns of integration and modularity between cranial modules (Rohlf and Corti 2000). The cranial landmarks were subdivided into developmental modules based on structures derived from neural-crest, that is, elements of the facial skeleton (block 1), and mesoderm, specific components of the neurocranium (block 2) (Figure 5.1a, b). 2B-PLS analysis uses a singular value decomposition to find pairs of axes, one axis per block, which account for the maximum amount of co-variation between the two sets of variables examined, in this case the landmark coordinates that delimit the two cranial blocks. The axis in one block is only correlated to the corresponding axis in the other block, making it easy to examine co-variation one pair of PLS axes at a time (Mitteroecker and Bookstein 2007). In this study, following the partitioning of our modules based on distinct tissue origin, the landmarks for each block were subjected to separate Procrustes fits, and the covariation pattern between-blocks was analyzed across groups. Principal component analyses (PCAs) were also conducted on the cranial and mandibular landmarks to compare the effects of increased SHH signaling on the two regions (Figure 5.1c). Last, mean shapes were computed in Avizo from group-specific Procrustes coordinates to illustrate the overall morphology of each group in our sample. All other morphometric and statistical analyses, including construction of the wireframe diagrams, were performed in MorphoJ and R programming software (version 4.0.1) (Klingenberg 2011; Team 2013).

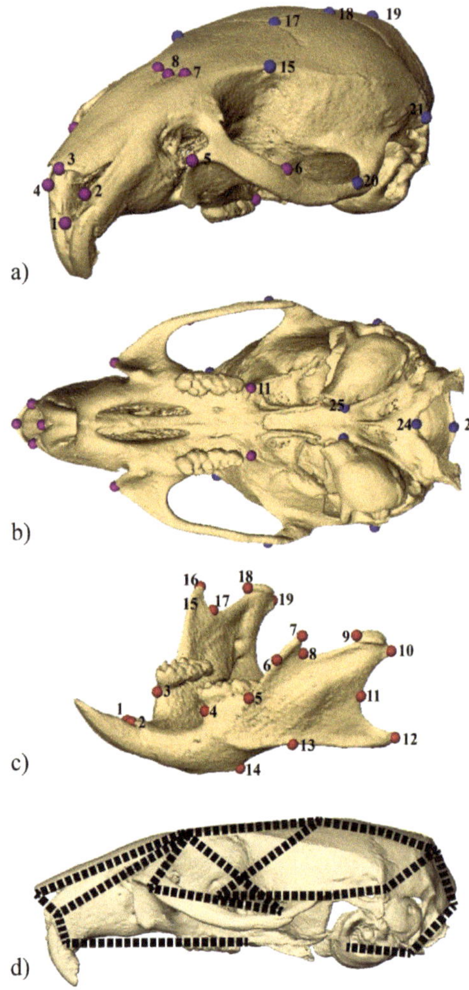

FIGURE 5.1 (a) Lateral view of cranium; (b) Inferior view of cranium; (c) Lateral view of mandible; (d) Illustration of wireframe diagram in lateral view of the cranium. Facial landmarks 1–13; Neurocranial landmarks 13–26; Mandibular landmarks 1–26 (only 19 landmarks are visible in the lateral oblique view).

5.2.2 Results and Discussion

5.2.2.1 SHH and Craniofacial Modularity and Integration

A 2B-PLS analysis of the Procrustes coordinates was performed to examine the pattern of integration between the facial skeleton (block 1) and parts of the neurocranium (block 2) in a sample of SAG-treated and vehicle-treated mice. PLS 1 (95%) accounts for the bulk of the covariation in the sample and shows a deviation in integration pattern between SAG- and vehicle-treated mice in aspects of the face and neurocranium along this axis (Figure 5.2a). As illustrated in the PLS plot, the treated and untreated mice do not overlap and are largely separated along block 1. The relative positions of the SAG-treated and vehicle groups also suggest differences in the degree of within-module changes in the face. The mice treated with 40 μg/g of SAG occupy the negative scores on PLS 1 and show the most pronounced cranial shape changes along the antero-posterior and mediolateral dimensions, exhibiting a dramatically retracted and mediolaterally expanded snout and depressed nasal bones associated with a slightly rounded posterior neurocranium (Figure 5.2b). The vehicle-treated mice present an overall elongated snout and superior-inferiorly flat neurocranium compared to the SAG-treated groups (Figure 5.2b). The SAG-treated

a)

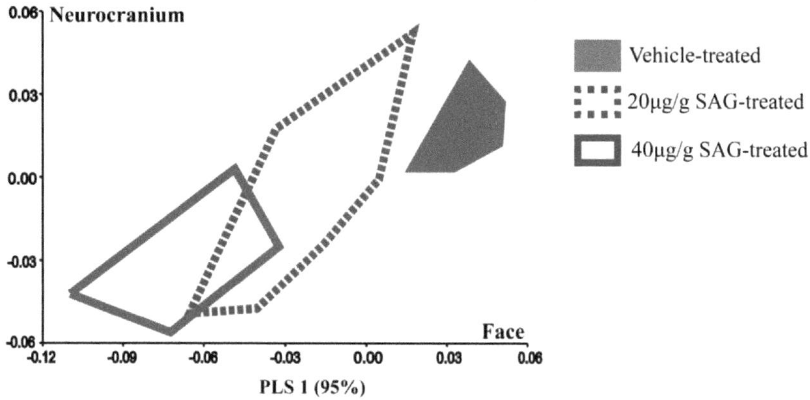

b)

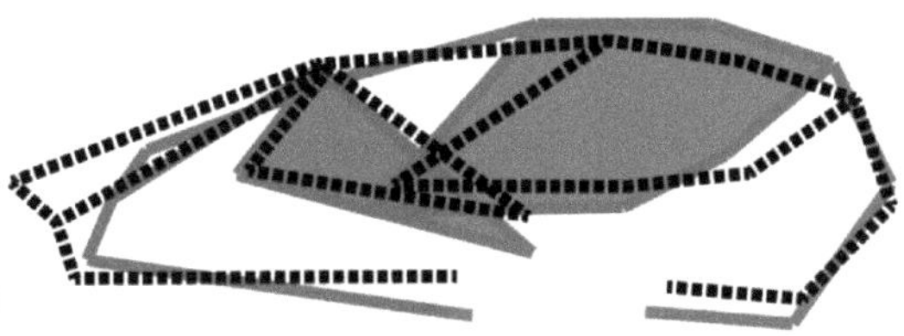

FIGURE 5.2 (a) Plot of 2B-PLS analysis showing the pattern of integration between face (block 1) and neurocranium (block 2) along PLS 1. The solid gray convex hull represents the vehicle-treated mice on the positive end of PLS 1; the dashed convex hull represents the 20 μg/g SAG-treated mice; the solid convex hull represents the 40 μg/g SAG-treated mice on the negative end of PLS 1. The vehicle-treated are separated from both groups of SAG-treated mice. There is also less within-group variation in the former than in the latter two groups, the vehicle-treated mice showing the largest amount of within-group variation in the sample. The general pattern of integration between the vehicle- and SAG-treated mice seems to be different along PLS 1. (b) The dashed wireframe represents the positive end of PLS 1 occupied by the vehicle-treated mice and the solid wireframe represents the negative end of PLS 1 occupied by the 40 μg/g SAG-treated mice. The wireframes are in lateral view. The mice treated with 40 μg/g of SAG show the most severe changes in cranial shape, exhibiting a dramatically retracted and mediolaterally expanded snout and depressed nasal bones associated with a slightly rounded posterior neurocranium. In contrast, the vehicle-treated mice present an overall elongated cranial shape.

groups also show more within-group variation than the unaffected mice, with the lower-dosed mice being the most variable in aspects of shape covariation. The distribution of groups along PLS 1 also suggests that 20 μg/g of SAG administered at birth is enough to cause marked variation in the association between the face and elements of the neurocranium during postnatal growth. Overall, results along PLS 1 indicate that upregulation of the SHH pathway at P0 causes perturbations that partially alter the structural interaction of the facial skeleton with the neurocranium during postnatal growth.

To further explore the within-module variation across the groups, I examined PLS 1 and PLS 2 separately in the two blocks (Figure 5.3a, b). Block 1 (facial skeleton) clearly distinguishes the vehicle-treated mice from the SAG-treated groups (Figure 5.3a). The SAG-treated mice also show considerable within-group variation in the face along both PLS axes; the vehicle-treated mice are more constrained in their group scatter. Shape variation in block 2 (neurocranium) also separates the vehicle-treated from the SAG-treated mice (Figure 5.3b), but the distinction among the groups is not as marked as it is in block 1, and mice treated with 20 μg/g of SAG overlap in aspects of neurocranial shape with both the 40 μg/g SAG-treated and vehicle-treated specimens (Figure 5.3a, b). It is apparent that shape changes in the neurocranium do not distinguish the groups as obviously as the shape changes in the facial skeleton. This supports the idea that upregulation of SHH has a more severe effect on the facial skeleton than parts of the neurocranium during postnatal development. Higher doses of SAG do alter the shape of the cranial vault but to a lesser degree than the facial skeleton. Together, the analyses of covariation between and within the blocks demonstrate that the amount of modular change in the facial skeleton compared to the neurocranium could be driving the deviations in the pattern of integration between the groups, particularly in the double-dosed mice. That is, the differences in within-block patterns of variation can greatly influence covariation patterns between blocks.

It is also important to note the effects of dosage. The 40 μg/g mice are the most distinct from the vehicle-treated specimens in both facial and neurocranial shape (Figure 5.3a, b). These results demonstrate that the higher the dosage of SAG, the more severe the effects on different cranial regions, including parts that are not directly responsive to SHH signaling. The latter reveals the underlying interconnectedness among modules and bring to light the possible role of another HH pathway ligand, Indian Hedgehog (IHH). During cranial development, IHH is involved in the proliferation and differentiation of chondrocytes and endochondral bone development (St-Jacques et al. 1999; Young et al. 2006; Pan et al. 2013). The overexpression of SHH increases the IHH inhibitor, PTHrP, affecting normal growth of the cranial base in the later stages of development (Mak et al. 2008). While the majority of the morphological outcomes of SAG are concentrated in SHH-responsive regions, the role of IHH in overall cranial development and potential pleiotropic effects of artificial perturbations of a key signaling pathway cannot be discounted and needs to be further evaluated.

These results additionally support previous findings from studies on experimental manipulation of SHH signaling that demonstrated that an increase in SHH produces expansion of the frontonasal region, whereas a decrease in SHH signaling causes narrowing of the frontonasal process (Young et al. 2010, 2014; Marcucio et al. 2015). Postnatal upregulation of SHH via SAG at P0 also targets midline facial structures causing mediolateral widening of the snout and indentation in the nasal bones that further manifest as a broad and depressed snout in adult specimens (Figure 5.4). These morphological changes represent localized patterns of variation in the cranium

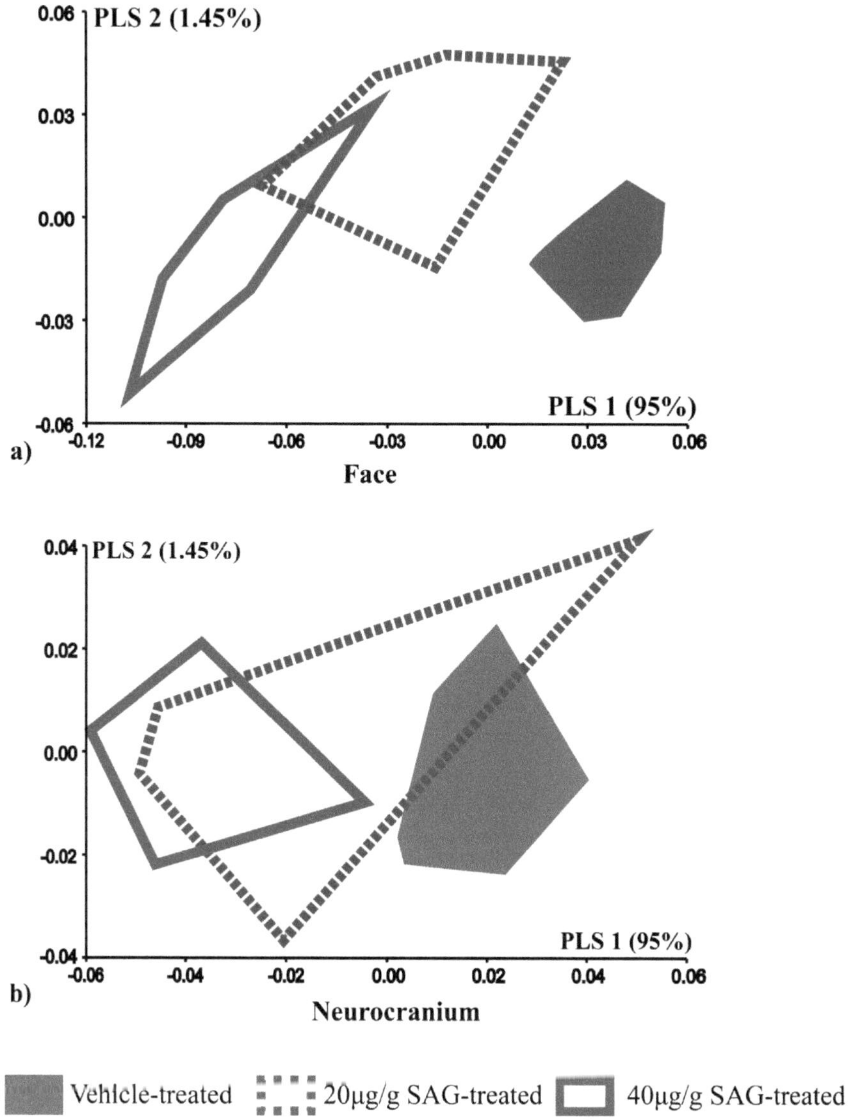

FIGURE 5.3 (a) Plot of PLS 1 vs PLS 2 of the facial landmarks; (b) Plot of PLS 1 vs PLS 2 of the neurocranial landmarks. The solid convex hull represents the vehicle-treated mice at the positive score along PLS 1; the dashed convex represents the 20 μg/g SAG-treated mice; the unfilled convex hull represents the 40 μg/g SAG-treated mice on the negative ends of PLS 1.

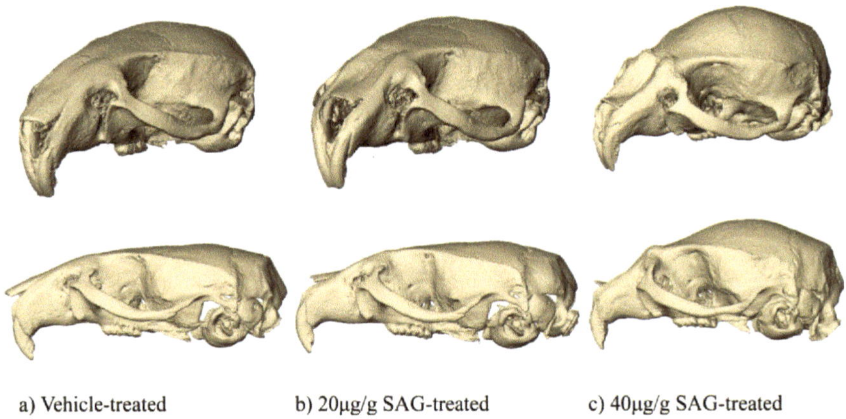

FIGURE 5.4 (a) Vehicle-treated mice; (b) 20 μg/g SAG-treated mice; (c) 40 μg/g SAG-treated mice. Isosurfaces of mean shapes computed from Procrustes shape coordinates.

delimiting the midfacial skeleton as a module based on the differential effects of SHH signaling. This in turn highlights developmental avenues by which novel phenotypic variation and subsequent novel phenotypic changes might emerge in species through evolutionary time.

5.2.2.2 SHH and Cranio-Mandibular Modularity

The effects of elevated SHH signaling were also examined in the mandible (relative to the cranium). Mouse models have shown that reduction of *Shh* expression in the pharyngeal endoderm causes depletion of CNCC in PA1, affecting development of Meckel's cartilage and consequently mandibular morphogenesis (Jeong et al. 2004; Billmyre and Klingensmith 2015). Therefore, considering the importance of *Shh* in both maxillary and mandibular development, a PCA of mandibular and cranial landmarks, respectively, was conducted to evaluate the effects of *Shh* upregulation in these structures across our groups (Figures 5.5 and 5.6). The PCA of mandibular landmarks shows a separation of all three groups of mice along PC 1 (46.9%), with the 40 μg/g specimens being the most distinct in mandibular morphology (Figure 5.5). The 20 μg/g-dosed mice fall between the range of variation of the other two groups. PC 2 (11.6%) captures the within-group variation in the 20 μg/g and vehicle-treated mice (Figure 5.5). Shape variation along PC 1 shows that the higher-dosed mice have an anterior-posteriorly retracted and medio-laterally wide mandible compared to the overall elongated shape of the vehicle-treated mice (Figure 5.5). PC 2 captures shape changes in aspects of the coronoid and condylar processes of the ramus.

SHH and IHH ligands of the HH pathway are involved in mandibular embryogenesis (Nie et al. 2005; Sugito et al. 2011; Billmyre and Klingensmith 2015; Xu et al. 2019). However, little is known about the role of these pathways in postnatal development of the mandible. *Shh* expression has been found to regulate development of Meckel's cartilage by E10.5 and controls growth of the oral-aboral axis of the

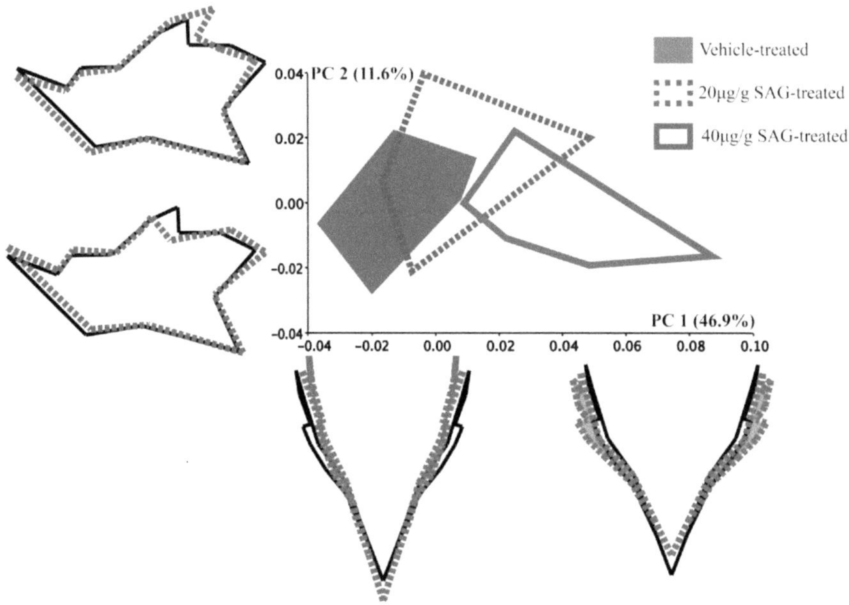

FIGURE 5.5 PCA plot, PC 1 vs PC 2 of mandibular landmarks. The solid convex hull represents the vehicle-treated mice on the negative end of PC 1, the dashed convex represents the 20 μg/g SAG-treated mice and the unfilled convex hull represents the 40 μg/g SAG-treated mice on the positive end of PC 1. The wireframe diagrams long PC 1 represent the shape changes in superior view from the negative to the positive scores of the axis. The wireframe diagrams on PC 2 capture the shape variation in lateral view from the negative to the positive end of the axis. The solid black wireframe represents the mean shape of the entire sample, and the dashed gray wireframe depicts the variation along the axes.

mandible (Billmyre and Klingensmith 2015; Xu et al. 2019). In later stages of embryonic development (E18.5), *Ihh* is shown to be essential for organizing and mediating growth of the secondary cartilage in the mandibular symphysis (Sugito et al. 2011). Inhibition of *Ihh*-expression in chondrocytes of secondary cartilage affects endochondral ossification, which in turn truncates the development of the mandibular symphysis (Sugito et al. 2011). Though only select portions of the mandible form endochondrally (symphysis and coronoid process), the mandibular shape differences seen between the SAG-treated and vehicle-treated mice, particularly the mice that were given 40 μg/g of SAG, suggest that SAG affects both SHH and IHH responsive regions in the skull but to varying degrees. Further exploration of the HH ligands in mandibular development might clarify the influence of HH signaling in mandibular evolution.

Compared to the mandible, PCA of the cranial landmarks more clearly distinguish the affected and unaffected groups (Figure 5.6). PC 1 (57.4%) separates the SAG mice from the unaffected group. The double-dosed SAG and vehicle-treated mice occupy the positive and negative ends of PC 1, respectively. The mice that were administered 20 μg/g of SAG show maximum within-group variation on

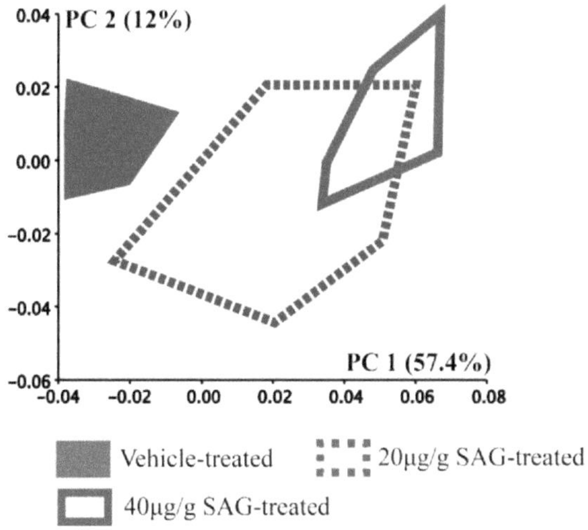

FIGURE 5.6 PCA plot of cranial landmarks. The solid convex hull represents the vehicle-treated mice on the negative end of PC 1; the dashed convex represents the 20 μg/g SAG-treated mice; the unfilled convex hull represents the 40 μg/g SAG-treated mice on the positive end of PC 1. The vehicle-treated mice are distinct along PC 1 from both the SAG-treated groups. There is considerable within-group variation among in the lower-dosed SAG mice compared to both vehicle-treated and the double-dosed SAG mice, as captured along PC 2.

PC 2 (12%), with the vehicle-treated mice showing the least. PC 1 shows the SAG-treated mice, mainly driven by the mice that were given 40 μg/g of SAG, to have anterior-posteriorly retracted snouts and superior-inferiorly taller neurocrania than the vehicle-treated mice. PC 2 mainly captures changes in the superior-inferior dimension of the cranium. These shape changes in the cranium follow the pattern seen in the mandible. Interestingly, the vehicle-treated mice show less within-group variation in the cranium than in the mandible. It is worth considering how functional demands of mastication possibly influence the degree and structure of variation between the two structures.

These results reveal that higher the dosage of SAG, the more severe the effects on both cranial and mandibular morphology. The original dose of 20 μg/g of SAG is associated with increased variation in cranio-mandibular phenotype relative to variation in the higher-dosed animals, possibly due to incomplete penetrance in some specimens. Though the facial skeleton shows more impact of elevated SHH signaling than the mandible, SHH affects the development of many tissues and bones of the skull. SHH-mediated signals contribute to a multifaceted set of interactions among genes and regulatory networks that drive communication among cells of the head. Complex craniofacial traits composed of many tissue types vary together by exploiting patterns of gene-to-cell and cell-to-cell signaling, resulting in soft tissue complexes developing in close association and synchrony with skeletal tissues (Richtsmeier and Flaherty 2013). Though the synchrony is clearly based on the coordinated genetic regulation of development, physical associations among tissues can lead to biomechanical impacts of one growing tissue on another that can affect the regulation of gene expression (Butler et al. 2009). Together these effects can modify the pattern of covariation within and between structures.

Modularity arises when genetic and developmental processes influence some structures more than others and sometimes not at all. Our findings confirm that SHH activity has differential effects on the postnatal growth of the cranium and mandible, implicating SHH as a mechanism that generates individualized, yet interconnected, modules in the skull.

5.3 SHH AND CRANIOFACIAL EVOLUTION IN PRIMATES

Phenotypic integration and modularity have been widely studied in the context of primate morphological evolution (Cheverud 1995; Ackermann and Cheverud 2004; Bastir and Rosas 2005; Mitteroecker and Bookstein 2008; Ackermann 2009; Rolian 2009; Porto et al. 2009; Lewton 2012; Neaux et al. 2018; Powell et al. 2018). Primates display a tremendous variety in skull form, body size, locomotor patterns, and environmental adaptations. In particular, the range of craniofacial tendencies include changes in cranial capacity, orbital morphology, shape of the midface, dentition, orientation of the basicranium relative to the face, and mandibular shape (Ackermann 2007). As mentioned earlier, the most dramatic craniofacial changes among primates occurred along the *Hominini* line, which includes modern humans and their extinct relatives. The most significant morphological changes in humans comprise the dramatic expansion of brain size, accompanied by marked orthognathia, and overall reduction of the facial skeleton. *SHH* is an evolutionarily conserved gene critical for normal development of the mammalian face. The results presented here provide insight into the differential and modular effects of SHH signaling on the craniofacial skeleton. Moreover, rodent models serve as a useful proxy for human and nonhuman primates to understand signaling vital for morphogenesis.

The study of mechanism and process is particularly important to understand the developmental determinants that contribute to the emergence of novel phenotypes, such as pronounced retraction of the midface in humans. Modules are regarded as being more evolvable, that is, plastic, than traits that are constrained by strong integration across multiple morphological regions (Wagner and Altenberg 1996; Klingenberg 2004). In human and nonhuman primates, the face has experienced repeated evolutionary transformations and is considered less integrated and semi-independent from components of the neurocranium that are tied to brain growth (Lieberman et al. 2000). So, how can we apply information from experimental mouse models to understand cellular processes that organize modules in the primate skull?

The function of pathways such as SHH is highly conserved across mammals (Young et al. 2010). In particular, SHH signaling is critical for the patterning and outgrowth of the upper and mid face—regions that have undergone tremendous change in human evolution, as mentioned. Dorus et al. (2006) have demonstrated that molecular evolution of SHH is accelerated in all primates compared to other mammals. In particular, humans show a rampant addition of serines or threonines in the C-terminus of SHH potentially leading to post-translational modifications and processing of SHH-N production. Post-translational processing of SHH-N is essential during embryonic development. This along with the polymorphisms at the human SHH locus suggest that SHH has experienced adaptive evolution in human and nonhuman primates (Dorus et al. 2006). Considering that elevation of SHH causes drastic retraction and mediolateral expansion of the snout in SAG-treated mice, it is reasonable to propose that SHH signaling played a critical evolutionary role in patterning the primate midface. However, the suggestion here is not that variation in SHH signaling is the primary mechanism responsible for producing human-specific traits, but that subtle species-specific alterations in such highly conserved pathways might produce diverse phenotypes across an evolutionary time scale, that is, within the primate phylogeny, and specifically among hominins.

An increase in dosage of SAG from 20 μg/g to 40 μg/g not only caused further changes to midline structures of the snout (e.g., depressed nasal bone) but also made parts of the neurocranium more rounded compared to the unaffected and lower-dosed mice. The latter finding exemplifies the hierarchical ways in which modularity and integration manifest in the cranium. These two concepts are interrelated and speak to how biological systems, and consequently skeletal structures, are organized. Within the cranium, because SHH is primarily responsible for morphogenesis of the midface, it unites/integrates the components that make up the midface more so than aspects of the cranial vault or basicranium cranium. However, an amplified increase is SHH signaling has a more pervasive effect on the cranium, concomitantly causing changes in structures other than the midface and confirming the hierarchical ways in which the cranium is integrated. Thus, to reiterate, the significant trends towards orthognathia and globularity in hominin evolution, culminating in modern human craniofacial morphology, could, at least in part, be a result of subtle changes in the magnitude, timing, and range of action of SHH and other related pathways within the primate phylogeny (Dorus et al. 2006).

5.4 CONCLUSIONS

As dictated by the availability of data, most questions on primate morphological evolution are addressed by quantifying patterns of variation and covariation in skeletal remains. Mouse models and other experimental animals have facilitated research into underlying processes that generate variation and covariation. Modularity and integration are interrelated concepts in evolutionary developmental biology and denote that there is more interaction within subparts or ‘modules’ of an organism than there is between modules, resulting in elements of the phenotype that display a certain degree of developmental and/or functional independence from contiguous units. In exploring the effects of a SHH-based agonist, SAG, on postnatal craniofacial morphology, I found that elevation in SHH signaling at birth targets the face unequivocally, causing widening and shortening of the midface in adult mice. Concentrated and patterned effects of SHH provide insight into how developmental networks structure variation in the skull, revealing the intimate connection between the genotype and phenotype. However, this result does not suggest that traits from different developmental components are autonomous or unaltered. Numerous studies have demonstrated pleiotropy between regions, and it is also shown here that neurocranial elements are impacted when the dosage of SAG is doubled. The overall findings of this study are meant to underscore the role of SHH signaling in organizing modules in the skull. Investigating perturbations in other signaling pathways and growth factors such as BMPs, FGFs, and FGFRs can further enhance our understanding of the relationship between the genotype and phenotypic modularity and integration.

5.5 ACKNOWLEDGMENTS

I thank Joan T. Richtsmeier and M. Kathleen Pitirri for the invitation to contribute to this volume. I also thank Roger Reeves and two anonymous reviewers whose comments greatly improved the manuscript. Joan T. Richtsmeier was involved in the initial discussions that led to this chapter. I am grateful to Roger Reeves for providing the data used in this study. Funding for the original project was provided by the LuMind Foundation, Research Down syndrome, and the National Institutes of Health (5R01 HD38384–16).

5.6 REFERENCES

Ackermann, R. R. 2005. Ontogenetic integration of the hominoid face. *J Hum Evol* 48:175–97.

Ackermann, R. R. 2007. Craniofacial variation and developmental divergence in primate and human evolution. In *Tinkering: The Microevolution of Development: Novartis Foundation Symposium*, ed. G. Bock and J. Goode, 284:262. Chichester, NY: John Wiley.

Ackermann, R. R. 2009. Morphological integration and the interpretation of fossil hominin diversity. *Evol Biol* 36:149–56.

Ackermann, R. R., and J. M. Cheverud. 2004. Morphological integration in primate evolution. In *Phenotypic Integration: Studying the Ecology and Evolution of Complex Phenotypes*, ed. M. Pigliucci and K. Preston, 302–19. Oxford: Oxford University Press.

Adams, D. C., F. J. Rohlf, and D. E. Slice. 2004. Geometric morphometrics: ten years of progress following the 'revolution'. *Ital J Zool* 71:5–16.

Ahlgren, S. C., and M. Bronner-Fraser. 1999. Inhibition of Sonic Hedgehog signaling in vivo results in craniofacial neural crest cell death. *Curr Biol* 9:1304–14.

Amano, O., T. Doi, T. Yamada, A. Sasaki, et al. 2010. Meckel's cartilage: discovery, embryology and evolution. *J Oral Biosci* 52:125–35.

Atchley, W. R. 1993. Genetic and developmental aspects of variability in the mammalian mandible. In *The Skull*, vol. I, ed. J. Hanken and B. K. Hall, 207–47. Chicago: Chicago University Press.

Baker, V. H., and M. Bronner-Fraser. 2001. Vertebrate cranial placodes I. Embryonic induction. *Dev Biol* 232:1–61.

Balk, K., and L. G. Biesecker. 2008. The clinical atlas of Greig cephalopolysyndactyly syndrome. *Am J Hum Genet A* 146:548–57.

Bastir, M., and A. Rosas. 2005. Hierarchical nature of morphological integration and modularity in the human posterior face. *Am J Physl Anthropol* 128:26–34.

Billmyre, K. K., and J. Klingensmith. 2015. Sonic Hedgehog from pharyngeal arch 1 epithelium is necessary for early mandibular arch cell survival and later cartilage condensation differentiation. *Dev Dyn* 244:564–76.

Bolker, J. A. 2000. Modularity in development and why it matters to evo-devo. *Am Zool* 40:770–6.

Bookstein, F. L., P. Gunz, P. Mitteroecker, H. Prossinger, et al. 2003. Cranial integration in *Homo*: singular warps analysis of the midsagittal plane in ontogeny and evolution. *J Hum Evol* 44:167–87.

Bush, J. O., and R. Jiang. 2012. Palatogenesis: morphogenetic and molecular mechanisms of secondary palate development. *Development* 139:231–43.
Butler, D. L., S. A. Goldstein, R. E. Guldberg, X. G. Guo, et al. 2009. The impact of biomechanics in tissue engineering and regenerative medicine. *Tissue Eng Part B Rev* 15:477–84.
Cardini, A., and S. Elton. 2008. Does the skull carry a phylogenetic signal? Evolution and modularity in the Guenons. *Biol J of the Linn Soc* 93:813–34.
Chai, Y., and R. E. Jr. Maxson. 2006. Recent advances in craniofacial morphogenesis. *DevDyn* 235:2353–75.
Chen, J. K., J. Taipale, K. E. Young, T. Maiti, et al. 2002. Small molecule modulation of smoothened activity. *PNAS* 99:14071–6.
Cheverud, J. M. 1982. Phenotypic, genetic, and environmental morphological integration in the cranium. *Evolution* 36:499–516.
Cheverud, J. M. 1995. Morphological integration in the saddle-back tamarin (*Saguinus Fuscicollis*) cranium. *Am Nat* 145:63–89.
Chiang, C., Y. Litingtung, E. Lee, K. E. Young, et al. 1996. Cyclopia and defective axial patterning in mice lacking Sonic Hedgehog gene function. *Nature* 383:407–13.
Das, I., J.-M. Park, J. H. Shin, S. K. Jeon, et al. 2013. Hedgehog agonist therapy corrects structural and cognitive deficits in a down syndrome mouse model. *Sci Transl Med* 5:201ra120–201ra120.
Dorus, S., J. R. Anderson, E. J. Vallender, S. L. Gilbert, et al. 2006. Sonic Hedgehog, a key development gene, experienced intensified molecular evolution in primates. *Hum Mol Genet* 15:2031–7.
Geng, X., and G. Oliver. 2009. Pathogenesis of holoprosencephaly. *J Clin Investig* 119:1403–13.
Goswami, A., and D. P. Polly. 2010. Methods for studying morphological integration and modularity. *Paleontological Society* 16:213–43.
Hallgrímsson, B., and D. E. Lieberman. 2008. Mouse models and the evolutionary developmental biology of the skull. *Integr Comp Biol* 48:373–84.
Hallgrimsson, B., D. E. Lieberman, N. M. Young, T. Parsons, et al. 2007. Evolution of covariance in the mammalian skull. In *Novartis Foundation Symposium*, 284:164–85. Chichester and New York: John Wiley.
Hallgrímsson, B., K. Willmore, C. Dorval, and D. M. L. Cooper. 2004. Craniofacial variability and modularity in macaques and mice. *J Exp Zool B Mol Dev Evol* 302:207–25.
Helms, J. A. 2005. New insights into craniofacial morphogenesis. *Development* 132:851–61.
Hu, D., and J. A. Helms. 1999. The role of Sonic Hedgehog in normal and abnormal craniofacial morphogenesis. *Development* 126:4873–84.
Hu, D., and R. S. Marcucio. 2009a. A SHH-responsive signaling center in the forebrain regulates craniofacial morphogenesis via the facial ectoderm. *Development* 136:107–16.
Hu, D., and R. S. Marcucio. 2009b. Unique organization of the frontonasal ectodermal zone in birds and mammals. *Dev Biol* 325:200–10.
Hu, D., R. S. Marcucio, and J. A. Helms. 2003. A zone of frontonasal ectoderm regulates patterning and growth in the face. *Development* 130:1749–58.
Jeong, J., J. Mao, T. Tenzen, A. H. Kottmann, et al. 2004. Hedgehog signaling in the neural crest cells regulates the patterning and growth of facial primordia. *Genes Dev* 18:937–51.
Jiang, R., J. O. Bush, and A. C. Lidral. 2006. Development of the upper lip: morphogenetic and molecular mechanisms. *Dev Dyn* 235:1152–66.
Jiang, X., S. Iseki, R. E. Maxson, H. M. Sucov, et al. 2002. Tissue origins and interactions in the mammalian skull vault. *Dev Biol* 241:106–16.
Klingenberg, C. P. 2004. Integration, modules, and development. In *Phenotypic Integration: Studying the Ecology and Evolution of Complex Phenotypes*, ed. M. Pigliucci and K. Preston, 213–30. New York: Oxford University Press.

Klingenberg, C. P. 2005. Developmental constraints, modules, and evolvability. In *Variation*, ed. B. Hallgrímsson and B. K. Hall, 219–47. San Diego: Academic Press.

Klingenberg, C. P. 2008. Morphological integration and developmental modularity. *Annu Rev Ecol Evol Syst* 39:115–32.

Klingenberg, C. P. 2009. Morphometric integration and modularity in configurations of landmarks: tools for evaluating a priori hypotheses. *Evol Dev* 11:405–21.

Klingenberg, C. P. 2011. MorphoJ: an integrated software package for geometric morphometrics. *Mol Ecol Resour* 11:353–7.

Klingenberg, C. P. 2016. Size, shape, and form: concepts of allometry in geometric morphometrics. *Dev Genes Evol* 226:113–37.

Klingenberg, C. P., K. Mebus, and J.-C. Auffray. 2003. Developmental integration in a complex morphological structure: how distinct are the modules in the mouse mandible? *Evol Dev* 5:522–31.

Klingenberg, C. P., and N. Navarro. 2012. Development of the mouse mandible: a model system for complex morphological structures. In *Evolution of the House Mouse*, ed. M. Macholan, S. Baird, P. Munclinger, and J. Pialek, 135–49. Cambridge: Cambridge University Press.

Kronmiller, J. E., T. Nguyen, W. Berndt, and A. Wickson. 1995. Spatial and temporal distribution of Sonic Hedgehog MRNA in the embryonic mouse mandible by reverse transcription/polymerase chain reaction and *in situ* hybridization analysis. *Arch Oral Biol* 40:831–8.

Lacruz, R. S., C. B. Stringer, W. H. Kimbel, B. Wood, et al. 2019. The evolutionary history of the human face. *Nat Ecol Evol* 3:726–36.

Lewton, K. L. 2012. Evolvability of the primate pelvic girdle. *Evol Biol* 39:126–39.

Lieberman, D. E., C. F. Ross, and M. J. Ravosa. 2000. The primate cranial base: ontogeny, function, and integration. *Am J Phys Anthropol* 113:117–69.

Mak, K. K., Y. Bi, C. Wan, P.-T. Chuang, et al. 2008. Hedgehog signaling in mature osteoblasts regulates bone formation and resorption by controlling PTHrP and RANKL expression. *Dev Cell* 14:674–88.

Marcucio, R. S., D. R. Cordero, D. Hu, and J. A. Helms. 2005. Molecular interactions coordinating the development of the forebrain and face. *Dev Biol* 284:48–61.

Marcucio, R. S., B. Hallgrimsson, and N. M. Young. 2015. Facial morphogenesis: physical and molecular interactions between the brain and the face. *Curr Top Dev Biol* 115:299–320.

Martínez-Abadías, N., Y. Heuzé, Y. Wang, E. W. Jabs, et al. 2011. FGF/FGFR signaling coordinates skull development by modulating magnitude of morphological integration: evidence from Apert syndrome mouse models. *PLoS One* 6:e26425.

Mayr, E. 1982. *The Growth of Biological Thought: Diversity, Evolution, and Inheritance.* Cambridge, MA: Harvard University Press.

McBratney-Owen, B., S. Iseki, S. D. Bamforth, B. R. Olsen, et al. 2008. Development and tissue origins of the mammalian cranial base. *Dev Biol* 322:121–32.

Mitteroecker, P., and F. L. Bookstein. 2007. The conceptual and statistical relationship between modularity and morphological integration. *Syst Biol* 56:818–36.

Mitteroecker, P., and F. L. Bookstein. 2008. The evolutionary role of modularity and integration in the hominoid cranium. *Evolution* 62:43–58

Moss, M. L. 1968. A theoretical analysis of the functional matrix. *Acta Biotheor* 18:195–202.

Moss, M. L., and R. W. Young. 1960. A functional approach to craniology. *Am J Phys Anthropol* 18:281–92.

Moss-Salentijn, L. 1997. Melvin L. Moss and the functional matrix. *J Dent Res* 76:1814–17.

Muenke, M., and M. M. Jr. Cohen. 2000. Genetic approaches to understanding brain development: holoprosencephaly as a model. *Dev Disabil Res Rev* 6:15–21.

Neaux, D., G. Sansalone, J. A. Ledogar, S. H. Ledogar, et al. 2018. Basicranium and face: assessing the impact of morphological integration on primate evolution. *J Hum Evol* 118:43–55.

Neaux, D., S. Wroe, J. A. Ledogar, S. H. Ledogar, et al. 2019. Morphological integration affects the evolution of midline cranial base, lateral basicranium, and face across primates. *Am J Phys Anthropol* 170:37–47.

Nie, X., K. Luukko, I. H. Kvinnsland, and P. Kettunen. 2005. Developmentally regulated expression of Shh and Ihh in the developing mouse cranial base: comparison with Sox9 expression. *Anat Rec* 286:891–8.

Odent, S., T. Attié-Bitach, M. Blayau, M. Mathieu, et al. 1999. Expression of the Sonic Hedgehog (SHH) gene during early human development and phenotypic expression of new mutations causing holoprosencephaly. *Hum Mol Genet* 8:1683–9.

Olson, E. C., and R. L. Miller. 1999. *Morphological Integration*. Chicago: University of Chicago Press.

Osumi-Yamashita, N., Y. Ninomiya, and K. Eto. 2002. Mammalian craniofacial embryology in vitro. *Int J Dev Biol* 41:187–94.

Pan, A., L. Chang, A. Nguyen, and A. W. James. 2013. A review of hedgehog signaling in cranial bone development. *Front Physiol* 4:1–14.

Pavlicev, M., J. M. Cheverud, and G. P. Wagner. 2009. Measuring morphological integration using eigenvalue variance. *Evol Biol* 36:157–70.

Porto, A., F. B. de Oliveira, L. T. Shirai, V. D. Conto, et al. 2009. The evolution of modularity in the mammalian skull I: morphological integration patterns and magnitudes. *Evol Biol* 36:118–35.

Powell, V., B. Esteve-Altava, J. Molnar, B. Villmoare, et al. 2018. Primate modularity and evolution: first anatomical network analysis of primate head and neck musculoskeletal system. *Sci Rep* 8:1–10.

Richtsmeier, J. T., and K. Flaherty. 2013. Hand in glove: brain and skull in development and dysmorphogenesis. *Acta Neuropathol* 125:469–89.

Roessler, E., E. Belloni, K. Gaudenz, P. Jay, et al. 1996. Mutations in the human Sonic Hedgehog gene cause holoprosencephaly. *Nat Genet* 14:357–60.

Rohlf, F. J., and M. Corti. 2000. Use of two-block partial least-squares to study covariation in shape. *Syst Biol* 49:740–53.

Rohlf, F. J., and D. Slice. 1990. Extensions of the Procrustes method for the optimal superimposition of landmarks. *Syst Biol* 39:40–59.

Rolian, C. 2009. Integration and evolvability in primate hands and feet. *Evol Biol* 36:100–17.

Senders, C. W., E. C. Peterson, A. G. Hendrickx, and M. A. Cukierski. 2003. Development of the upper lip. *JAMA Facial Plast Surg* 5:16–25.

Singh, N., T. Dutka, B. M. Devenney, K. Kawasaki, et al. 2015. Acute upregulation of hedgehog signaling in mice causes differential effects on cranial morphology. *Dis Models Mech* 8:271–9.

Singh, N., K. Harvati, J.-J. Hublin, and C. P. Klingenberg. 2012. Morphological evolution through integration: a quantitative study of cranial integration in *Homo*, *Pan*, *Gorilla* and *Pongo*. *J Hum Evol* 62:155–64.

Slice, D. E. 2007. Geometric morphometrics. *Annu Rev Anthropol* 36:261–81.

St-Jacques, B., M. Hammerschmidt, and A. P. McMahon. 1999. Indian hedgehog signaling regulates proliferation and differentiation of chondrocytes and is essential for bone formation. *Genes Dev* 13:2072–86.

Sugito, H., Y. Shibukawa, T. Kinumatsu, T. Yasuda, et al. 2011. Ihh signaling regulates mandibular symphysis development and growth. *J Dent Res* 90:625–31.

Szabo-Rogers, H. L., L. E. Smithers, W. Yakob, and K. J. Liu. 2010. New directions in craniofacial morphogenesis. *Dev Biol* 341:84–94.

Team, R. C. 2013. *R: A Language and Environment for Statistical Computing.* Vienna, Austria. http://www.R-project.org/.

Wagner, G. P., and L. Altenberg. 1996. Perspective: complex adaptations and the evolution of evolvability. *Evolution* 50:967–76.

Xavier, G. M., M. Seppala, W. Barrell, A. A. Birjandi, et al. 2016. Hedgehog receptor function during craniofacial development. *Dev Biol* 415:198–215.

Xu, J., H. Liu, Y. Lan, M. Adam, et al. 2019. Hedgehog signaling patterns the oral-aboral axis of the mandibular arch. *eLife* 8:e40315.

Young, B., N. Minugh-Purvis, T. Shimo, B. St-Jacques, et al. 2006. Indian and Sonic Hedgehogs regulate synchondrosis growth plate and cranial base development and function. *Dev Biol* 299:272–82.

Young, N. M., H. J. Chong, D. Hu, B. Hallgrímsson, et al. 2010. Quantitative analyses link modulation of Sonic Hedgehog signaling to continuous variation in facial growth and shape. *Development* 137:3405–9.

Young, N. M., D. Hu, A. L. Lainoff, F. J. Smith, et al. 2014. Embryonic bauplans and the developmental origins of facial diversity and constraint. *Development* 141:1059–63.

6 Brown Adipose Tissue, Nonshivering Thermogenesis, and Energy Availability

Maureen J. Devlin

CONTENTS

6.1 Introduction 132
  6.1.1 Brown Adipose Tissue 132
  6.1.2 Key Questions about BAT Energetics 134
6.2 Quantification of BAT Volume and Metabolic Activity 135
  6.2.1 BAT, Glucose, and Fatty Acids 135
  6.2.2 Infrared Skin Temperature Measurement 136
6.3 BAT and Body Composition in Subadults 137
  6.3.1 BAT in Fetal Stage and Infancy 137
  6.3.2 Childhood BAT 138
  6.3.3 Adolescence and Puberty 138
  6.3.4 Adulthood and Aging 138
  6.3.5 Candidate Mechanisms 139
6.4 Energy Restriction and Chronic Stress 139
  6.4.1 Experimental Effects of Energy Restriction on BAT 140
  6.4.2 Chronic Stressors and Glucocorticoids 141
6.5 Obesity and Nonshivering versus Diet-Induced Thermogenesis 141
  6.5.1 BAT and Obesity 142
  6.5.2 Diet-Induced Thermogenesis and BAT 143
  6.5.3 Diet and Obesity in UCP1 Knockout Mice 144
6.6 Pregnancy and Lactation 144
  6.6.1 Pregnancy 144
  6.6.2 Lactation 145
6.7 Exercise and Bone 146
  6.7.1 BAT and Bone 146
  6.7.2 Exercise, Irisin, and BAT 146
6.8 Implications for Hominin Evolutionary Biology 147
6.9 Summary and Future Directions 148
6.10 Acknowledgments 149
6.11 References 149

6.1 INTRODUCTION

Primates evolved in the tropics, and even today, most extant taxa live in tropical forests, woodlands, and savannas (Martin 1986). Hominins are a notable exception to this pattern, successfully expanding out of Africa and colonizing high and low latitudes despite a major Pliocene cooling trend and the Ice Ages of the Pleistocene. Multiple taxa, including Neanderthals, Denisovans, and modern humans, succeeded in harsh environments despite chronic cold stress and sometimes scarce food resources. Cultural and environmental buffering such as fire, shelter, and clothing aided survival in the cold (Wales 2012), but hominins must also have depended on biological mechanisms of temperature homeostasis, including changes in body size and shape (known as Bergmann's law and Allen's rule), muscle contractions from shivering or exercise, and nonshivering thermogenesis in order to maintain body temperature (Steegmann et al. 2002; Steegmann 2007). Latitudinal variation in body breadth and limb proportions for improved heat retention are well recognized in humans (Ruff 1993). As a source of heat, muscle contractions are effective but energetically expensive (Eyolfson et al. 2001; Krustrup et al. 2003). Nonshivering thermogenesis (NST) is less well studied in humans, in part because it primarily occurs in a specific type of fat called brown adipose tissue that until recently was thought to be scarce in adults (Heaton 1972). However, the potential of NST as a mechanism for hominin cold adaptation is particularly interesting for three reasons. First, as discussed in the following, NST occurs in specialized fat cells called brown adipocytes, whose capacity for heat production rises and falls in parallel with cold exposure, so this mechanism is energetically flexible. Second, NST involves a direct tradeoff of calories between ATP and heat; from a life history perspective, conservation of such a metabolically expensive tissue suggests its benefits outweighed its costs. Third, data from humans and animal models show that cold exposure can induce bone loss, which may be partly ameliorated by NST. Thus, understanding NST in extant humans is essential for developing hypotheses about its role in hominin evolution and its effects on skeletal phenotype.

6.1.1 Brown Adipose Tissue

Adipocytes, cells that are specialized for the storage of fat, are an essential depot for energy storage, endocrine signaling, and other crucial functions in mammals (Zwick et al. 2018). Most research has focused on white adipose tissue (WAT), which is increasing rapidly in human populations and contributing to a worrisome rise in metabolic disease (Hruby and Hu 2015). Less well known is bone marrow adipose tissue (BMAT), a complex depot whose function may relate to surviving starvation (Fazeli et al. 2013; Devlin and Rosen 2015; Scheller et al. 2015; Suchacki and Cawthorn 2018). A third depot, brown adipose tissue (BAT), is common in mammals, especially in small-bodied and hibernating species that rely on NST. Its characteristic brown color comes from a high density of mitochondria uniquely expressing uncoupling protein-1 (UCP1) in the inner mitochondrial membrane of the cell. During NST, UCP1 uncouples cellular respiration (ATP production) and allows energy to instead be dissipated as heat (Cannon and Nedergaard 2004) (Figure 6.1). Brown fat

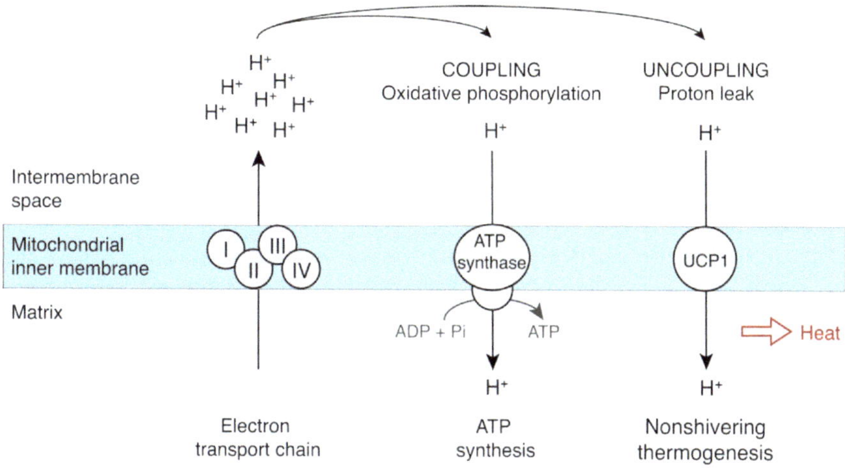

FIGURE 6.1 Uncoupling protein in the inner mitochondrial membrane diverts protons away from ATP synthase so that the proton motive force is released as heat. (Adapted from Collins, S., and Surwit, R. S., *Recent Prog Horm Res*, 56, 309–28, 2001.)

contributes to temperature homeostasis in human newborns but was thought to have little role in adults because this depot regresses with age. However, a second type of functional UCP1-expressing adipocyte was identified in adults in 2009 (Au-Yong et al. 2009; Cypess et al. 2009; Saito et al. 2009; van Marken Lichtenbelt et al. 2009; Virtanen et al. 2009). Like brown adipocytes, these so-called beige adipocytes express UCP1, contain multilocular lipid droplets and numerous mitochondria, and generate heat via NST when stimulated by norepinephrine (Wu et al. 2012). Distinct from classical brown adipocytes, beige adipocytes are facultatively rather than constitutively expressed and can be induced to proliferate by environmental stimuli, particularly cold (Wu et al. 2012, 2013). Furthermore, while the classical brown adipocytes of newborns and rodents are predominantly found in the interscapular region of the body, inducible beige adipocytes are located within white fat depots, particularly in the cervical, supraclavicular, and axillary regions and sometimes in perirenal, pericardial, and paravertebral depots (Sacks and Symonds 2013) (Figure 6.2). Finally, and most importantly, beige adipocytes are derived from a different stem cell progenitor lineage than classical brown adipocytes (Harms and Seale 2013). Lineage tracing studies in mice showed that brown adipocytes derive from pluripotent embryonic dermomyotome cells that also contribute to the skeletal muscle and dermis of the back, while beige adipocytes derive from progenitors related to adipogenic and smooth muscle cells (Long et al. 2014; Wang and Seale 2016). Herein, both brown and beige adipocytes will be collectively referred to as BAT. BAT is more often detected in winter than in summer and in women than in men and tends to be less common with advancing age and/or obesity, although the direction of causation remains unclear (Pfannenberg et al. 2010; Ouellet et al. 2011; Vijgen et al. 2011; Yoneshiro et al. 2011, 2016).

6.1.2 Key Questions about BAT Energetics

The unique characteristic of BAT is that it expends rather than stores energy, taking up glucose and free fatty acids during NST (Hoeke et al. 2016; Schilperoort et al. 2016). The discovery that humans retain beige adipocytes into adulthood raised the possibility that stimulation of BAT could be used to increase energy expenditure for weight loss and for blood glucose control, either physiologically (e.g., cold exposure) or pharmacologically (e.g., sympathetic nervous system upregulation) (Harms and Seale 2013; Sidossis and Kajimura 2015; Betz and Enerback 2018; Chouchani et al. 2019). From an evolutionary perspective, it is more compelling to think about the role of BAT and its integration into overall energy metabolism in a life history framework.

Here I will review current issues in measuring BAT metabolic activity, its function through ontogeny (e.g., childhood and adolescent growth, pregnancy and lactation, and aging) and interactions between BAT and energy availability (e.g., starvation and obesity), exercise, and chronic stress. Finally, I will propose approaches to better understand its evolutionary significance. Life history theory predicts that organisms' energy availability is constrained, requiring tradeoffs between competing biological functions (e.g., maintenance, reproduction, activity). In this context, if BAT energy consumption is a significant component of an organism's energy budget, then BAT thermogenesis should be reduced in the absence of cold stress and/or during negative energy balance. If its energy usage is not significant, then we would expect BAT activity to be unchanged during energetic stress. Thus, the general hypothesis is that BAT thermogenesis should be decreased during times of high energy expenditure,

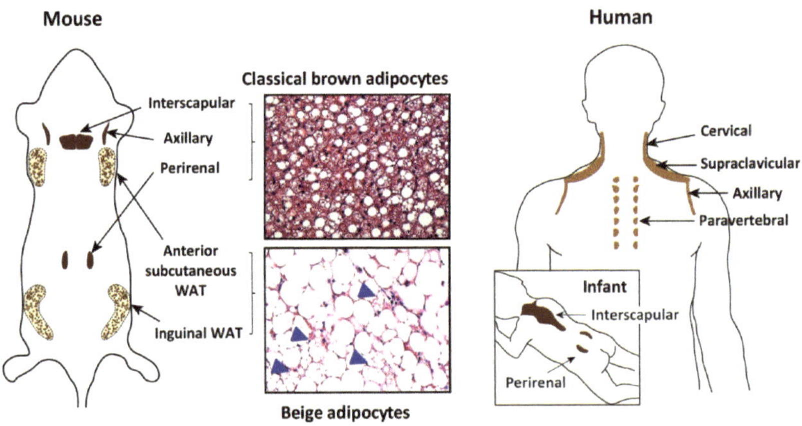

FIGURE 6.2 Human infants have constitutive brown adipocytes in interscapular and perirenal depots. In adults, inducible beige adipocytes are located within white adipose tissue depots in the supraclavicular, deep cervical, axillary, and paravertebral regions. (Adapted from *Trends Endocrinol Metab*, 29, Ikeda, K., Maretich, P., and Kajimura, S., The Common and Distinct Features of Brown and Beige Adipocytes, 191–200, 2018, an open access article under the CC BY-NC-ND license [http://creativecommons.org/licenses/by-nc-nd/4.0/], with permission from Elsevier.)

including rapid growth, pregnancy and lactation, and exercise, and during times of low energy intake, including starvation. Cases that do not follow this pattern might provide interesting life history insights. Data will be drawn from human studies whenever possible, but the difficulty of measuring BAT activity in vivo necessitates incorporation of data from animal models, primarily rodents. Although such experimental data cannot be used to make conclusions about BAT function or energy consumption in humans, they are critical for developing mechanistic hypotheses.

6.2 QUANTIFICATION OF BAT VOLUME AND METABOLIC ACTIVITY

Adult BAT was rediscovered through analyses of 18^{F}-FDG-PET-CT scans (18F-FDG), in which radiolabeled glucose is preferentially taken up by highly metabolically active tissues (typically cancer cells) (Cypess et al. 2009; Saito et al. 2009; van Marken Lichtenbelt et al. 2009; Virtanen et al. 2009). This method became the gold standard for identifying BAT in vivo because its glucose uptake increases during active NST. Briefly, sympathetic nervous system release of norepinephrine in response to cold or other stimuli activates beta-adrenergic receptors on brown and beige adipocytes, leading to increased blood flow and glucose consumption by BAT (Cannon and Nedergaard 2004). However, it has become clear that the preferred metabolic substrate for BAT is fatty acids rather than glucose (Hoeke et al. 2016; Schilperoort et al. 2016), and 18F-FDG is costly and involves radiation exposure, reducing its practicality for research. These issues have led to exploration of alternative options for BAT quantification.

6.2.1 BAT, Glucose, and Fatty Acids

Although sympathetic activation causes a marked increase in BAT blood glucose uptake, the emerging consensus is that intracellular triglycerides (fatty acids) rather than circulating blood glucose are its preferred energy source during cold-stimulated NST (Hoeke et al. 2016; Schilperoort et al. 2016). Sympathetic activation of beta-adrenergic receptors causes release of intracellular triglycerides (Cannon and Nedergaard 2004), and supraclavicular BAT fat content declines within 10 minutes of cold exposure, indicating rapid fatty acid metabolism (Oreskovich et al. 2019). Further, there is evidence that the presence of long chain fatty acids is sufficient to trigger uncoupling and that, conversely, chemically blocking fatty acid release inhibits uncoupling (Cannon and Nedergaard 2004; Shabalina et al. 2008; Blondin et al. 2017a).

The finding that fatty acids, rather than glucose, are the preferred substrate for BAT following cold challenge means that 18F-FDG likely underestimates BAT energy usage and perhaps even presence/absence and tissue volume of BAT (Carpentier et al. 2018). Glucose uptake by BAT is impaired in insulin resistance and type 2 diabetes, but fatty acid uptake is not (Orava et al. 2011), so 18F-FDG based measurements of BAT may be particularly inaccurate in this population (Hoeke et al. 2016; Schilperoort et al. 2016). Glucose uptake does increase in BAT during NST, and depleted intracellular fatty acids are replenished through a combination of blood glucose and fatty acids (Hoeke et al. 2016; Schilperoort et al. 2016). Interestingly, BAT has metabolic activity and takes up glucose even in warm conditions, perhaps to increase or replenish triglyceride stores for future NST (Weir et al. 2018). However, the existing reliance on glucose-based measurements suggests some existing conclusions about BAT, such as the observation that BAT prevalence and function are negatively correlated with age and obesity, could be due in part to decreased insulin sensitivity rather than to a complete absence of BAT. This point should be kept in mind while evaluating the studies reported in the following.

6.2.2 Infrared Skin Temperature Measurement

To address the limited practicality of 18F-FDG in vulnerable populations such as children, non-clinical studies, and field settings, several groups have investigated infrared measurement of supraclavicular skin temperature as a non-invasive, radiation-free index of BAT activity (Figure 6.3). In general, these studies show that the decrease in skin temperature after cold challenge is blunted in the supraclavicular region compared to nearby areas (e.g., chest and deltoid), particularly in subjects with confirmed BAT activity, suggesting infrared measurement is detecting BAT thermogenesis (Symonds et al. 2012; Salem et al. 2016; van der Lans et al. 2016; Yoneshiro et al. 2016; Levy 2019). Interestingly, in some cases overall metabolic rate declines despite higher skin surface temperatures, which could reflect a reduced metabolic response to cold due to habituation and/or an increase in central metabolism combined with a greater decrease in peripheral metabolism due to vasoconstriction (reviewed in Levy et al. 2018) The same pattern was seen in an indigenous circumpolar population, the Yakut (Sakha), in whom sternal temperature decreased following cold challenge but supraclavicular temperature did not, consistent with BAT thermogenesis (Levy et al. 2018). In studies including both supraclavicular skin temperature as well as 18F-FDG-based measurement of BAT volume and glucose

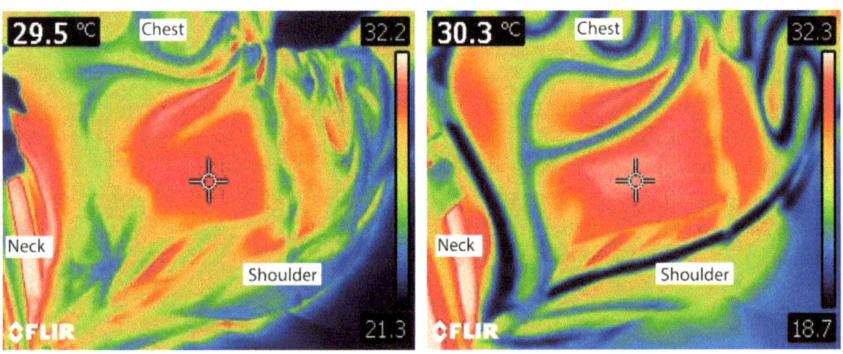

FIGURE 6.3 Thermal imaging (FLIR E60BX) shows that skin temperature of the right supraclavicular region, where BAT is commonly found, increased by 0.8°C from room temperature (left) to cold challenge (right). The subject is an adult Finnish male who is supine and at rest during the measurement. (Figure courtesy of Dr. Cara Ocobock.)

uptake, the correlations between these measures are often modest (r2 ~0.2–0.35) (Boon et al. 2014; Chondronikola et al. 2016; van der Lans et al. 2016). This finding is unsurprising, since skin surface temperature depends on blood circulation and subcutaneous white fat as well as brown fat (Jimenez-Pavon et al. 2019). However, infrared thermal imaging is an essential tool for obtaining data on cold-induced thermogenesis in field settings, and error can be minimized by following careful protocols, as outlined in Levy (2019).

6.3 BAT AND BODY COMPOSITION IN SUBADULTS

Although the classical brown adipocytes present in newborns regress in infancy, inducible beige adipocytes are common in children and adolescents, and BAT mass actually increases from childhood into puberty. These findings raise interesting questions about the factors regulating BAT proliferation in subadults and BAT influence on body mass homeostasis. Given that BAT is metabolically expensive, its increase in volume during peak somatic growth is particularly puzzling from a life history perspective. Studies of subadult BAT are understandably limited and include retrospective analyses of 18F-FDG imaging performed for cancer surveillance and postmortem histological studies, as well as MRI-based imaging. Even these limited data generally support the hypothesis that BAT is a phenotypically plastic depot, whose size and metabolic activity are influenced by energy availability.

6.3.1 BAT in Fetal Stage and Infancy

MRI-based imaging of fetal BAT volume and body composition showed that BAT is detectable by 26–30 weeks of gestation and that infants born premature have less BAT compared to full-term peers (Lean et al. 1986; Baumgart 2008). Candidate stimuli include angiopoietin-like protein 8 (ANGPTL8), a newly identified potential moderator of fetal BAT development. ANGPTL8 is inversely related to neonatal size and fat mass in vivo and increases brown adipocyte markers in vitro, suggesting it may influence the balance between BAT and WAT (Martinez-Perez et al. 2016). Another candidate is insulin-like growth factor, which increases in late gestation (Symonds et al. 2003); in rodents and in vitro, inactivation of the IGF-1 receptor disrupts brown adipocyte function (Valverde et al. 2004; Boucher et al. 2012).

Historical studies showed that neonatal NST in BAT is energetically expensive, increasing oxygen consumption and plasma glycerol by nearly twofold after 20 minutes at 25–26°C vs. 34–35°C (Dawkins and Scopes 1965). A postmortem histological assessment of periadrenal BAT depots from birth–age 16 showed that for the first 2 months of life, multilocular lipid droplet size and number decreased, followed by progressive infiltration of large monolocular lipid droplets in older children (Emery and Dinsdale 1978), indicating an increasingly WAT-like phenotype. An MRI study of infants quantified the relative fat fraction of the supraclavicular region, with a lower fat fraction indicating a higher proportion of brown adipocytes (Entringer et al. 2017). Infants with more BAT (a lower fat fraction) gained significantly less body fat from 0–6 months and had lower body fat at 6 months of age (r^2 ~24%) (Entringer et al. 2017). However, another study of infants found no relationship between BAT and subcutaneous fat gain in the trunk but did find that infants who had smaller decreases in BAT from 0–6 months of age had greater gains in paraspinous muscle mass (Ponrartana et al. 2016).

6.3.2 Childhood BAT

BAT is relatively common in children, found in 10–60% of healthy subjects if measured at or below room temperature (~21°C) (Yeung et al. 2003; Zukotynski et al. 2009, 2010; Hong et al. 2011; Gilsanz et al. 2012b; Rockstroh et al. 2015). Supraclavicular temperature increase in response to cold challenge did not differ in prepubertal boys and girls (Malpique et al. 2019) or young adults (Kistner et al. 2018) who had been small for gestational age (SGA) compared to individuals who were appropriate for gestational age (AGA). However, BAT variation among children is correlated with body composition. Several studies reported lower body fat in children who had active BAT compared to children who did not (Chalfant et al. 2012; Gilsanz et al. 2012a; Ponrartana et al. 2013). Histological analyses of subcutaneous, visceral, and perirenal adipose tissue from children undergoing elective surgery found that none of the 44 subjects with high BMI had evidence of active BAT, while 10% of 87 subjects at normal BMI did; most of the BAT-positive subjects were under 6 years of age (Rockstroh et al. 2015). Similarly, 18F-FDG scans from pediatric cancer patients showed that during treatment, only 10% of subjects had active BAT, but in subsequent cancer-free scans, 80% of subjects had active BAT (Gilsanz et al. 2012a), perhaps because TNF, a cytokine produced by tumors, is a major suppressor of BAT (Salles et al. 1996).

6.3.3 Adolescence and Puberty

Both BAT volume and glucose uptake increase across the pubertal transition (Gelfand et al. 2005; Drubach et al. 2011; Gilsanz et al. 2012b), with the incidence of metabolically active BAT increasing from about 15% in Tanner stage 1 to 75% in Tanner stages 2–5 (Gilsanz et al. 2012b). Tissue volume of BAT increased by 8–10-fold in late puberty vs. early puberty in both sexes, in parallel with increased muscle volume (although BAT volume was not scaled by body size), leading the authors to conclude that BAT was responsive to the hormonal milieu of puberty (Gilsanz et al. 2012b). A lower supraclavicular fat fraction on MRI (a more BAT-like depot) in adolescents was negatively correlated with total body adiposity and positively correlated with thigh muscle volume and with osteocalcin, a marker of bone formation (Andersson et al. 2019). Adolescents with a higher supraclavicular fat fraction (a less BAT-like depot) on MRI had higher adiposity and higher glucose following glucose tolerance test (Lundstrom et al. 2019).

6.3.4 Adulthood and Aging

BAT mass typically declines beginning in middle age (Heaton 1972; Saito et al. 2009), with women retaining higher BAT mass and activity than men do at all ages (Pfannenberg et al. 2010). The frequency of active BAT after cold exposure declined from >50% in subjects under age 30 to only one of eight subjects aged 50–60 and zero of eight over age 60 (Yoneshiro et al. 2011). Body fat did not increase with age in BAT-positive subjects but did in BAT-negative subjects (Yoneshiro et al. 2011). In elderly subjects (mean age 80 years), there was no evidence of metabolically active BAT (Franz et al. 2015). Finally, data from the circumpolar Sakha showed that younger adults had a higher metabolic rate in winter than in summer, but older adults did not show seasonal changes (Snodgrass et al. 2005; Leonard et al. 2014), consistent with an age-related decline in BAT thermogenesis.

6.3.5 Candidate Mechanisms

BAT is correlated with lower obesity, improved glucose tolerance, and increased muscle and bone mass in subadults, but the direction of causality remains unclear. It is possible that individuals with less BAT are predisposed to obesity through decreased energy expenditure and/or that individuals with higher BMI may have decreased need for NST, causing involution of BAT. It should be noted that most studies have recruited subjects from industrialized economies with cultural buffering (e.g., indoor heating, warm clothing), so children may have less need for NST to maintain temperature homeostasis compared to prior generations. Beyond cold exposure, physiological candidates that might induce BAT formation include endogenous hormones, such as growth hormone, the sex steroids, leptin, and insulin, although data from children and adolescents are limited. DHEA increases in children at adrenarche, and DHEA or DHEAS treatment has been shown to stimulate BAT in rats (Lea-Currie et al. 1997; Ryu et al. 2003).

Estrogen and testosterone are low during childhood but peak during adolescence. Although there are no human data directly relating estrogen to BAT, a recent mechanistic study in female rats found estrogen increased BAT metabolic activity via suppression of AMP-activated protein kinase (AMPK), an enzyme that regulates energy balance in the ventromedial hypothalamus of the brain (Martinez de Morentin et al. 2014; Huynh et al. 2016). Interestingly, testosterone had suppressive effects on BAT cells in vitro (Rodriguez et al. 2002; Monjo et al. 2003). Overall, these findings provide potential explanations for the high prevalence of BAT in adolescents and are consistent with the observation that BAT is less abundant in men than in women, in whom body fat, leptin, and estrogen tend to be higher.

6.4 ENERGY RESTRICTION AND CHRONIC STRESS

It is critical to understand interconnections between BAT, undernutrition, and overnutrition both to think about this fat depot from a life history perspective and to understand how pharmacological manipulation of energy expenditure by BAT might work. During negative energy balance (when energy intake is chronically below energy expenditure), mammals adapt by decreasing energy expenditure for nonessential functions, including immune function, reproduction, and body temperature (Prentice 2005; McCue 2010). The prediction is that sympathetic stimulation of BAT as well as its metabolic activity should decrease during negative energy balance. This question has not been studied directly in humans, although there is evidence that catecholamine metabolites in urine and circulating thyroid hormones decrease during undernutrition, which would reduce sympathetic outflow to BAT (Jung et al. 1980; Prentice 2005). To date, the majority of mechanistic data comes from animal models.

6.4.1 Experimental Effects of Energy Restriction on BAT

In mice and hamsters, single or repeated bouts of fasting decreased BAT mass, mRNA, and protein content, which rebounded after refeeding (Levin and Trayhurn 1987; Desautels and Dulos 1988; Trayhurn and Jennings 1988; Champigny and Ricquier 1990), suggesting energy conservation. Levels of the T4–5' deiodinase enzyme (DIO2), which converts the thyroid hormone thyroxine (T4) to the active form triiodothyronine (T3) in BAT, also decreased with fasting and increased with refeeding (Champigny and Ricquier 1990). A longer-term study of BAT thermogenesis in mice after up to 1 year of 20% caloric restriction (CR) showed that 1-year-old mice had less WAT infiltration of BAT depots, higher expression of BAT marker genes including Prdm16 and Pgc1a, and higher T3 and DIO2 levels (Corrales et al. 2019). In other words, BAT morphology and gene expression in the 1-year-old CR mice resembled that of 3-month-old controls rather than 1-year-old controls. However, the 1-year-old CR mice had lower body temperature and lower glucose uptake in BAT following cold challenge vs. 3-month-old controls, indicating lower BAT metabolic activity (Corrales et al. 2019). In ad libitum–fed rats, females had higher BAT activity than males, and 40% CR markedly decreased UCP1 protein content and triglyceride levels in BAT of females but not males (Valle et al. 2005; Valle et al. 2007), suggesting sex differences in the metabolic response to CR.

These experiments support the overall hypothesis that NST in BAT is suppressed in negative energy balance. Food intake and body temperature are centrally regulated, and the same hypothalamic orexigenic signals that spur food-seeking behavior also suppress sympathetic outflow to BAT (Nakamura and Nakamura 2018). Prime candidates for central energy-dependent regulation of BAT activity include ghrelin, known as the "hunger hormone", and adiponectin, a hormone produced in WAT that increases during negative energy balance (Huynh et al. 2016). Starvation increases ghrelin, which in turn increases orexigenic neuropeptide Y (NPY) and agouti-related peptide (AgRP), stimulating hunger, decreasing activity level, and blunting NST in BAT (Nakamura and Nakamura 2018). Ghrelin administered intravenously or directly into the brain suppressed BAT noradrenaline release, but only if the vagus nerve was intact, indicating ghrelin can reduce energy expenditure by suppressing sympathetic stimulation of BAT via the vagus nerve (Mano-Otagiri et al. 2009). Ghrelin and adiponectin also upregulate hypothalamic AMPK kinase, which suppresses NST (Huynh et al. 2016). Estrogen, leptin, and insulin, all of which fall in negative energy balance, downregulate hypothalamic AMPK, so low levels of these hormones allow greater suppression of NST (Huynh et al. 2016). In other words, hormonal signals of nutritional sufficiency or deficiency can increase or decrease NST via the hypothalamus, consistent with the hypothesis that BAT metabolic activity responds to energy status.

6.4.2 Chronic Stressors and Glucocorticoids

As with energy restriction, the expectation is that chronic physiological and/or psychological stressors would decrease NST in BAT, consistent with an overall downregulation of energy expenditure for non-essential functions. Thus, while short-term stress leads to catecholamine release that upregulates BAT (Ramage et al. 2016), the dysregulation of diurnal changes in glucocorticoids seen in long-term stress is more likely to downregulate BAT. Data from in vitro studies generally support this prediction. In cells from a BAT tumor called a hibernoma, corticosterone blunts cellular responses to norepinephrine (van den Beukel et al. 2014), and dexamethasone blunts *UCP1* transcription in response to norepinephrine (Soumano et al. 2000). Glucocorticoids also suppress NST by upregulating hypothalamic AMPK kinase (Huynh et al. 2016) and by shunting lipids to fat storage and reducing norepinephrine activation of UCP1 (Feve et al. 1992; Kiely et al. 1994; Strack et al. 1995). In rats, corticosterone treatment inhibits BAT response to cold and to norepinephrine (Moriscot et al. 1993).

In humans, the data are more complex and depend on study duration. Infusion of the glucocorticoid prednisolone during cold exposure increased glucose uptake by BAT (Ramage et al. 2016) and did not reduce BAT activity in response to the sympathetic agonist isoproterenol (Scotney et al. 2017). One week of prednisolone caused significant decreases in 18F-FDG glucose uptake and supraclavicular skin temperature, along with increased lipid synthesis and energy expenditure, perhaps due to the increased cost of storing energy as lipid (Thuzar et al. 2018). A larger retrospective 18F-FDG study showed patients treated with glucocorticoids or with hypercortisolemia were less likely to have metabolically active BAT, and in those who did, BAT volume and glucose uptake were lower vs. controls (Ramage et al. 2016). In vitro, short-term prednisolone exposure increased BAT activity in response to neurotransmitters, but in longer-term cultures, glucocorticoid suppressed UCP1 in human cells (Ramage et al. 2016).

Overall, these studies support the general hypothesis that BAT is downregulated during energy restriction and chronic exposure to stressors, consistent with a shift in energy availability toward essential functions and away from temporarily expendable functions. The next question is whether overnutrition causes the opposite pattern, that is, increased thermogenesis.

6.5 OBESITY AND NONSHIVERING VERSUS DIET-INDUCED THERMOGENESIS

The persistent observation that people with active BAT have lower body fat and better metabolic health compared to people who do not has led to the hypothesis that BAT is protective against obesity, perhaps by using energy for nonshivering thermogenesis or (more speculatively) by burning off excess calories as part of diet-induced thermogenesis. These hypotheses spurred efforts to identify potential weight loss interventions, including testing the responsiveness of BAT to chronic cold exposure and to diets of varying calorie content and macronutrient balance.

6.5.1 BAT and Obesity

Experimental efforts to increase BAT abundance and energy expenditure through cold exposure have been most effective in young subjects with normal BMI. In men, 4 weeks of 2 hours/day at 10°C was sufficient to increase active BAT volume by 50% (Blondin et al. 2014), and short-term cold exposure increased circulating norepinephrine, improved insulin sensitivity, and altered circulating fatty acid concentration (Iwen et al. 2017). Overnight exposure to 19°C elevated BAT glucose intake and increased energy expenditure by about 5%, although individual responses varied substantially (Chen et al. 2013). In a study of ten men and women combining 18F-FDG, magnetic resonance imaging (MRI), and proton magnetic resonance spectroscopy (1H MRS) to measure tissue temperature, heat production in BAT correlated positively with BAT metabolic activity and negatively with supraclavicular fat fraction (Koskensalo et al. 2017). Even in metabolically obese subjects ($N = 10$), glucose uptake increased in BAT and in skeletal muscle after 10 days of cold exposure, although the response was inversely correlated with age and body fat (Hanssen et al. 2016).

These studies consistently show that BAT energy consumption can be increased by cold (assuming subjects are willing to undergo it), but tend to be short term; have very small sample sizes; often include only one sex; and often focus on younger, leaner subjects who are already more likely to have active BAT, so the results may not hold in longer-term analyses with larger sample sizes. For example, analysis of retrospective 18F-FDG scans showed that adults who were BAT-positive had lower BMI and visceral and subcutaneous fat but were also more likely to be young and female (Brendle et al. 2018). Also, as discussed, fatty acids are preferentially metabolized by BAT over glucose, and these substrates may have different responses to cold. Fatty acid clearance by BAT as measured using a radiolabeled long-chain fatty acid called 14(R,S)-[^{18}F]-fluoro-6-thia-heptadecanoic acid (^{18}FTHA) did not increase even after a 4-week cold acclimation, whereas BAT oxidative metabolism increased more than twofold (Blondin et al. 2017b). Finally, it remains unclear whether industrialized humans have sufficient amounts of BAT to make a meaningful difference in energy metabolism. Across studies, estimates for the amount of BAT in adults ranged from as low as 4 cc to over 1500 cc, and the proportion of adults with active BAT following cold stimulation ranged from 20–100% (Blondin et al. 2015a). Leitner et al. (2017) found that BAT constituted up to 1.5% of body mass in young men, although the depots of lean subjects were larger (334 vs. 130 ml) and had lower fat fraction compared to obese counterparts. In another study in men, daily energy expenditure did not differ between BAT+ and BAT– subjects, but the former exhibited more diet-induced thermogenesis and had a lower respiratory quotient, suggesting preferential fat metabolism (Hibi et al. 2016). However, the small sample size and short study durations of these studies make it difficult to extrapolate over longer time frames. Under typical physiological conditions, NST is likely not maximally stimulated, and brown adipocytes in adult humans attenuate with age, so it remains difficult to estimate actual calorie expenditure.

Similar to the human data, a study of surgical removal of BAT in mice followed by cold exposure for 9 days at 4°C found increased oxygen consumption, fat loss, changes in muscle sarcoplasmic reticulum and mitochondria, and increased sarcolipin levels, all suggestive of a shift to NST in muscle via calcium cycling (Bal et al. 2016). However, in *Ucp1KO* and WT mice housed at 18°C or 30°C, cold-acclimated WT mice showed about a 50% increase in norepinephrine-induced O2 consumption compared to warm-acclimated mice, but cold-acclimated *Ucp1KO* did not (Golozoubova et al. 2006). The authors concluded that although UCP1 in BAT is not the only thermogenic mechanism in mammals, it is the only mechanism that allows adaptive acclimatization to the cold (Golozoubova et al. 2006). Although it is difficult to resolve these conflicting results, the role of muscle in nonshivering thermogenesis—involving mechanisms other than UCP1—is an active area of investigation.

6.5.2 Diet-Induced Thermogenesis and BAT

Diet-induced thermogenesis (DIT) is the thermic effect of food, or the heat generated by digestion. A recent meta-analysis found that DIT for a mixed diet was about 5–15% of daily energy expenditure (Westerterp 2004), although intra-individual variation was as high as 40–50% when assessed by indirect calorimetry, even after correction for BMR and activity (Ravussin et al. 1986; Tataranni et al. 1995). DIT varies by macronutrient, with protein (20–30%) and alcohol (10–30%) inducing more DIT than carbohydrate (5–10%) or fat (<3%) (Westerterp 2004). Even during caloric restriction, protein intake is positively correlated with daily energy expenditure. In adults, a crossover study of 1000-kCal high-protein, high-fat, and high-carbohydrate diets found similar weight loss on each but a smaller decline in daily energy expenditure on the high protein diet (Whitehead et al. 1996). In overfeeding, DIT increased by 2.4–2.8-fold more in subjects on high-protein vs. low- and normal-protein diets, but after 6 weeks, there were no significant differences among groups in DIT following a standard meal, indicating that DIT did not adapt to changes in macronutrient composition (Sutton et al. 2016).

Beyond this obligatory DIT, which is an inherent cost of digestion, some early studies suggested the existence of facultative DIT to expend excess calories during positive energy balance (Rothwell and Stock 1979; Acheson et al. 1984; Jequier and Schutz 1988). Such a mechanism would partly explain why BAT-positive individuals are resistant to obesity but also raises evolutionary questions: it is difficult to envisage circumstances in which it would be advantageous to burn off extra calories rather than storing them as fat. It is unclear whether overnutrition influences human DIT. Earlier meta-analyses of human overfeeding studies found that 59% (29/49) reported lower DIT in obese compared to lean subjects (de Jonge and Bray 1997), but only 31% (5/16) reported possible evidence for facultative DIT (Joosen and Westerterp 2006). More recently, a 24-hour overfeeding study found energy expenditure and overall DIT increased, but BAT metabolic activity did not, even in subjects who were BAT+ in response to cold (Schlogl et al. 2013). Another found post-meal BAT glucose uptake increased following a high-calorie, high-fat meal, although less than following cold exposure (Vosselman et al. 2013). Finally, U Din et al. (2018) found increased BAT thermogenesis following a mixed meal equivalent to that of cold exposure. The latter studies also showed that cold exposure induced greater fatty acid uptake, and food induced greater glucose uptake, indicating BAT substrate preference differs by stimulus.

Studies in rodents also found evidence of UCP1-dependent DIT (Bachman et al. 2002; Lowell and Bachman 2003). Although the mechanism has not been identified (Kontani et al. 2005; Feldmann et al. 2009; Rowland et al. 2016), it appears limited to classical brown rather than beige adipocytes (von Essen et al. 2017), and varies depending on macronutrient composition. In mice pair-fed to controls on three different high-fat diets (low carbohydrate and low protein; low carbohydrate and normal protein; moderate carbohydrate and normal protein), high fat intake did increase sympathetic outflow but did not alter *Ucp1* mRNA or UCP1 protein expression, so although mice gained less weight on the isocaloric high fat diets, the authors concluded that increased DIT was not the cause (Betz et al. 2012). In contrast, a high-protein, carbohydrate-free, normal-fat diet decreased sympathetic outflow to BAT and reduced its thermogenic capacity, although thermogenic response to acute cold remained intact, with no differences in weight gain between groups (Brito et al. 1998).

6.5.3 Diet and Obesity in UCP1 Knockout Mice

Another line of evidence to understand the role of UCP1 in body mass homeostasis comes from diet experiments in *Ucp1KO* mice. At standard vivarium housing temperature, 20–22C, *Ucp1KO* mice showed no differences in obesity from wildtype mice (Enerback et al. 1997; Liu et al. 2003). However, when housed at thermoneutrality (~27–30°C), *Ucp1KO* mice became obese (Liu et al. 2003; Feldmann et al. 2009), and there was no evidence for facultative DIT (von Essen et al. 2017). This pattern holds in both obesity-prone and obesity-resistant strains (Luijten et al. 2019), further supporting that the underlying mechanism is the lack of *Ucp1*. The implication is that without functional BAT, *Ucp1KO* mice must keep warm via metabolically expensive alternatives such as shivering, and the extra calories burned help to prevent obesity. Interestingly, *Ucp1KO* mice develop greater obesity in response to high fat diet compared to wildtype controls at thermoneutrality, suggesting there is some energy expenditure in BAT even in the absence of cold (Feldmann et al. 2009).

6.6 PREGNANCY AND LACTATION

Gestation and lactation are energetically expensive, with energy requirements increasing across the first (90 kCal/day), second (287 kCal/day), and third (466 kCal/day) trimesters and in full (626 kCal/day) and partial lactation (461 kCal/day) (Butte and King 2005). These demands necessitate reallocation of maternal metabolic expenditure when energy is constrained, such as reductions in maternal BMR and activity level (Prentice and Whitehead 1987). It is reasonable to hypothesize that during negative energy balance, NST in BAT would also be reduced, particularly during late gestation and lactation. However, BAT dynamics in human pregnancy have not been well studied, and most data come from rodent models.

6.6.1 Pregnancy

Little is known about BAT metabolic activity in human pregnancy, and the increased body mass and metabolic rate of pregnancy may produce sufficient heat to reduce the need for NST regardless of energy availability. One study reported maternal core body temperature fell after the first trimester of pregnancy, reaching its lowest point 12 weeks postpartum (Hartgill et al. 2011), but did not assess the mechanism.

In pregnant and lactating mice, food intake increased in the last 2–3 days of the 21-day gestation and remained high through lactation (Speakman 2008), while BAT metabolic activity fell despite hypertrophy of the tissue itself (Trayhurn et al. 1982b; Andrews et al. 1986). Similarly, pregnant hamsters showed a decrease in BAT tissue metabolic activity, protein content, and weight from late pregnancy through lactation (Wade et al. 1986; Schneider and Wade 1987). These data are consistent with the hypothesis that energy would be reallocated from maternal thermogenesis to gestation when energetic costs peak in late pregnancy and lactation but also with the alternative hypothesis that pregnancy and lactation are exothermic, and the reduction in BAT activity reflects the decreased need for thermogenesis (Krol and Speakman 2019).

6.6.2 Lactation

In rodent models of lactation, there is clear evidence that BAT thermogenic activity, gene expression, norepinephrine content, and turnover are reduced, perhaps driven by high prolactin and low leptin (Trayhurn et al. 1982a; Trayhurn and Wusteman 1987; Trayhurn 1989; Krol et al. 2011). Two models proposed to explain this reduction in NST are the central limitation hypothesis, which holds that the organism's central capacity for energy generation sets the ceiling on energy expenditure, and the heat dissipation limit theory, which holds that maternal hyperthermia induced by milk production reduces the need for NST. Lactation data provide some support for each model. A series of studies tracking milk production and weight gain of pups in cold, standard, and warm housing showed that mothers housed in the cold (8°C) consumed more food, produced more milk, and had larger pups at weaning compared to mothers housed at thermoneutrality (30°C), despite ad lib access to food for all mice, consistent with the heat dissipation limit theory (Johnson and Speakman 2001; Johnson et al. 2001; Krol and Speakman 2003). Gerbil mothers also produced less milk at 30°C (thermoneutrality) than at 10°C or 21°C, consistent with the expectations of the heat dissipation theory (Yang et al. 2013). Interestingly, pup mass was smaller at both 10°C and 30°C vs. 21°C, and milk production decreased when lactating females were transferred from 21°C to 30°C but did not increase in transfers from 21°C to 10°C (Yang et al. 2013).

However, not all studies find evidence for the heat dissipation limit theory. In female Swiss mice housed at 23°C or 5°C, cold mice did not wean larger pups, implying hyperthermia was not a constraint (Zhao 2012). Female mice gestating in cold (5°C) vs standard (21°C) temperatures similarly showed no differences in gestation length or mean pup mass at birth; although cold mice ate 40% more than warm mice, pregnant females had similar increases in calorie consumption (+20%) and similar weight gains in both temperatures (Marsteller and Lynch 1987). Taken together, these observations support a U-shaped curve for milk production: at low temperatures, maternal milk production may be limited by the need for nonshivering thermogenesis, particularly when energy availability is constrained, and at high temperatures, milk production may be limited by hyperthermia, with some variation across rodent strains and/or species.

6.7 EXERCISE AND BONE

6.7.1 BAT and Bone

Multiple studies have shown that BAT volume is correlated with higher bone mineral density (BMD), greater bone size, and higher muscle mass in children (Gilsanz et al. 2011; Ponrartana et al. 2012) and in young adults (Bredella et al. 2012, 2013; Oelkrug et al. 2013; Bredella et al. 2014), although one study of young adults found no association between BAT volume and BMD (Sanchez-Delgado et al. 2019). There is also evidence that older adults treated with "beta blockers", which prevent catecholamine binding to beta-adrenergic receptors, have lower fracture rates (Bonnet et al. 2007; Yang et al. 2012). The observations that BAT volume increases during the phase of rapid adolescent skeletal acquisition and that BAT is positively correlated with bone and muscle mass in adolescents and adults have led to speculation about interactions between the sympathetic nervous system, BAT, and bone. Although temperature homeostasis and skeletal homeostasis may seem unrelated, there is a fundamental connection between BAT and bone: cold-induced catecholamine release by the sympathetic nervous system, which stimulates NST, also leads to bone loss (Elefteriou 2005; Devlin 2015). By maintaining temperature homeostasis, BAT might protect the skeleton from chronic sympathetic activation that would lead to bone loss (Devlin 2015).

In experimental models, young and older mice housed below thermoneutrality exhibited accelerated trabecular bone loss compared to mice housed at thermoneutrality (Iwaniec et al. 2016; Robbins et al. 2018). In young mice, impaired trabecular bone acquisition occurred at 16°C vs. 26°C despite marked increases in BAT UCP1 protein content, suggesting NST alone was insufficient to protect the skeleton (Robbins et al. 2018). In contrast, both beta-adrenergic receptor knockout mice (Bonnet et al. 2008; Kajimura et al. 2011) and mice treated with the beta blocker propranolol (Motyl et al. 2013) had reduced bone loss. Taken together, these data support the hypothesis that BAT reduces but cannot completely prevent sympathetic tone-induced bone loss.

6.7.2 Exercise, Irisin, and BAT

Recent studies reported the unexpected discovery that BAT is upregulated by exercise, but data from experimental models and humans are inconsistent. Several myokines, proteins secreted by muscle during exercise, have been linked to beiging of WAT (Stanford and Goodyear 2018). For example, exercise induces skeletal muscle to produce a hormone-like peptide called irisin from its precursor, FNDC5, in mice and humans (Bostrom et al. 2012). In mice, irisin induced browning in white adipose depots, leading to more BAT, and improved glucose homeostasis (Bostrom et al. 2012). Intermittent doses of irisin in mice led to improved cortical bone mass and strength (Colaianni et al. 2015) and reversed bone and muscle loss during experimental unloading (Colaianni et al. 2017), demonstrating direct effects of irisin on bone. Although the data are persuasive, this mechanism is puzzling from both energetic and physiological perspectives. Exercise is energetically expensive and generates heat, so both the calorie expenditure and resulting heat seem unlikely to be necessary. Nevertheless, irisin has generated significant interest as a potential pharmacological target for obesity and glucose intolerance (Arhire et al. 2019).

In contrast to the data in rodents, a study of young women found BAT volume and activity tended to be lower in trained athletes vs. controls, and BAT volume was correlated positively with irisin but negatively with lean mass (Singhal et al. 2015). The same pattern was seen in young men, with lower BAT activity in athletes vs. controls, although plasma irisin did not differ (Vosselman et al. 2015). A study comparing the effects of aerobic vs. anaerobic exercise on myokine levels found no effect of exercise type on irisin, although other myokines were higher after anaerobic exercise (He et al. 2019). Although more data are needed, these equivocal results may reflect training duration; the human studies enrolled trained athletes, whereas laboratory rodents are typically naïve to exercise.

6.8 IMPLICATIONS FOR HOMININ EVOLUTIONARY BIOLOGY

Active BAT is common in extant humans, even in industrialized populations with significant cultural buffering against cold, so it is reasonable to infer its presence in earlier hominins as well (Steegmann et al. 2002; Steegmann 2007). In support of BAT's importance in human evolution, *UCP1* polymorphisms are correlated with variation in basal metabolic rate, NST, and body fat (Yoneshiro et al. 2013; Nishimura et al. 2017; Chathoth et al. 2018; Sellayah 2019), and there is evidence of positive selection on haplotypes linked to higher NST in Europeans and East Asians as well as in high latitude-dwelling populations (Hancock et al. 2011; Bakker et al. 2014; Bigham 2019). Polymorphisms in genes regulating BAT proliferation and abundance also show latitudinal patterns, including PR/SET domain 16 (*PRDM16*) in northern vs. southern Europeans (Quagliariello et al. 2017) and in the T-box transcription factor 15/mitochondrial tryptophan tRNA synthetase 2 (*TBX15/WARS2*) locus that may result from archaic introgression from Denisovans and is most common in Inuit and East Asians and least common in Africans (Gburcik et al. 2012; Racimo et al. 2017). In the future, it would be particularly interesting to look for evidence of selection around 2 Ma, during *H. erectus* expansion into colder climates. It would also be interesting to investigate selection for climatic adaptation during the peopling of the Americas, since the first Americans were likely cold adapted and encountered progressively hotter and then colder climates as they moved southward through the continent.

Both brown and beige fat would have aided survival during expansions to high and low latitudes. Brown fat would have been crucial for thermoregulation in infants and young children, particularly if born in winter. Beyond early childhood, inducible beige fat would have been essential, as it provides flexibility to moderate BAT tissue volume and thermogenic activity depending on metabolic conditions. Modern humans in cold environments exhibit high energy expenditure and elevated metabolic rates, some of which may be due to NST in BAT (Levy et al.

2013; Leonard et al. 2014; Ocobock 2016; Levy et al. 2018; Ocobock et al. 2020). This cost should be incorporated into models of hominin energy requirements, but in order to do so, we need better data from living humans on (1) BAT energy expenditure during cold exposure, including measurement of both glucose and fatty acid uptake, and (2) BAT functional changes during metabolic stressors in humans, particularly during development, pregnancy and lactation, caloric restriction, and elite athletic training. Once these data are available, it will be possible to more precisely estimate BAT energy expenditure in cold-dwelling hominins, such as the cost of moving from the tropics to the temperate and polar zones. A better understanding of the evolutionary role of BAT across primates is also needed and will require assessment of interspecific BAT prevalence and comparison of key BAT genes in nonhuman primates, Neanderthals, Denisovans, and modern human populations. In living human populations, better energetic models would also allow estimation of the decrease in BAT calorie expenditure in industrialized humans due to indoor heating, insulated clothing, and more subcutaneous white fat, which might be contributing to the obesity epidemic.

6.9 SUMMARY AND FUTURE DIRECTIONS

In the past 10 years, scientific perspective on human brown adipose tissue has transformed from an ephemeral tissue of the newborn to a unique fat depot that persists in childhood, peaks in adolescence, and remains present in adults until at least the 5th decade of life. BAT is correlated with body composition, glucose metabolism, muscle mass, and bone mass. There is general support for the overall hypothesis that BAT metabolic activity decreases during energy restriction and chronic exposure to stressors, although there is less evidence for upregulation during overnutrition. The primary environmental stimulus for BAT is cold, but emerging studies indicate exercise also upregulates BAT via the protein irisin. Hormones and growth factors can alter BAT sensitivity to sympathetic stimulus by signaling energy status to the hypothalamus to up- or downregulate sympathetic tone. Key upregulators include estrogen, leptin, and insulin; downregulators include glucocorticoids, prolactin, ghrelin, and adiponectin. All of these observations of the interactions of BAT with internal and environmental complexes result from activities that occur at the level of the cell. Further understanding of cellular mechanisms will clarify these interactions, and better data pertaining to the interactions will provide more detailed hypotheses about BAT cell function.

Current research on BAT continues to seek pharmacological stimuli that might target obesity and the metabolic syndrome. For example, a new study indicates that after 4 weeks of treatment with the β3-adrenergic receptor agonist Mirabegron, which is used to control overactive bladder, women had higher BAT metabolic activity, total body energy expenditure, and insulin sensitivity, although body mass did not change (Rowe and Troen 1980). Key research areas for the future include developing more precise methods to measure BAT volume and metabolic activity and delineating its metabolism of fatty acids vs. glucose. Understanding the emerging role of skeletal muscle in nonshivering thermogenesis and its interaction with BAT is also a high priority, since some skeletal muscle remains throughout aging. An elegant study of interactions of WAT, BAT, and skeletal muscle during mild cold exposure (below the shivering threshold) in young, healthy men found that glucose disposal was about 50-fold higher in muscle than in BAT (Blondin et al. 2015b). Finally, longer-term studies are needed to determine the persistence of changes seen in short-term interventions. A better understanding of BAT interactions with systemic energy metabolism will make it possible to test hypotheses about its role in evolutionary history as well as in living humans.

6.10 ACKNOWLEDGMENTS

I am grateful to Joan T. Richtsmeier and Kathy Pitirri for inviting me to contribute to this volume, and to two anonymous reviewers whose excellent feedback improved the paper. Funding for this project was provided by NSF BCS-1638553 to MD.

6.11 REFERENCES

Acheson, K. J., E. Ravussin, J. Wahren, and E. Jequier. 1984. Thermic effect of glucose in man: obligatory and facultative thermogenesis. *J Clin Invest* 74:1572–80. doi:10.1172/JCI111573.

Andersson, J., J. Roswall, E. Kjellberg, H. Ahlstrom, et al. 2019. MRI estimates of brown adipose tissue in children—associations to adiposity, osteocalcin, and thigh muscle volume. *Magn Reson Imaging* 58:135–42. doi:10.1016/j.mri.2019.02.001.

Andrews, J. F., D. Richard, G. Jennings, and P. Trayhurn. 1986. Brown adipose tissue thermogenesis during pregnancy in mice. *Ann Nutr Metab* 30:87–93. doi:10.1159/000177180.

Arhire, L. I., L. Mihalache, and M. Covasa. 2019. Irisin: a hope in understanding and managing obesity and metabolic syndrome. *Front Endocrinol (Lausanne)* 10:524. doi:10.3389/fendo.2019.00524.

Au-Yong, I. T., N. Thorn, R. Ganatra, A. C. Perkins, et al. 2009. Brown adipose tissue and seasonal variation in humans. *Diabetes* 58:2583–7. doi:10.2337/db09-0833.

Bachman, E. S., H. Dhillon, C. Y. Zhang, S. Cinti, et al. 2002. betaAR signaling required for diet-induced thermogenesis and obesity resistance. *Science* 297:843–5.

Bakker, L. E., M. R. Boon, R. A. van der Linden, L. P. Arias-Bouda, et al. 2014. Brown adipose tissue volume in healthy lean south Asian adults compared with white Caucasians: a prospective, case-controlled observational study. *Lancet Diabetes Endocrinol* 2:210–17. doi:10.1016/S2213-8587(13)70156-6.

Bal, N. C., S. K. Maurya, S. Singh, X. H. Wehrens, et al. 2016. Increased reliance on muscle-based thermogenesis upon acute minimization of brown adipose tissue function. *J Biol Chem* 291:17247–57. doi:10.1074/jbc.M116.728188.

Baumgart, S. 2008. Iatrogenic hyperthermia and hypothermia in the neonate. *Clin Perinatol* 35:183–97. doi:10.1016/j.clp.2007.11.002.

Betz, M. J., M. Bielohuby, B. Mauracher, W. Abplanalp, et al. 2012. Isoenergetic feeding of low carbohydrate-high fat diets does not increase brown adipose tissue thermogenic capacity in rats. *PLoS One* 7:e38997. doi:10.1371/journal.pone.0038997.

Betz, M. J., and S. Enerback. 2018. Targeting thermogenesis in brown fat and muscle to treat obesity and metabolic disease. *Nat Rev Endocrinol* 14:77–87. doi:10.1038/nrendo.2017.132.

Bigham, A. W. 2019. Natural selection and adaptation to extreme environments: high latitudes and altitudes. In *A Companion to Anthropological Genetics*, ed. D. H. O'Rourke, 219–32. Hoboken, NJ: Wiley-Blackwell.

Blondin, D. P., F. Frisch, S. Phoenix, B. Guerin, et al. 2017a. Inhibition of intracellular triglyceride lipolysis suppresses cold-induced brown adipose tissue metabolism and increases shivering in humans. *Cell Metab* 25:438–47. doi:10.1016/j.cmet.2016.12.005.

Blondin, D. P., S. M. Labbé, S. Phoenix, B. Guerin, et al. 2015b. Contributions of white and brown adipose tissues and skeletal muscles to acute cold-induced metabolic responses in healthy men. *J Physiol* 593:701–14. doi:10.1113/jphysiol.2014.283598.

Blondin, D. P., S. M. Labbé, H. C. Tingelstad, C. Noll, et al. 2014. Increased brown adipose tissue oxidative capacity in cold-acclimated humans. *J Clin Endocrinol Metab* 99:E438–46. doi:10.1210/jc.2013-3901.

Blondin, D. P., S. M. Labbé, E. E. Turcotte, F. Haman, et al. 2015a. A critical appraisal of brown adipose tissue metabolism in humans. *Clinical Lipidology* 10:259–80.

Blondin, D. P., H. C. Tingelstad, C. Noll, F. Frisch, et al. 2017b. Dietary fatty acid metabolism of brown adipose tissue in cold-acclimated men. *Nat Commun* 8:14146. doi:10.1038/ncomms14146.

Bonnet, N., C. Gadois, E. McCloskey, G. Lemineur, et al. 2007. Protective effect of beta blockers in postmenopausal women: influence on fractures, bone density, micro and macroarchitecture. *Bone* 40:1209–16. doi:10.1016/j.bone.2007.01.006.

Bonnet, N., D. D. Pierroz, and S. L. Ferrari. 2008. Adrenergic control of bone remodeling and its implications for the treatment of osteoporosis. *J Musculoskelet Neuronal Interact* 8:94–104.

Boon, M. R., L. E. Bakker, R. A. van der Linden, L. Pereira Arias-Bouda, et al. 2014. Supraclavicular skin temperature as a measure of 18F-FDG uptake by BAT in human subjects. *PLoS One* 9:e98822. doi:10.1371/journal.pone.0098822.

Bostrom, P., J. Wu, M. P. Jedrychowski, A. Korde, et al. 2012. A PGC1-alpha-dependent myokine that drives brown-fat-like development of white fat and thermogenesis. *Nature* 481:463–8. doi:10.1038/nature10777.

Boucher, J., M. A. Mori, K. Y. Lee, G. Smyth, et al. 2012. Impaired thermogenesis and adipose tissue development in mice with fat-specific disruption of insulin and IGF-1 signalling. *Nat Commun* 3:902. doi:10.1038/ncomms1905.

Bredella, M. A., P. K. Fazeli, L. M. Freedman, G. Calder, et al. 2012. Young women with cold-activated brown adipose tissue have higher bone mineral density and lower Pref-1 than women without brown adipose tissue: a study in women with anorexia nervosa, women recovered from anorexia nervosa, and normal-weight women. *J Clin Endocrinol Metab* 97:E584–90. doi:10.1210/jc.2011-2246.

Bredella, M. A., P. K. Fazeli, B. Lecka-Czernik, C. J. Rosen, et al. 2013. IGFBP-2 is a negative predictor of cold-induced brown fat and bone mineral density in young non-obese women. *Bone* 53:336–9. doi:10.1016/j.bone.2012.12.046.

Bredella, M. A., C. M. Gill, C. J. Rosen, A. Klibanski, et al. 2014. Positive effects of brown adipose tissue on femoral bone structure. *Bone* 58:55–8. doi:10.1016/j.bone.2013.10.007.

Brendle, C., M. K. Werner, M. Schmadl, C. la Fougere, et al. 2018. Correlation of brown adipose tissue with other body fat compartments and patient characteristics: a retrospective analysis in a large patient cohort using PET/CT. *Acad Radiol* 25:102–10. doi:10.1016/j.acra.2017.09.007.

Brito, M. N., N. A. Brito, M. A. Garofalo, I. C. Kettelhut, et al. 1998. Sympathetic activity in brown adipose tissue from rats adapted to a high protein, carbohydrate-free diet. *J Auton Nerv Syst* 69:1–5. doi:10.1016/s0165-1838(97)00132-x.

Butte, N. F., and J. C. King. 2005. Energy requirements during pregnancy and lactation. *Public Health Nutr* 8:1010–27. doi:10.1079/phn2005793.

Cannon, B., and J. Nedergaard. 2004. Brown adipose tissue: function and physiological significance. *Physiol Rev* 84:277–359. doi:10.1152/physrev.00015.2003.

Carpentier, A. C., D. P. Blondin, K. A. Virtanen, D. Richard, et al. 2018. Brown adipose tissue energy metabolism in humans. *Front Endocrinol (Lausanne)* 9:447. doi:10.3389/fendo.2018.00447.

Chalfant, J. S., M. L. Smith, H. H. Hu, F. J. Dorey, et al. 2012. Inverse association between brown adipose tissue activation and white adipose tissue accumulation in successfully treated pediatric malignancy. *Am J Clin Nutr* 95:1144–9. doi:10.3945/ajcn.111.030650.

Champigny, O., and D. Ricquier. 1990. Effects of fasting and refeeding on the level of uncoupling protein mRNA in rat brown adipose tissue: evidence for diet-induced and cold-induced responses. *J Nutr* 120:1730–6. doi:10.1093/jn/120.12.1730.

Chathoth, S., M. H. Ismail, C. Vatte, C. Cyrus, et al. 2018. Association of uncoupling protein 1 (UCP1) gene polymorphism with obesity: a case-control study. *BMC Med Genet* 19:203. doi:10.1186/s12881-018-0715-5.

Chen, K. Y., R. J. Brychta, J. D. Linderman, S. Smith, et al. 2013. Brown fat activation mediates cold-induced thermogenesis in adult humans in response to a mild decrease in ambient temperature. *J Clin Endocrinol Metab* 98:E1218–23. doi:10.1210/jc.2012-4213.

Chondronikola, M., E. Volpi, E. Borsheim, T. Chao, et al. 2016. Brown adipose tissue is linked to a distinct thermoregulatory response to mild cold in people. *Front Physiol* 7:129. doi:10.3389/fphys.2016.00129.

Chouchani, E. T., L. Kazak, and B. M. Spiegelman. 2019. New advances in adaptive thermogenesis: UCP1 and beyond. *Cell Metab* 29:27–37. doi:10.1016/j.cmet.2018.11.002.

Colaianni, G., C. Cuscito, T. Mongelli, P. Pignataro, et al. 2015. The myokine irisin increases cortical bone mass. *Proc Natl Acad Sci U S A* 112:12157–62. doi:10.1073/pnas.1516622112.

Colaianni, G., T. Mongelli, C. Cuscito, P. Pignataro, et al. 2017. Irisin prevents and restores bone loss and muscle atrophy in hind-limb suspended mice. *Sci Rep* 7:2811. doi:10.1038/s41598-017-02557-8.

Collins, S., and R. S. Surwit. 2001. The beta-adrenergic receptors and the control of adipose tissue metabolism and thermogenesis. *Recent Prog Horm Res* 56:309–28. doi:10.1210/rp.56.1.309.

Corrales, P., Y. Vivas, A. Izquierdo-Lahuerta, D. Horrillo, et al. 2019. Long-term caloric restriction ameliorates deleterious effects of aging on white and brown adipose tissue plasticity. *Aging Cell* 18:e12948. doi:10.1111/acel.12948.

Cypess, A. M., S. Lehman, G. Williams, I. Tal, et al. 2009. Identification and importance of brown adipose tissue in adult humans. *N Engl J Med* 360:1509–17. doi:10.1056/NEJMoa0810780.

Dawkins, M. J., and J. W. Scopes. 1965. Non-shivering thermogenesis and brown adipose tissue in the human new-born infant. *Nature* 206:201–2. doi:10.1038/206201b0.

de Jonge, L., and G. A. Bray. 1997. The thermic effect of food and obesity: a critical review. *Obes Res* 5:622–31. doi:10.1002/j.1550-8528.1997.tb00584.x.

Desautels, M., and R. A. Dulos. 1988. Effects of repeated cycles of fasting-refeeding on brown adipose tissue composition in mice. *Am J Physiol* 255:E120–8. doi:10.1152/ajpendo.1988.255.2.E120.

Devlin, M. J. 2015. The 'skinny' on brown fat, obesity, and bone. *Am J Phys Anthropol* 156(Suppl 59):98–115. doi:10.1002/ajpa.22661.

Devlin, M. J., and C. J. Rosen. 2015. The bone-fat interface: basic and clinical implications of marrow adiposity. *Lancet Diabetes Endocrinol* 3:141–7. doi:10.1016/S2213-8587(14)70007-5.

Drubach, L. A., E. L. Palmer, L. P. Connolly, A. Baker, D. Zurakowski, and A. M. Cypess. 2011. Pediatric brown adipose tissue: detection, epidemiology, and differences from adults. *J Pediatr* 159:939–44. doi:10.1016/j.jpeds.2011.06.028.

Elefteriou, F. 2005. Neuronal signaling and the regulation of bone remodeling. *Cell Mol Life Sci* 62:2339–49.

Emery, J. L., and F. Dinsdale. 1978. Structure of periadrenal brown fat in childhood in both expected and cot deaths. *Arch Dis Child* 53:154–8.

Enerback, S., A. Jacobsson, E. M. Simpson, C. Guerra, et al. 1997. Mice lacking mitochondrial uncoupling protein are cold-sensitive but not obese. *Nature* 387:90–4. doi:10.1038/387090a0.

Entringer, S., J. Rasmussen, D. M. Cooper, S. Ikenoue, et al. 2017. Association between supraclavicular brown adipose tissue composition at birth and adiposity gain from birth to 6 months of age. *Pediatr Res* 82:1017–21. doi:10.1038/pr.2017.159.

Eyolfson, D. A., P. Tikuisis, X. Xu, G. Weseen, et al. 2001. Measurement and prediction of peak shivering intensity in humans. *Eur J Appl Physiol* 84:100–6. doi:10.1007/s004210000329.

Fazeli, P. K., M. C. Horowitz, O. A. MacDougald, E. L. Scheller, et al. 2013. Marrow fat and bone—new perspectives. *J Clin Endocrinol Metab* 98:935–45. doi:10.1210/jc.2012-3634.

Feldmann, H. M., V. Golozoubova, B. Cannon, and J. Nedergaard. 2009. UCP1 ablation induces obesity and abolishes diet-induced thermogenesis in mice exempt from thermal stress by living at thermoneutrality. *Cell Metab* 9:203–9. doi:10.1016/j.cmet.2008.12.014.

Feve, B., B. Baude, S. Krief, A. D. Strosberg, et al. 1992. Inhibition by dexamethasone of beta 3-adrenergic receptor responsiveness in 3T3-F442A adipocytes: evidence for a transcriptional mechanism. *J Biol Chem* 267:15909–15.

Franz, D., D. C. Karampinos, E. J. Rummeny, M. Souvatzoglou, et al. 2015. Discrimination between brown and white adipose tissue using a 2-point Dixon water-fat separation method in simultaneous PET/MRI. *J Nucl Med* 56:1742–7. doi:10.2967/jnumed.115.160770.

Gburcik, V., W. P. Cawthorn, J. Nedergaard, J. A. Timmons, et al. 2012. An essential role for TBX15 in the differentiation of brown and 'brite' but not white adipocytes. *Am J Physiol Endocrinol Metab* 303:E1053–60. doi:10.1152/ajpendo.00104.2012.

Gelfand, M. J., M. O'Hara, L. A. Curtwright, and J. R. Maclean. 2005. Pre-medication to block [(18)F]FDG uptake in the brown adipose tissue of pediatric and adolescent patients. *Pediatr Radiol* 35:984–90. doi:10.1007/s00247-005-1505-8.

Gilsanz, V., S. A. Chung, H. Jackson, F. J. Dorey, et al. 2011. Functional brown adipose tissue is related to muscle volume in children and adolescents. *J Pediatr* 158:722–6. doi:10.1016/j.jpeds.2010.11.020.

Gilsanz, V., H. H. Hu, M. L. Smith, F. Goodarzian, et al. 2012a. The depiction of brown adipose tissue is related to disease status in pediatric patients with lymphoma. *Am J Roentgenol* 198:909–13. doi:10.2214/AJR.11.7488.

Gilsanz, V., M. L. Smith, F. Goodarzian, M. Kim, et al. 2012b. Changes in brown adipose tissue in boys and girls during childhood and puberty. *J Pediatr* 160:604–9e1. doi:10.1016/j.jpeds.2011.09.035.

Golozoubova, V., B. Cannon, and J. Nedergaard. 2006. UCP1 is essential for adaptive adrenergic nonshivering thermogenesis. *Am J Physiol Endocrinol Metab* 291:E350–7. doi:10.1152/ajpendo.00387.2005.

Hancock, A. M., V. J. Clark, Y. Qian, and A. Di Rienzo. 2011. Population genetic analysis of the uncoupling proteins supports a role for UCP3 in human cold resistance. *Mol Biol Evol* 28:601–14. doi:10.1093/molbev/msq228.

Hanssen, M. J., A. A. van der Lans, B. Brans, J. Hoeks, et al. 2016. Short-term cold acclimation recruits brown adipose tissue in obese humans. *Diabetes* 65:1179–89. doi:10.2337/db15-1372.

Harms, M., and P. Seale. 2013. Brown and beige fat: development, function and therapeutic potential. *Nat Med* 19:1252–63. doi:10.1038/nm.3361.

Hartgill, T. W., T. K. Bergersen, and J. Pirhonen. 2011. Core body temperature and the thermoneutral zone: a longitudinal study of normal human pregnancy. *Acta Physiol (Oxf)* 201:467–74. doi:10.1111/j.1748-1716.2010.02228.x.

He, Z., Y. Tian, P. L. Valenzuela, C. Huang, et al. 2019. Myokine/adipokine response to 'aerobic' exercise: is it just a matter of exercise load? *Front Physiol* 10:691. doi:10.3389/fphys.2019.00691.

Heaton, J. M. 1972. The distribution of brown adipose tissue in the human. *J Anat* 112:35–9.

Hibi, M., S. Oishi, M. Matsushita, T. Yoneshiro, et al. 2016. Brown adipose tissue is involved in diet-induced thermogenesis and whole-body fat utilization in healthy humans. *Int J Obes (Lond)* 40:1655–61. doi:10.1038/ijo.2016.124.

Hoeke, G., S. Kooijman, M. R. Boon, P. C. Rensen, et al. 2016. Role of brown fat in lipoprotein metabolism and atherosclerosis. *Circ Res* 118:173–82. doi:10.1161/CIRCRESAHA.115.306647.

Hong, T. S., A. Shammas, M. Charron, K. A. Zukotynski, et al. 2011. Brown adipose tissue 18F-FDG uptake in pediatric PET/CT imaging. *Pediatr Radiol* 41:759–68. doi:10.1007/s00247-010-1925-y.

Hruby, A., and F. B. Hu. 2015. The epidemiology of obesity: a big picture. *Pharmacoeconomics* 33:673–89. doi:10.1007/s40273-014-0243-x.

Huynh, M. K., A. W. Kinyua, D. J. Yang, and K. W. Kim. 2016. Hypothalamic AMPK as a regulator of energy homeostasis. *Neural Plast* 2016:2754078. doi:10.1155/2016/2754078.

Ikeda, K., P. Maretich, and S. Kajimura. 2018. The common and distinct features of brown and beige adipocytes. *Trends Endocrinol Metab* 29:191–200. doi:10.1016/j.tem.2018.01.001.

Iwaniec, U. T., K. A. Philbrick, C. P. Wong, J. L. Gordon, et al. 2016. Room temperature housing results in premature cancellous bone loss in growing female mice: implications for the mouse as a preclinical model for age-related bone loss. *Osteoporos Int* 27:3091–101. doi:10.1007/s00198-016-3634-3.

Iwen, K. A., J. Backhaus, M. Cassens, M. Waltl, et al. 2017. Cold-induced brown adipose tissue activity alters plasma fatty acids and improves glucose metabolism in men. *J Clin Endocrinol Metab* 102:4226–34. doi:10.1210/jc.2017-01250.

Jequier, E., and Y. Schutz. 1988. Energy expenditure in obesity and diabetes. *Diabetes Metab Rev* 4:583–93. doi:10.1002/dmr.5610040604.

Jimenez-Pavon, D., J. Corral-Perez, D. Sanchez-Infantes, F. Villarroya, et al. 2019. Infrared thermography for estimating supraclavicular skin temperature and BAT activity in humans: a systematic review. *Obesity (Silver Spring)* 27:1932–49. doi:10.1002/oby.22635.

Johnson, M. S., and J. R. Speakman. 2001. Limits to sustained energy intake. V. Effect of cold-exposure during lactation in Mus musculus. *J Exp Biol* 204:1967–77.

Johnson, M. S., S. C. Thomson, and J. R. Speakman. 2001. Limits to sustained energy intake. I. Lactation in the laboratory mouse *Mus musculus*. *J Exp Biol* 204:1925–35.

Joosen, A. M., and K. R. Westerterp. 2006. Energy expenditure during overfeeding. *Nutr Metab (Lond)* 3:25. doi:10.1186/1743-7075-3-25.

Jung, R. T., P. S. Shetty, and W. P. James. 1980. Nutritional effects on thyroid and catecholamine metabolism. *Clin Sci (Lond)* 58:183–91. doi:10.1042/cs0580183.

Kajimura, D., E. Hinoi, M. Ferron, A. Kode, et al. 2011. Genetic determination of the cellular basis of the sympathetic regulation of bone mass accrual. *J Exp Med* 208:841–51. doi:10.1084/jem.20102608.

Kiely, J., J. R. Hadcock, S. W. Bahouth, and C. C. Malbon. 1994. Glucocorticoids down-regulate beta 1-adrenergic-receptor expression by suppressing transcription of the receptor gene. *Biochem J* 302(Pt 2):397–403. doi:10.1042/bj3020397.

Kistner, A., H. Ryden, B. Anderstam, A. Hellstrom, et al. 2018. Brown adipose tissue in young adults who were born preterm or small for gestational age. *J Pediatr Endocrinol Metab* 31:641–7. doi:10.1515/jpem-2017-0547.

Kontani, Y., Y. Wang, K. Kimura, K. I. Inokuma, et al. 2005. UCP1 deficiency increases susceptibility to diet-induced obesity with age. *Aging Cell* 4:147–55. doi:10.1111/j.1474-9726.2005.00157.x.

Koskensalo, K., J. Raiko, T. Saari, V. Saunavaara, et al. 2017. Human brown adipose tissue temperature and fat fraction are related to its metabolic activity. *J Clin Endocrinol Metab* 102:1200–7. doi:10.1210/jc.2016-3086.

Krol, E., S. A. Martin, I. T. Huhtaniemi, A. Douglas, et al. 2011. Negative correlation between milk production and brown adipose tissue gene expression in lactating mice. *J Exp Biol* 214:4160–70. doi:10.1242/jeb.061382.

Krol, E., and J. R. Speakman. 2003. Limits to sustained energy intake. VI. Energetics of lactation in laboratory mice at thermoneutrality. *J Exp Biol* 206:4255–66. doi:10.1242/jeb.00674.

Krol, E., and J. R. Speakman. 2019. Switching off the furnace: brown adipose tissue and lactation. *Mol Aspects Med* 68:18–41. doi:10.1016/j.mam.2019.06.003.

Krustrup, P., R. A. Ferguson, M. Kjaer, and J. Bangsbo. 2003. ATP and heat production in human skeletal muscle during dynamic exercise: higher efficiency of anaerobic than aerobic ATP resynthesis. *J Physiol* 549:255–69. doi:10.1113/jphysiol.2002.035089.

Lea-Currie, Y. R., S. M. Wu, and M. K. McIntosh. 1997. Effects of acute administration of dehydroepiandrosterone-sulfate on adipose tissue mass and cellularity in male rats. *Int J Obes Relat Metab Disord* 21:147–54. doi:10.1038/sj.ijo.0800382.

Lean, M. E., W. P. James, G. Jennings, and P. Trayhurn. 1986. Brown adipose tissue uncoupling protein content in human infants, children and adults. *Clin Sci (Lond)* 71:291–7. doi:10.1042/cs0710291.

Leitner, B. P., S. Huang, R. J. Brychta, C. J. Duckworth, et al. 2017. Mapping of human brown adipose tissue in lean and obese young men. *Proc Natl Acad Sci USA* 114:8649–54. doi:10.1073/pnas.1705287114.

Leonard, W. R., S. B. Levy, L. A. Tarskaia, T. M. Klimova, et al. 2014. Seasonal variation in basal metabolic rates among the Yakut (Sakha) of northeastern Siberia. *Am J Hum Biol* 26:437–45. doi:10.1002/ajhb.22524.

Levin, I., and P. Trayhurn. 1987. Thermogenic activity and capacity of brown fat in fasted and refed golden hamsters. *Am J Physiol* 252:R987–93. doi:10.1152/ajpregu.1987.252.5.R987.

Levy, S. B. 2019. Field and laboratory methods for quantifying brown adipose tissue thermogenesis. *Am J Hum Biol* 31:e23261. doi:10.1002/ajhb.23261.

Levy, S. B., T. M. Klimova, R. N. Zakharova, A. I. Federov, et al. 2018. Brown adipose tissue, energy expenditure, and biomarkers of cardio-metabolic health among the Yakut (Sakha) of northeastern Siberia. *Am J Hum Biol* 30:e23175. doi:10.1002/ajhb.23175.

Levy, S. B., W. R. Leonard, L. A. Tarskaia, T. M. Klimova, et al. 2013. Seasonal and socioeconomic influences on thyroid function among the Yakut (Sakha) of eastern Siberia. *Am J Hum Biol* 25:814–20. doi:10.1002/ajhb.22457.

Liu, X., M. Rossmeisl, J. McClaine, M. Riachi, et al. 2003. Paradoxical resistance to diet-induced obesity in UCP1-deficient mice. *J Clin Invest* 111:399–407. doi:10.1172/JCI15737.

Long, J. Z., K. J. Svensson, L. Tsai, X. Zeng, et al. 2014. A smooth muscle-like origin for beige adipocytes. *Cell Metab* 19:810–20. doi:10.1016/j.cmet.2014.03.025.

Lowell, B. B., and E. S. Bachman. 2003. Beta-adrenergic receptors, diet-induced thermogenesis, and obesity. *J Biol Chem* 278:29385–8. doi:10.1074/jbc.R300011200.

Luijten, I. H. N., H. M. Feldmann, G. von Essen, B. Cannon, et al. 2019. In the absence of UCP1-mediated diet-induced thermogenesis, obesity is augmented even in the obesity-resistant 129S mouse strain. *Am J Physiol Endocrinol Metab* 316:E729–40. doi:10.1152/ajpendo.00020.2019.

Lundstrom, E., J. Ljungberg, J. Andersson, H. Manell, et al. 2019. Brown adipose tissue estimated with the magnetic resonance imaging fat fraction is associated with glucose metabolism in adolescents. *Pediatr Obes* 14:e12531. doi:10.1111/ijpo.12531.

Malpique, R., J. M. Gallego-Escuredo, G. Sebastiani, J. Villarroya, et al. 2019. Brown adipose tissue in prepubertal children: associations with sex, birthweight, and metabolic profile. *Int J Obes (Lond)* 43:384–91. doi:10.1038/s41366-018-0198-7.

Mano-Otagiri, A., H. Ohata, A. Iwasaki-Sekino, T. Nemoto, et al. 2009. Ghrelin suppresses noradrenaline release in the brown adipose tissue of rats. *J Endocrinol* 201:341–9. doi:10.1677/JOE-08-0374.

Marsteller, F. A., and C. B. Lynch. 1987. Reproductive responses to variation in temperature and food supply by house mice: II. Lactation. *Biol Reprod* 37:844–50. doi:10.1095/biolreprod37.4.844.

Martin, R. D. 1986. Primates: a definition. In *Major Topics in Primate and Human Evolution*, ed. B. Wood, L. Martin, and P. Andrews, 1–31. Cambridge: Cambridge University Press.

Martinez de Morentin, P. B., I. Gonzalez-Garcia, L. Martins, R. Lage, et al. 2014. Estradiol regulates brown adipose tissue thermogenesis via hypothalamic AMPK. *Cell Metab* 20:41–53. doi:10.1016/j.cmet.2014.03.031.

Martinez-Perez, B., M. Ejarque, C. Gutierrez, C. Nunez-Roa, et al. 2016. Angiopoietin-like protein 8 (ANGPTL8) in pregnancy: a brown adipose tissue-derived endocrine factor with a potential role in fetal growth. *Transl Res* 178:1–12. doi:10.1016/j.trsl.2016.06.012.

McCue, M. D. 2010. Starvation physiology: reviewing the different strategies animals use to survive a common challenge. *Comp Biochem Physiol A Mol Integr Physiol* 156:1–18. doi:10.1016/j.cbpa.2010.01.002.

Monjo, M., A. M. Rodriguez, A. Palou, and P. Roca. 2003. Direct effects of testosterone, 17 beta-estradiol, and progesterone on adrenergic regulation in cultured brown adipocytes: potential mechanism for gender-dependent thermogenesis. *Endocrinology* 144:4923–30. doi:10.1210/en.2003-0537.

Moriscot, A., R. Rabelo, and A. C. Bianco. 1993. Corticosterone inhibits uncoupling protein gene expression in brown adipose tissue. *Am J Physiol* 265:E81–7. doi:10.1152/ajpendo.1993.265.1.E81.

Motyl, K. J., K. A. Bishop, V. E. DeMambro, S. A. Bornstein, et al. 2013. Altered thermogenesis and impaired bone remodeling in Misty mice. *J Bone Miner Res* 28:1885–97. doi:10.1002/jbmr.1943.

Nakamura, Y., and K. Nakamura. 2018. Central regulation of brown adipose tissue thermogenesis and energy homeostasis dependent on food availability. *Pflugers Arch* 470:823–37. doi:10.1007/s00424-017-2090-z.

Nishimura, T., T. Katsumura, M. Motoi, H. Oota, et al. 2017. Experimental evidence reveals the UCP1 genotype changes the oxygen consumption attributed to non-shivering thermogenesis in humans. *Sci Rep* 7:5570. doi:10.1038/s41598-017-05766-3.

Ocobock, C. 2016. Human energy expenditure, allocation, and interactions in natural temperate, hot, and cold environments. *Am J Phys Anthropol* 161:667–75. doi:10.1002/ajpa.23071.

Ocobock, C., P. Soppela, M. T. Turunen, V. Stenback, et al. 2020. Elevated resting metabolic rates among female, but not male, reindeer herders from subarctic Finland. *Am J Hum Biol*:e23432. doi:10.1002/ajhb.23432.

Oelkrug, R., N. Goetze, C. Exner, Y. Lee, et al. 2013. Brown fat in a protoendothermic mammal fuels eutherian evolution. *Nat Commun* 4:2140. doi:10.1038/ncomms3140.

Orava, J., P. Nuutila, M. E. Lidell, V. Oikonen, et al. 2011. Different metabolic responses of human brown adipose tissue to activation by cold and insulin. *Cell Metab* 14:272–9. doi:10.1016/j.cmet.2011.06.012.

Oreskovich, S. M., F. J. Ong, B. A. Ahmed, N. B. Konyer, et al. 2019. MRI reveals human brown adipose tissue is rapidly activated in response to cold. *J Endocr Soc* 3:2374–84. doi:10.1210/js.2019-00309.

Ouellet, V., A. Routhier-Labadie, W. Bellemare, L. Lakhal-Chaieb, et al. 2011. Outdoor temperature, age, sex, body mass index, and diabetic status determine the prevalence, mass, and glucose-uptake activity of 18F-FDG-detected BAT in humans. *J Clin Endocrinol Metab* 96:192–9. doi:10.1210/jc.2010-0989.

Pfannenberg, C., M. K. Werner, S. Ripkens, I. Stef, et al. 2010. Impact of age on the relationships of brown adipose tissue with sex and adiposity in humans. *Diabetes* 59:1789–93. doi:10.2337/db10-0004.

Ponrartana, S., P. C. Aggabao, T. A. Chavez, N. L. Dharmavaram, et al. 2016. Changes in brown adipose tissue and muscle development during infancy. *J Pediatr* 173:116–21. doi:10.1016/j.jpeds.2016.03.002.

Ponrartana, S., P. C. Aggabao, H. H. Hu, G. M. Aldrovandi, et al. 2012. Brown adipose tissue and its relationship to bone structure in pediatric patients. *J Clin Endocrinol Metab* 97:2693–8. doi:10.1210/jc.2012-1589.

Ponrartana, S., H. H. Hu, and V. Gilsanz. 2013. On the relevance of brown adipose tissue in children. *Ann N Y Acad Sci* 1302:24–9. doi:10.1111/nyas.12195.

Prentice, A. M. 2005. Starvation in humans: evolutionary background and contemporary implications. *Mech Ageing Dev* 126:976–81. doi:10.1016/j.mad.2005.03.018.

Prentice, A. M., and R. G. Whitehead. 1987. The energetics of human reproduction. *Symposia of the Zoological Society of London* 57:275–304.

Quagliariello, A., S. De Fanti, C. Giuliani, P. Abondio, et al. 2017. Multiple selective events at the PRDM16 functional pathway shaped adaptation of western European populations to different climate conditions. *J Anthropol Sci* 95:235–47. doi:10.4436/JASS.95011.

Racimo, F., D. Gokhman, M. Fumagalli, A. Ko, et al. 2017. Archaic adaptive introgression in TBX15/WARS2. *Mol Biol Evol* 34:509–24. doi:10.1093/molbev/msw283.

Ramage, L. E., M. Akyol, A. M. Fletcher, J. Forsythe, et al. 2016. Glucocorticoids acutely increase Brown adipose tissue activity in humans, revealing species-specific differences in UCP-1 regulation. *Cell Metab* 24:130–41. doi:10.1016/j.cmet.2016.06.011.

Ravussin, E., S. Lillioja, T. E. Anderson, L. Christin, et al. 1986. Determinants of 24-hour energy expenditure in man: methods and results using a respiratory chamber. *J Clin Invest* 78:1568–78. doi:10.1172/JCI112749.

Robbins, A., Catmb Tom, M. N. Cosman, C. Moursi, et al. 2018. Low temperature decreases bone mass in mice: implications for humans. *Am J Phys Anthropol* 167:557–68. doi:10.1002/ajpa.23684.

Rockstroh, D., K. Landgraf, I. V. Wagner, J. Gesing, et al. 2015. Direct evidence of brown adipocytes in different fat depots in children. *PLoS One* 10:e0117841. doi:10.1371/journal.pone.0117841.

Rodriguez, A. M., M. Monjo, P. Roca, and A. Palou. 2002. Opposite actions of testosterone and progesterone on UCP1 mRNA expression in cultured brown adipocytes. *Cell Mol Life Sci* 59:1714–23. doi:10.1007/pl00012499.

Rothwell, N. J., and M. J. Stock. 1979. A role for brown adipose tissue in diet-induced thermogenesis. *Nature* 281:31–5.

Rowe, J. W., and B. R. Troen. 1980. Sympathetic nervous system and aging in man. *Endocr Rev* 1:167–79. doi:10.1210/edrv-1-2-167.

Rowland, L. A., S. K. Maurya, N. C. Bal, L. Kozak, et al. 2016. Sarcolipin and uncoupling protein 1 play distinct roles in diet-induced thermogenesis and do not compensate for one another. *Obesity (Silver Spring)* 24:1430–3. doi:10.1002/oby.21542.

Ruff, C. B. 1993. Climatic adaptation and hominid evolution: the thermoregulatory imperative. *Evol Anthropol* 2:53–60.

Ryu, J. W., M. S. Kim, C. H. Kim, K. H. Song, et al. 2003. DHEA administration increases brown fat uncoupling protein 1 levels in obese OLETF rats. *Biochem Biophys Res Commun* 303:726–31. doi:10.1016/s0006-291x(03)00409-1.

Sacks, H., and M. E. Symonds. 2013. Anatomical locations of human brown adipose tissue: functional relevance and implications in obesity and type 2 diabetes. *Diabetes* 62:1783–90. doi:10.2337/db12-1430.

Saito, M., Y. Okamatsu-Ogura, M. Matsushita, K. Watanabe, et al. 2009. High incidence of metabolically active brown adipose tissue in healthy adult humans: effects of cold exposure and adiposity. *Diabetes* 58:1526–31. doi:10.2337/db09-0530.

Salem, V., C. Izzi-Engbeaya, C. Coello, D. B. Thomas, et al. 2016. Glucagon increases energy expenditure independently of brown adipose tissue activation in humans. *Diabetes Obes Metab* 18:72–81. doi:10.1111/dom.12585.

Salles, G., J. Bienvenu, Y. Bastion, Y. Barbier, et al. 1996. Elevated circulating levels of TNFalpha and its p55 soluble receptor are associated with an adverse prognosis in lymphoma patients. *Br J Haematol* 93:352–9. doi:10.1046/j.1365-2141.1996.5181059.x.

Sanchez-Delgado, G., B. Martinez-Tellez, Y. Garcia-Rivero, F. M. Acosta, et al. 2019. Association between brown adipose tissue and bone mineral density in humans. *Int J Obes (Lond)* 43:1516–25. doi:10.1038/s41366-018-0261-4.

Scheller, E. L., C. R. Doucette, B. S. Learman, W. P. Cawthorn, et al. 2015. Region-specific variation in the properties of skeletal adipocytes reveals regulated and constitutive marrow adipose tissues. *Nat Commun* 6:7808. doi:10.1038/ncomms8808.

Schilperoort, M., G. Hoeke, S. Kooijman, and P. C. Rensen. 2016. Relevance of lipid metabolism for brown fat visualization and quantification. *Curr Opin Lipidol* 27:242–8. doi:10.1097/MOL.0000000000000296.

Schlogl, M., P. Piaggi, P. Thiyyagura, E. M. Reiman, et al. 2013. Overfeeding over 24 hours does not activate brown adipose tissue in humans. *J Clin Endocrinol Metab* 98:E1956–60. doi:10.1210/jc.2013-2387.

Schneider, J. E., and G. N. Wade. 1987. Body composition, food intake, and brown fat thermogenesis in pregnant djungarian hamsters. *Am J Physiol* 253:R314–20. doi:10.1152/ajpregu.1987.253.2.R314.

Scotney, H., M. E. Symonds, J. Law, H. Budge, et al. 2017. Glucocorticoids modulate human brown adipose tissue thermogenesis in vivo. *Metabolism* 70:125–32. doi:10.1016/j.metabol.2017.01.024.

Sellayah, D. 2019. The impact of early human migration on brown adipose tissue evolution and its relevance to the modern obesity pandemic. *J Endocr Soc* 3:372–86. doi:10.1210/js.2018-00363.

Shabalina, I. G., E. C. Backlund, J. Bar-Tana, B. Cannon, et al. 2008. Within brown-fat cells, UCP1-mediated fatty acid-induced uncoupling is independent of fatty acid metabolism. *Biochim Biophys Acta* 1777:642–50. doi:10.1016/j.bbabio.2008.04.038.

Sidossis, L., and S. Kajimura. 2015. Brown and beige fat in humans: thermogenic adipocytes that control energy and glucose homeostasis. *J Clin Invest* 125:478–86. doi:10.1172/JCI78362.

Singhal, V., G. D. Maffazioli, N. Cano Sokoloff, K. E. Ackerman, et al. 2015. Regional fat depots and their relationship to bone density and microarchitecture in young oligo-amenorrheic athletes. *Bone* 77:83–90. doi:10.1016/j.bone.2015.04.005.

Snodgrass, J. J., W. R. Leonard, L. A. Tarskaia, V. P. Alekseev, et al. 2005. Basal metabolic rate in the Yakut (Sakha) of Siberia. *Am J Hum Biol* 17:155–72. doi:10.1002/ajhb.20106.

Soumano, K., S. Desbiens, R. Rabelo, E. Bakopanos, et al. 2000. Glucocorticoids inhibit the transcriptional response of the uncoupling protein-1 gene to adrenergic stimulation in a brown adipose cell line. *Mol Cell Endocrinol* 165:7–15. doi:10.1016/s0303-7207(00)00276-8.

Speakman, J. R. 2008. The physiological costs of reproduction in small mammals. *Philos Trans R Soc Lond B Biol Sci* 363:375–98. doi:10.1098/rstb.2007.2145.

Stanford, K. I., and L. J. Goodyear. 2018. Muscle-adipose tissue cross talk. *Cold Spring Harb Perspect Med* 8. doi:10.1101/cshperspect.a029801.

Steegmann, A. T., Jr. 2007. Human cold adaptation: an unfinished agenda. *Am J Hum Biol* 19:218–27. doi:10.1002/ajhb.20614.

Steegmann, A. T., Jr., F. J. Cerny, and T. W. Holliday. 2002. Neandertal cold adaptation: physiological and energetic factors. *Am J Hum Biol* 14:566–83. doi:10.1002/ajhb.10070.

Strack, A. M., C. J. Horsley, R. J. Sebastian, S. F. Akana, et al. 1995. Glucocorticoids and insulin: complex interaction on brown adipose tissue. *Am J Physiol* 268:R1209–16. doi:10.1152/ajpregu.1995.268.5.R1209.

Suchacki, K. J., and W. P. Cawthorn. 2018. Molecular interaction of bone marrow adipose tissue with energy metabolism. *Curr Mol Biol Rep* 4:41–9. doi:10.1007/s40610-018-0096-8.

Sutton, E. F., G. A. Bray, J. H. Burton, S. R. Smith, et al. 2016. No evidence for metabolic adaptation in thermic effect of food by dietary protein. *Obesity (Silver Spring)* 24:1639–42. doi:10.1002/oby.21541.

Symonds, M. E., K. Henderson, L. Elvidge, C. Bosman, et al. 2012. Thermal imaging to assess age-related changes of skin temperature within the supraclavicular region co-locating with brown adipose tissue in healthy children. *J Pediatr* 161:892–8. doi:10.1016/j.jpeds.2012.04.056.

Symonds, M. E., A. Mostyn, S. Pearce, H. Budge, et al. 2003. Endocrine and nutritional regulation of fetal adipose tissue development. *J Endocrinol* 179:293–9. doi:10.1677/joe.0.1790293.

Tataranni, P. A., D. E. Larson, S. Snitker, and E. Ravussin. 1995. Thermic effect of food in humans: methods and results from use of a respiratory chamber. *Am J Clin Nutr* 61:1013–19. doi:10.1093/ajcn/61.4.1013.

Thuzar, M., W. P. Law, J. Ratnasingam, C. Jang, et al. 2018. Glucocorticoids suppress brown adipose tissue function in humans: a double-blind placebo-controlled study. *Diabetes Obes Metab* 20:840–8. doi:10.1111/dom.13157.

Trayhurn, P. 1989. Thermogenesis and the energetics of pregnancy and lactation. *Can J Physiol Pharmacol* 67:370–5. doi:10.1139/y89-060.

Trayhurn, P., J. B. Douglas, and M. M. McGuckin. 1982a. Brown adipose tissue thermogenesis is 'suppressed' during lactation in mice. *Nature* 298:59–60. doi:10.1038/298059a0.

Trayhurn, P., and G. Jennings. 1988. Nonshivering thermogenesis and the thermogenic capacity of brown fat in fasted and/or refed mice. *Am J Physiol* 254:R11–16. doi:10.1152/ajpregu.1988.254.1.R11.

Trayhurn, P., P. M. Jones, M. M. McGuckin, and A. E. Goodbody. 1982b. Effects of overfeeding on energy balance and brown fat thermogenesis in obese (ob/ob) mice. *Nature* 295:323–5. doi:10.1038/295323a0.

Trayhurn, P., and M. C. Wusteman. 1987. Sympathetic activity in brown adipose tissue in lactating mice. *Am J Physiol* 253:E515–20. doi:10.1152/ajpendo.1987.253.5.E515.

U Din, M., T. Saari, J. Raiko, N. Kudomi, et al. 2018. Postprandial oxidative metabolism of human brown fat indicates thermogenesis. *Cell Metab* 28:207–16 e3. doi:10.1016/j.cmet.2018.05.020.

Valle, A., A. Catala-Niell, B. Colom, F. J. Garcia-Palmer, et al. 2005. Sex-related differences in energy balance in response to caloric restriction. *Am J Physiol Endocrinol Metab* 289:E15–22. doi:10.1152/ajpendo.00553.2004.

Valle, A., F. J. Garcia-Palmer, J. Oliver, and P. Roca. 2007. Sex differences in brown adipose tissue thermogenic features during caloric restriction. *Cell Physiol Biochem* 19:195–204. doi:10.1159/000099207.

Valverde, A. M., C. Mur, M. Brownlee, and M. Benito. 2004. Susceptibility to apoptosis in insulin-like growth factor-I receptor-deficient brown adipocytes. *Mol Biol Cell* 15:5101–17. doi:10.1091/mbc.e03-11-0853.

van den Beukel, J. C., A. Grefhorst, C. Quarta, J. Steenbergen, et al. 2014. Direct activating effects of adrenocorticotropic hormone (ACTH) on brown adipose tissue are attenuated by corticosterone. *FASEB J* 28:4857–67. doi:10.1096/fj.14-254839.

van der Lans, A. A., M. J. Vosselman, M. J. Hanssen, B. Brans, et al. 2016. Supraclavicular skin temperature and BAT activity in lean healthy adults. *J Physiol Sci* 66:77–83. doi:10.1007/s12576-015-0398-z.

van Marken Lichtenbelt, W. D., J. W. Vanhommerig, N. M. Smulders, J. M. Drossaerts, et al. 2009. Cold-activated brown adipose tissue in healthy men. *N Engl J Med* 360:1500–8. doi:10.1056/NEJMoa0808718.

Vijgen, G. H., N. D. Bouvy, G. J. Teule, B. Brans, et al. 2011. Brown adipose tissue in morbidly obese subjects. *PLoS One* 6:e17247. doi:10.1371/journal.pone.0017247.

Virtanen, K. A., M. E. Lidell, J. Orava, M. Heglind, et al. 2009. Functional brown adipose tissue in healthy adults. *N Engl J Med* 360:1518–25. doi:10.1056/NEJMoa0808949.

von Essen, G., E. Lindsund, B. Cannon, and J. Nedergaard. 2017. Adaptive facultative diet-induced thermogenesis in wild-type but not in UCP1-ablated mice. *Am J Physiol Endocrinol Metab* 313:E515–27. doi:10.1152/ajpendo.00097.2017.

Vosselman, M. J., B. Brans, A. A. van der Lans, R. Wierts, et al. 2013. Brown adipose tissue activity after a high-calorie meal in humans. *Am J Clin Nutr* 98:57–64. doi:10.3945/ajcn.113.059022.

Vosselman, M. J., J. Hoeks, B. Brans, H. Pallubinsky, et al. 2015. Low brown adipose tissue activity in endurance-trained compared with lean sedentary men. *Int J Obes (Lond)* 39:1696–702. doi:10.1038/ijo.2015.130.

Wade, G. N., G. Jennings, and P. Trayhurn. 1986. Energy balance and brown adipose tissue thermogenesis during pregnancy in Syrian hamsters. *Am J Physiol* 250:R845–50. doi:10.1152/ajpregu.1986.250.5.R845.

Wales, N. 2012. Modeling Neanderthal clothing using ethnographic analogues. *J Hum Evol* 63:781–95. doi:10.1016/j.jhevol.2012.08.006.

Wang, W., and P. Seale. 2016. Control of brown and beige fat development. *Nat Rev Mol Cell Biol* 17:691–702. doi:10.1038/nrm.2016.96.

Weir, G., L. E. Ramage, M. Akyol, J. K. Rhodes, et al. 2018. Substantial metabolic activity of human brown adipose tissue during warm conditions and cold-induced lipolysis of local triglycerides. *Cell Metab* 27:1348–55 e4. doi:10.1016/j.cmet.2018.04.020.

Westerterp, K. R. 2004. Diet induced thermogenesis. *Nutr Metab (Lond)* 1:5. doi:10.1186/1743-7075-1-5.

Whitehead, J. M., G. McNeill, and J. S. Smith. 1996. The effect of protein intake on 24-h energy expenditure during energy restriction. *Int J Obes Relat Metab Disord* 20:727–32.

Wu, J., P. Bostrom, L. M. Sparks, L. Ye, et al. 2012. Beige adipocytes are a distinct type of thermogenic fat cell in mouse and human. *Cell* 150:366–76. doi:10.1016/j.cell.2012.05.016.

Wu, J., P. Cohen, and B. M. Spiegelman. 2013. Adaptive thermogenesis in adipocytes: is beige the new brown? *Genes Dev* 27:234–50. doi:10.1101/gad.211649.112.

Yang, D. B., L. Li, L. P. Wang, Q. S. Chi, et al. 2013. Limits to sustained energy intake. XIX. A test of the heat dissipation limitation hypothesis in Mongolian gerbils (*Meriones unguiculatus*). *J Exp Biol* 216:3358–68. doi:10.1242/jeb.085233.

Yang, S., N. D. Nguyen, J. A. Eisman, and T. V. Nguyen. 2012. Association between beta-blockers and fracture risk: a Bayesian meta-analysis. *Bone* 51:969–74. doi:10.1016/j.bone.2012.07.013.

Yeung, H. W., R. K. Grewal, M. Gonen, H. Schoder, et al. 2003. Patterns of (18)F-FDG uptake in adipose tissue and muscle: a potential source of false-positives for PET. *J Nucl Med* 44:1789–96.

Yoneshiro, T., S. Aita, M. Matsushita, Y. Okamatsu-Ogura, et al. 2011. Age-related decrease in cold-activated brown adipose tissue and accumulation of body fat in healthy humans. *Obesity (Silver Spring)* 19:1755–60. doi:10.1038/oby.2011.125.

Yoneshiro, T., M. Matsushita, S. Nakae, T. Kameya, et al. 2016. Brown adipose tissue is involved in the seasonal variation of cold-induced thermogenesis in humans. *Am J Physiol Regul Integr Comp Physiol* 310:R999–R1009. doi:10.1152/ajpregu.00057.2015.

Yoneshiro, T., T. Ogawa, N. Okamoto, M. Matsushita, et al. 2013. Impact of UCP1 and beta3AR gene polymorphisms on age-related changes in brown adipose tissue and adiposity in humans. *Int J Obes (Lond)* 37:993–8. doi:10.1038/ijo.2012.161.

Zhao, Z. J. 2012. Effect of cold exposure on energy budget and thermogenesis during lactation in Swiss mice raising large litters. *Biol Open* 1:397–404. doi:10.1242/bio.2012661.

Zukotynski, K. A., F. H. Fahey, S. Laffin, R. Davis, et al. 2009. Constant ambient temperature of 24 degrees C significantly reduces FDG uptake by brown adipose tissue in children scanned during the winter. *Eur J Nucl Med Mol Imaging* 36:602–6. doi:10.1007/s00259-008-0983-y.

Zukotynski, K. A., F. H. Fahey, S. Laffin, R. Davis, et al. 2010. Seasonal variation in the effect of constant ambient temperature of 24 degrees C in reducing FDG uptake by brown adipose tissue in children. *Eur J Nucl Med Mol Imaging* 37:1854–60. doi:10.1007/s00259-010-1485-2.

Zwick, R. K., C. F. Guerrero-Juarez, V. Horsley, and M. V. Plikus. 2018. Anatomical, physiological, and functional diversity of adipose tissue. *Cell Metab* 27:68–83. doi:10.1016/j.cmet.2017.12.002.

7 Interaction between Environmental Temperature and Craniofacial Morphology in Human Evolution

A Focus on Upper Airways

Laura Maréchal and Yann Heuzé

CONTENTS

7.1 Introduction 161
7.2 Upper Airway Morphology 164
  7.2.1 Nasal Anatomy and Physiology 164
  7.2.2 Climate-Related Variation of Nasal Structures 166
  7.2.3 Function of the Paranasal Sinuses 167
7.3 Effects of Environmental Temperature on Bone Formation 169
  7.3.1 Thermoregulation in Extreme Climatic Environments 170
  7.3.2 Mechanisms of Temperature Influence on Upper Airway Bone and Cartilage 171
  7.3.3 Temperature-Sensitive Developmental Pathways 172
7.4 Implications for Human Craniofacial Evolution 175
7.5 Conclusion 176
7.6 Acknowledgments 176
7.7 References 176

7.1 INTRODUCTION

The variation and evolution of craniofacial skeletal morphology in hominids is the result of a complex interplay between genetic factors (e.g., Adhikari et al. 2016; Pickrell et al. 2016; Shaffer et al. 2016; Zaidi et al. 2017; Claes et al. 2018) and biomechanical pressures related to brain growth, mastication, and respiration (e.g., Moss and Young 1960; Enlow 1990; Richtsmeier et al. 2006; Lieberman 2011; Bastir and Rosas 2013), as well as multiple environmental factors including diet, activity level, and ecogeographic variables such as temperature (e.g., Steegmann and Platner 1968; Roseman 2004; Rae et al. 2006; Evteev et al. 2014; Menéndez et al. 2014; Sardi 2018; Wroe et al. 2018; Martin et al. 2021). Although developmental shifts in brain ontogeny and selective pressures in response to the biomechanical forces related to mastication and respiration appear to be predominant in human skull evolution (for a recent comprehensive review, see Lesciotto and Richtsmeier 2019), environmental factors might still have a significant influence, though possibly more challenging to disentangle from other processes.

The influence of these environmental factors, particularly temperature, has been addressed in prior research, mainly focusing on their impact on infra- or post-crania morphology and body proportions (e.g., Holliday 1997; Ruff 2002; Holliday and Hilton 2010). The pioneering work of Bergmann (1847, translated in James 1970) states that homoeothermic organisms maintain stable internal body temperature by balancing the production of warmth within the volume of their body and the loss of warmth from its surface. In this thermoregulation process, the surface area-to-volume ratio of the body is therefore a predominant factor. According to Bergmann's rule, within a broadly distributed genus, species of larger size are found in colder environments, while species of smaller size are found in warmer environments. This rule, though still debated (e.g., Scholander 1955; Mayr 1963; McNab 1971, 2010, 2012; Crognier 1981a; Ruff 1994; Katzmarzyk and Leonard 1998; Ashton et al. 2000; Meiri and Dayan 2003; Ochocinska and Taylor 2003; Blackburn and Hawkins 2004; Meiri et al. 2004; Rodriguez et al. 2006; Clauss et al. 2013; Foster and Collard 2013; Alhajeri and Steppan 2015; Gohli and Voje 2016; Brown et al. 2017; Nunes et al. 2017; Sargis et al. 2018), also applies to the genus *Homo*. Allen's rule, based on another foundational work concerning the influence of environmental temperature on the morphology of homoeothermic organisms (Allen 1877), states that in colder climates individuals tend to possess shorter limbs and extremities, thus reducing the surface area-to-volume ratio and the associated heat dissipation. In warm climates, the opposite phenomenon is observed. However, Bergmann's and Allen's rules were based on observed variation of the postcranial skeleton and may not apply to the craniofacial skeleton. The impact of environmental temperature on the variation and evolution of skull phenotypes remains less clear, for example, in Neanderthals (e.g., Steegmann et al. 2002, but see Weaver 2009). Moreover, attempts to provide mechanistic hypotheses to explain craniofacial morphological variation attributed to environmental temperature remain scarce. Hence, in the present chapter, we will review research relevant to two critical questions, which, from our perspective, are related and need to be addressed together to increase our comprehension of the role of climate in the evolution of craniofacial morphology:

- What anatomical and functional units of the skull exhibit temperature-associated patterns of morphological variation?
- What developmental processes (genetic, molecular, and cellular) involved in craniofacial growth and development are sensitive to temperature and could contribute to the explanation of such variation?

Among the different anatomical and functional units of the skull, the facial skeleton shows clear signs of morphological variation related to environmental temperature. In particular, temperature-related variation of the shape and size of the nasal cavity has been a focus of research in paleoanthropology and physical anthropology for many years. The nasal cavity forms the gateway to the respiratory system (Enlow 1990). As such, the morphology of this interface region has long been considered a reliable proxy for studying the link between hominins and their environment (e.g., Davies 1932; Weiner 1954; Carey and Steegmann 1981; Yokley 2009; Noback et al. 2011). Two main hypotheses have been put forward to interpret how adaptive pressures impact nasal cavity morphology. These hypotheses are not mutually exclusive and relate to two factors that might simultaneously influence nasal morphology. The first hypothesis considers the morphology of the nasal cavity in relation to its air conditioning function and climatic adaptation (Charles 1930; Shea 1977; Cole 1982b; Churchill et al. 2004; Yokley 2009; Butaric et al. 2010; de Azevedo et al. 2017; Butaric and Klocke 2018; Evteev and Grosheva 2019; Heuzé 2019). The air conditioning function of the nasal cavity is the process by which the inspired air reaches body core temperature and a full saturation with water vapor to protect the alveolar lining in the lower airway (Elad et al. 2008; Wolf et al. 2004). The second hypothesis emphasizes the role of the nasal cavity as the upper part of the respiratory system regulating the amount of air inhaled and is thus closely related to the energetic demands of the body (Hall 2005; Froehle et al. 2013; Holton et al. 2014, 2016; Wroe et al. 2018).

Neanderthals, who lived in Eurasia until about 28 kya ago (Finlayson et al. 2006), have received much attention on these matters. The morphology of the Neanderthal appendicular and facial skeletons has often been interpreted as cold adapted (e.g., Steegmann et al. 2002) and/or as the result of genetic drift (Weaver 2009). In the facial region, this variation includes a larger nasal cavity in Neanderthal relative to anatomically modern *Homo*. This larger nasal cavity would allow an increased incoming airflow, associated with a larger volume of the ribcage to meet the high energetic demands of large-brained and heavy-bodied Neanderthals (Coon 1962; Franciscus and Churchill 2002; García-Martínez et al. 2018; Wroe et al. 2018) while providing an efficient way to condition air in cold climates (de Azevedo et al. 2017, but see Bastir 2019). Using computational fluid dynamics methods, one can quantify several airflow features characterizing respiration, as well as air conditioning efficiency in extant normal and pathological samples (Burgos et al. 2017; Kim et al. 2017). These methods have recently been used to study Neanderthals (de Azevedo et al. 2017; Wroe et al. 2018), though the virtual reconstruction of nasal mucosa of fossil specimens, achieved by morphing the modern human airway to fossil nasal cavities (see also Bourke et al. 2014), might be problematic (Evteev and Heuzé 2018). Indeed, based on a relatively small sample ($N = 30$), Heuzé (2019) reported a rather low correlation between the volume of the bony nasal cavity and the negative volume defined by the nasal mucosa, that is, the functioning nasal airway, thus preventing robust direct interpolation of nasal airway volume on the basis of nasal cavity volume.

In this chapter, we take a step back and address the question of the interaction between environmental temperature and facial skeletal morphology from a new perspective. Our purpose is to provide an overview of the current knowledge of the temperature-related morphological variation of nasal and paranasal structures and to explore the temperature-sensitive pathways that might have a role in this variation.

Genes sensitive to temperature are obviously of importance in temperature-related morphological variation, and examples are provided in this chapter. However, genes do not directly produce phenotypes (Cohen and McLean 2000; Richtsmeier and Lesciotto 2020). Rather, phenotypes are the products of complex interactions between the genome and the internal and external environment. A concept central to these interactions is developmental plasticity, which is the response of cells, tissues, organs, and/or an individual organism to environmental variation that occurs within the lifespan of an individual with a single genotype and results in the formation of more than one phenotype (Hall and Witten 2018). Acknowledging the key role of cellular processes in the making of phenotypes, this review focuses on the cellular response to temperature changes. In doing so, our purpose is to pave the way to a better understanding of the mechanisms that explain the relationship between climate variation and the morphology of the structures involved in respiration. We believe that such an approach will shed a new light on the role of temperature in human evolution.

7.2 UPPER AIRWAY MORPHOLOGY

As stated, the nasal cavity forms the gateway to the respiratory system (Enlow 1990). The covariation of environmental air temperature and humidity with morphology of the nasal structure has consequently been extensively addressed and these ecogeographic factors are considered a driving force in the expression of phenotypic variation and adaptation (Thomson 1913; Thomson and Buxton 1923; Davies 1932; Woo and Morant 1934; Negus 1952; Weiner 1954; Cottle 1955; Negus 1960; Wolpoff 1968; Hiernaux and Froment 1976; Carey and Steegmann 1981; Crognier 1981a, 1981b; Franciscus and Trinkaus 1988; Franciscus and Long 1991; Franciscus 1995; Roseman 2004; Roseman and Weaver 2004; Harvati and Weaver 2006a, 2006b; Márquez and Laitman 2008; Hubbe et al. 2009; Yokley 2009; Butaric et al. 2010; Bastir et al. 2011; Noback et al. 2011; Evteev et al. 2014; Jaskulska 2014; Butaric 2015; Maddux et al. 2016a; Zaidi et al. 2017; Marks et al. 2019). In the following, we discuss nasal anatomy and physiology and summarize what is currently known about climate-related phenotypic variation. Finally, we focus on the paranasal sinuses, their role in respiratory energetics, and current hypotheses about their covariation with nasal morphology and environmental factors.

7.2.1 Nasal Anatomy and Physiology

The nasal cavity is the area of the craniofacial skeleton that contains the nasal airway, the first anatomical region of the respiratory system involved in respiratory energetics and air conditioning. Conditioning of the inspired air in the nasal airway is achieved through contact with the respiratory mucosa producing heat exchange via convection and moisture exchange via evaporation (Cole 1982a; Naclerio et al. 2007; Yokley 2009). Skeletally, the nasal cavity is bounded and defined by maxillary, nasal, palatal, vomer, sphenoid, frontal, ethmoid, and lacrimal bones. The shape, size, and relative position of these bones affect the morphology of the nasal cavity that, at least indirectly, conditions the quantity of air that can be inhaled and

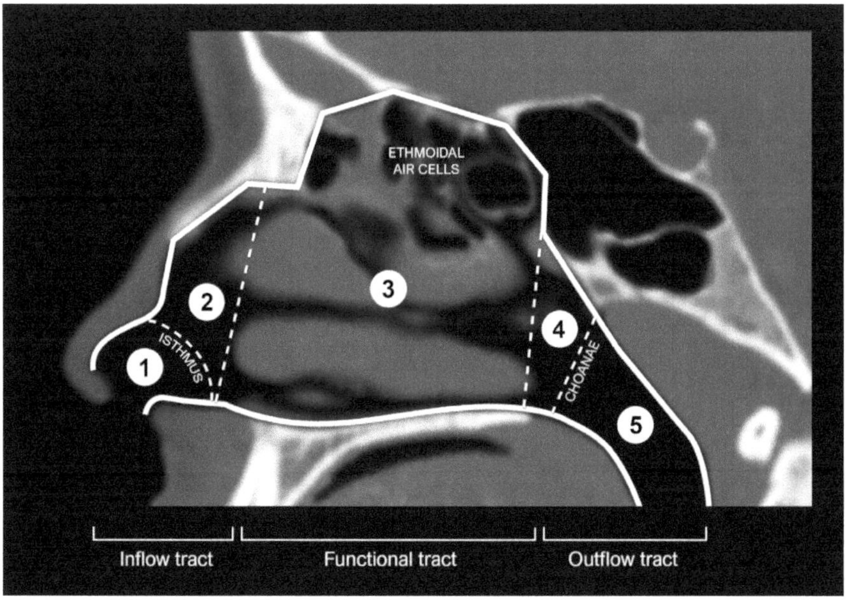

FIGURE 7.1 Sagittal cross-section from a CT scan showing the structural elements of the incoming nasal airflow pathway in lateral view. The inflow tract consists of the vestibulum (1) and the anterior cavum (2), which are separated by the isthmus. The functional tract is the area of the turbinates (3). The outflow tract is composed of the posterior cavum (4), choanae, and nasopharynx (5).

the air conditioning performances. The dimensions of the entry (piriform aperture) and exit (choanae) points significantly influence respiratory energetics (Swift and Proctor 1977; Bastir and Rosas 2013). Maddux et al. (2016b) argued that nasal cavity height was more likely to be associated with energetics, while nasal cavity width and length play an important role in air conditioning (Noback et al. 2011).

The mucous membrane lining the nasal cavity delimits the nasal airway that can be divided into three different units (Figure 7.1) (Mlynski et al. 2001; Bastir et al. 2020). The first part, the inflow tract, directs and diffuses the airflow, and contains the vestibulum, isthmus, and anterior cavum. The second part, the functional tract, includes the nasal turbinates. The third part, the outflow tract, directs warmed and humidified air toward the lower respiratory tract and is composed of the posterior cavum, choanae, and nasopharynx.

The thickness of the highly vascularized mucous membrane of the nasal airway fluctuates by contraction and expansion, depending on physiological factors such as blood pressure, nasal cycle, and nasal function (Cauna 1982; Elad et al. 2008; Yokley 2009; White et al. 2015). The variation in mucosa thickness causes congestion and decongestion of the nasal airway and affects its size and shape. Consequently, the speed, volume, and direction of airflow are also affected (Cauna 1982), as well as the efficiency of air conditioning processes (Churchill et al. 2004; Naftali et al. 2005; Zhao and Jiang 2014; Ma et al. 2018). Indeed, the nasal airway is the place where air conditioning takes place, which is necessary to optimize gas exchanges in the pulmonary alveoli and thus participate in global homeostatic thermoregulation (Havenith 2005; White 2006) while protecting the lungs from thermal damage, desiccation, and infection (Proetz 1951, 1953; Walker and Wells 1961; Cole 1982b; Proctor 1982; Keyhani et al. 1995; Williams 1998; Keck et al. 2000; Eccles 2002; Wolf et al. 2004; Yokley 2006; Doorly et al. 2008; Elad et al. 2008; Yokley 2009; Hildebrandt et al. 2013). A large mucosal surface and a narrow channel generally facilitate heat and moisture exchange (Schmidt-Nielsen et al. 1970; Collins et al. 1971; Hanna and Scherer 1986; Schroter and Watkins 1989; Lindemann et al. 2009).

The nasal turbinates (or conchae) are complex, curled structures that extend from the side and upper walls of each nostril and play a major role in respiratory processes. These structures are divided into lower, intermediate, and upper turbinates (Moore 1981; Smith et al. 2006; Maier and Ruf 2014). They are covered with an epithelium that is olfactory for the upper turbinate (Zhao 2004; Sahin-Yilmaz and Naclerio 2011) and respiratory for the intermediate and lower turbinates (Doorly et al. 2008; Wen et al. 2008; Xiong et al. 2008; Sommer et al. 2012; Kim et al. 2017; Marks et al. 2019). Beneath the turbinates lie the superior, middle, and inferior meatuses that communicate posteriorly with the outflow tract; the middle meatus also holds the opening of the maxillary sinus (ostium maxillare). Ontogenetically, the turbinates and the associated meatuses develop from the six furrows, resembling ethmoturbinals, appearing on each lateral branch of the cartilaginous nasal capsule during weeks 9 to 10 of human fetal development (Jankowski 2013). Postnatally, lower turbinates include numerous seromucous cells, providing a significant input of water vapor required in the air humidification process (Cole 1982b; Tos 1982; Keyhani et al. 1995; Naftali et al. 2005; Na et al. 2012). Studies using computational fluid dynamics methods have demonstrated the role of the two lower turbinates in mediating the velocity and direction of airflow (Keyhani et al. 1995; Wang et al. 2005; Inthavong et al. 2007; Doorly et al. 2008; Zhu et al. 2011; Na et al. 2012; Li et al. 2017; Inthavong et al. 2018).

7.2.2 Climate-Related Variation of Nasal Structures

Numerous studies focusing on the nasal area have shown an association between its morphology and environmental factors (e.g., air temperature, humidity, altitude). Two main proxies for nasal morphology have been used to study this relationship: the negative volume defined by bone (i.e., nasal cavity) and the negative volume defined by soft tissue (i.e., nasal airway).

The first studies that accurately demonstrated a relationship between nasal morphology and eco-geographical factors were conducted using dry skulls and based on measurements of the facial skeleton and nasal aperture (Thomson 1913; Thomson and Buxton 1923; Davies 1932; Weiner 1954; Hoyme 1965; Wolpoff 1968; Hiernaux and Froment 1976; Carey and Steegmann 1981; Crognier 1981a,b; St. Hoyme and Işcan 1989; Franciscus and Long 1991; Roseman 2004; Hubbe et al. 2009; Leong and Eccles 2009). Later studies addressed the morphology of the entire nasal cavity and confirmed this relationship (Yokley 2009; Noback et al. 2011; Evteev et al. 2014; Fukase et al. 2016). Comparative inter-population studies have demonstrated that the only area of the nasal complex affected by variation related to eco-geographic factors is the internal nasal cavity (Maddux et al. 2016b). This area is also the main site of heat and moisture exchange within the nasal complex (Ingelstedt 1956; Cole 1982b; Keck et al. 2000; Naftali et al. 2005; Elad et al. 2008).

The results of these studies show that, when humans live in cold environments, they appear to possess a nasal cavity that is reduced mediolaterally and increased anteroposteriorly and superoinferiorly (Churchill et al. 2004; Doorly et al. 2008; Yokley 2009; Holton et al. 2011, 2013; Maddux et al. 2016b). Narrow nasal passages facilitate heat and moisture exchange by increasing the mucosal surface area relative to air volume (SA/V) ratio. This configuration increases nasal resistance for conditioning incoming airflow but also increases the amount of heat and water recovered during exhalation, thereby improving the conditioning capacity of the inner nasal cavity (Schmidt-Nielsen et al. 1970; Collins et al. 1971; Hanna and Scherer 1986; Schroter and Watkins 1989; Lindemann et al. 2009). In addition, when the anteroposterior dimension of the nasal cavity increases, so does the time that airflow occurs in the nasal cavity, which also contains a larger volume of mucous membrane along this dimension, increasing efficiency of air conditioning (Inthavong et al. 2007; Noback et al. 2011). Last, the variation in nasal cavity height might be related to another aspect of climatic adaptation: energy demands. Several studies have shown that individuals with higher metabolic demands for oxygen consumption, which is generally the case in colder and/or drier environments, tend to have taller nasal cavities (Froehle 2008; Bastir and Rosas 2013, 2016; Holton et al. 2016). Furthermore, the increase in nasal height of individuals living in cold climates might also compensate for the reduction of nasal breadth that is also observed in these environmental conditions, thus maintaining a sufficient volumetric intake capacity (Maddux et al. 2016b).

While the nasal cavity has often been studied, the nasal airway has only become the focus of studies in the last few years. An in vivo study (Yokley 2009) measured the SA/V ratio of individuals from European and African ancestry and observed that this SA/V ratio was only higher in European individuals when the nasal airway was fully decongested (i.e., nasal mucosa fully contracted). When the nasal mucosa was not fully contracted, the SA/V ratio showed no significant differences between the individuals from European and African ancestry. This study underlines the importance of focusing not only on the nasal cavity but also on the volume delimited by the mucosa (i.e., nasal airway) and its morphology. An important issue when addressing nasal airway morphology is to take into consideration the nasal cycle, which consists of the alternative partial congestion and decongestion of the right and left sides of the nasal airway during breathing, thus optimizing respiratory air conditioning (Hasegawa and Kern 1977; Cauna 1982; Eccles 1982, 1996; Watelet and Cauwenberge 1999; White et al. 2015; Pendolino et al. 2018). The influence of the periodicity of these nasal cycles on airway morphology needs to be addressed by studying larger samples and measuring the right and left sides separately (Heuzé 2019).

7.2.3 Function of the Paranasal Sinuses

The paranasal sinuses are mucous membrane-lined cavities within bones that surround the nasal area (Figure 7.2). These include: the frontal sinuses, communicating

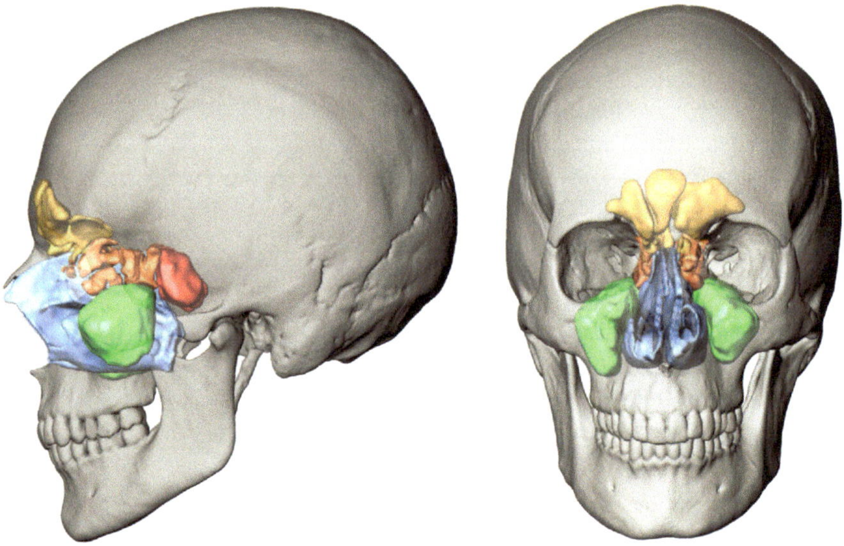

FIGURE 7.2 3D reconstruction of a human adult skull allowing the visualization (transparency) and localization of nasal and paranasal structures: nasal airway (blue), ethmoidal air cells (orange), frontal sinuses (yellow), sphenoid sinus (red), and maxillary sinuses (green).

with the nasal region through the meatus; the maxillary sinuses, communicating with the nasal region through the semilunar hiatus; and the sphenoid sinus. Ethmoidal air cells, which are thin-walled cavities located in and defined by the ethmoidal labyrinth, are also often considered paranasal sinuses. However, ethmoid bone development in humans starts during the first trimester of gestation, while the other paranasal sinuses develop entirely after birth in aerial conditions, which could ontogenetically grant them another status (Jankowski 2013). Furthermore, the mechanisms of paranasal sinus formation differ from those of the ethmoidal complex. One theory explaining paranasal sinus development—the epithelial theory—states that the maxillary, frontal, and sphenoid sinuses are produced via epithelial diverticula overflowing from the ethmoid labyrinth and causing a pneumatization of the surrounding bones (Zuckerkandl 1893; Zollikofer and Weissmann 2008).

Several functions have been assigned to paranasal sinuses without real consensus. One theory is that these structures were involved in reducing the weight and increasing pneumatization of the skull (O'Malley 1923; Tillier 1977). Another theory raised the hypothesis of a potential role in thermal insulation of the brain and the eyes or in voice resonance (Masuda 1992). A third hypothesis, the mechanistic theory, sees the paranasal sinuses as simple residual cavities that result from morphological changes in the surrounding structures that are constrained by biomechanical forces during craniofacial development (Rae et al. 2003; Holton et al. 2013; Jankowski 2013; Butaric 2015; Butaric and Maddux 2016; Noback et al. 2016; Maddux and Butaric 2017; Buck et al. 2019; Evteev and Grosheva 2019). An additional hypothesis focuses on the role of the paranasal sinuses in the warming and humidification of inspired air (Gannon et al. 1997).

Another function attributed to the paranasal sinuses is the production of nitric oxide by the inner membranes of the sinuses, under the action of the enzyme I-NOS (NOS-2) (Lundberg et al. 1995, 1996). Nitric oxide is a powerful vasodilator, particularly involved in the cellular functions of the respiratory, nervous, and immune systems (Lundberg 2008; Keir 2009; Márquez et al. 2014). It is produced by the endothelial cells of sinus blood vessels and relaxes the muscle fibers of the vascular nasal wall when pumped into incoming airflow, thus regulating the intranasal temperature (Lundberg et al. 1995; Holden et al. 1999). In addition, nitric oxide may help maintain a sterile environment in the respiratory tract through two mechanisms: (1) nitric oxide is toxic to many viruses and bacteria and may therefore play a role in protection against infections (Mancinelli and McKay 1983; Croen 1993), and (2) the level of nitric oxide affects the beat frequency of the cilia of airway epithelial cells that propel mucus-trapped debris and particles out of the lungs (Jain et al. 1993).

The proximity of the paranasal sinuses to the nasal structures and their role in respiratory processes has led researchers to study the relationship between the paranasal sinuses and climatic factors (e.g., Rae et al. 2011; Butaric 2015; Evteev and Grosheva 2019). While the sinuses show a high level of within- and between-group variation (Evteev and Grosheva 2019) and are strongly correlated with craniofacial size (Rae et al. 2011; Butaric 2015), a covariation has been measured between the maxillary sinus and nasal structures (Butaric 2015). Indeed, individuals from cold-dry climates tend to have a larger maxillary sinus volume associated with a medial displacement of the lateral nasal walls, thus causing a reduction of the internal nasal breadth (Holton et al. 2013; Butaric 2015; Butaric and Maddux 2016; Evteev and Grosheva 2019). Nevertheless, some confusion remains about the exact function of each of these sinuses, their relationship with neighboring structures, and how variation in paranasal sinuses relates to skeletal morphology of the entire face, not just the nasal aperture, and environmental factors.

7.3 EFFECTS OF ENVIRONMENTAL TEMPERATURE ON BONE FORMATION

The studies summarized previously focus on the correlation or covariation between ecogeographic factors (mainly temperature) and morphology of the nasal region and the hypothesized advantages of these phenotypes in cold or warm environments. Though it is widely acknowledged that the morphology of the nasal region depends in large part on the articulation of bones that surround it, the processes that contribute to bone formation that underlie this variation and generate its expression are rarely addressed. For instance, we do not yet understand the relative roles of heredity and developmental plasticity in the production of craniofacial phenotypes (Lovejoy et al. 2003). Here we present an overview of the manifestation of thermoregulation in extreme climatic environments and propose hypotheses on how stress caused by extreme temperature could modify bone formation processes and the resulting morphology of the nasal region. We then address the interaction between genes and the environment in the production of a phenotype through the identification of pathways or temperature-sensitive genes that could have an effect on the nasal and paranasal structures via cellular processes involved in bone and cartilage development.

We emphasize that trying to explain a phenotype as the direct result of temperature influence is unrealistic, as many other factors must be considered to explain craniofacial skeletal morphology. Diet, nutrition, and activity levels are examples of environmental inputs that can influence bone metabolism (e.g., Kiliaridis et al. 1985; Paschetta et al. 2010; Menéndez et al. 2014). Furthermore, temperature is not the only parameter defining ambient air, which also depends on the less acknowledged factor of humidity. Another major consideration is the integration among structures constituting the craniofacial skeleton that can induce a secondary variation in some specific area that is a response to the variation of its surrounding bony environment (Sardi et al. 2018; Scott et al. 2018). Age can also affect craniofacial morphology. Tooth loss and subsequent bone resorption in the maxillary area would induce a modification of palate morphology that could also affect the nasal and paranasal structures (Albert et al. 2007; Joganic and Heuzé 2019). Though not meant to be exhaustive, we present some temperature-sensitive genes, pathways, and cellular processes involved in bone and cartilage formation and that could influence aspects of craniofacial morphology.

7.3.1 Thermoregulation in Extreme Climatic Environments

Thermoregulation includes all the mechanisms used by an organism to control its body temperature and ensure optimal regulation of all metabolic processes (Iwen et al. 2018; Romanovsky 2018). In homeothermic species, the production of internal body heat must always be balanced with body surface heat loss in order to maintain homeostasis. To this end, various mechanisms have evolved, including skin vasodilatation, sweating, behavioral adaptations, insulation of the body by fur, clothing, and/or intradermal fat accumulation (Cannon and Nedergaard 2004; Kasza et al. 2014; Alexander et al. 2015; Fischer et al. 2016; Kasza et al. 2016). Biological responses such as vasoconstriction and vasodilatation also allow heat to be retained or lost by the extremities. For instance, cold temperature leads to vasoconstriction that directs the blood flow towards the trunk and vital organs to reduce the dispersion of blood heat in the extremities (Tansey and Johnson 2015). Indeed, the extremities of homeothermic species are characterized by their regional heterothermy, that is, the ability to drop temperature of the limbs while maintaining that of the trunk (Harrison and Clegg 1969; Ponganis et al. 2003; Serrat et al. 2008).

Experimental studies have measured an interaction between temperature, cell proliferation, and bone matrix production, which may affect cartilage growth and thus modify the morphology of endochondral bone (Serrat et al. 2008, 2010, 2015; Serrat 2014). Hence, temperature can affect vertebral number in ectothermic and homoeothermic vertebrates (Hall 2015), as well as limb and extremity length, which tend to shorten in response to a decrease in temperature (Allen 1877; Feldhamer 2007). One of the explanations of limb shortening proposed by Serrat (2014) is that the vasoconstriction induced by cold stress leads to a decreased blood flow in the extremities, altering the transport of important nutrients, oxygen, and hormones, and ultimately affecting endochondral ossification. Indeed, growth plates, though avascular, benefit from the nutritional support of the surrounding vasculature that transports solutes diffusing through the extracellular matrix to reach the cartilaginous cells (Brookes and Revell 1998). An alteration in blood flow might therefore affect this nutrient supply, thus affecting normal growth. Interestingly, the effects of temperature on limb length can be observed within a single generation of outbred mice reared at warm and cold temperatures during the postnatal growth period (Serrat et al. 2008). These observations underscore the potential role of phenotypic plasticity. When driven by environmental factors, phenotypic plasticity is considered a greater evolutionary force than random mutation (West-Eberhard 2005), but few studies have addressed the potential effect of phenotypic plasticity on morphological variation of the upper airway. Rae et al. (2006) contributed to this question in their study of the dry crania of cold- and warm-reared rats. Their results show that cold stress causes subtle but significant changes in facial shape as well as maxillary sinus and nasal cavity volumes, suggesting developmental plasticity of the craniofacial skeleton in response to climatic variation.

7.3.2 Mechanisms of Temperature Influence on Upper Airway Bone and Cartilage

Potential responses of the craniofacial skeleton to cold stress are not well understood, but some authors have proposed hypotheses that might explain how temperature influences craniofacial variation. For example, a recent study on nasal turbinate morphology (Marks et al. 2019) hypothesized that the modification of cartilage development in cold environments could also apply to nasal turbinate cartilage and all cartilages of the nasal capsule of the forming chondrocranium. If this change occurred, the morphology of the surrounding non-cartilaginous skeletal structures developed through intramembranous ossification (e.g., the maxillary, premaxillary and nasal bones) would also likely be affected (Chae et al. 2003; Egeli et al. 2004; Opperman et al. 2005; Wealthall and Herring 2006; Al Dayeh et al. 2013; Hall and Precious 2013; Hartman et al. 2016; Holton et al. 2018). By altering the expression of genes and pathways that affect cellular processes involved in cartilage metabolism, temperature could then have an indirect effect on growth of the facial bones.

Selective brain cooling (SBC) is the mechanism that keeps the brain at a cooler temperature than the rest of the body through the precooling of the blood supplying the brain and provides an example of complex interaction between blood flow intensity and morphological variation. This mechanism is achieved by dilation or constriction of the veins. For example, constriction of the veins returning blood from the nose and face precools that blood that will then be supplied arteriorly to the brain (Caputa 2004). Regulation of the evaporation of water at the mucosal surfaces of nasal turbinates also contributes to SBC (Irmak et al. 2004). Furthermore, ethmoidal air cells and the sphenoid sinus could potentially help cool the adjacent brain lobes and vessels by thermal conduction. Part of the morphological variation of craniofacial structures might reflect adaptive changes in growth patterns that produce morphology ensuring a more effective SBC specific to the environment. Finally, we know that cold stress implies greater oxygen consumption, due to the sympathetic activity associated with brown adipose tissue production and muscle activation during nonshivering thermogenesis (Lowell and Spiegelman 2000). The increase of incoming airflow, which is closely related to the energetic demands of the body, could also play a role in the morphological variation of nasal structures (see also Maddux et al. 2016b).

7.3.3 Temperature-Sensitive Developmental Pathways

Temperature not only affects the skeleton but almost every system in the body, causing many interactions among tissues that may experience different consequences of thermal stress (Tattersall et al. 2012). Although the number of genome-wide studies on the molecular basis of craniofacial morphology have been expanding in the last few years (e.g., Adhikari et al. 2016; Weinberg et al. 2018; Xiong et al. 2019), our understanding of the genetic basis for craniofacial variation is incomplete. It is therefore not an easy task to identify the temperature-sensitive genes that might directly and/or indirectly affect craniofacial morphology. Here, we highlight a few of these genes and associated pathways that were identified through in vivo and in vitro studies and which may explain some of the temperature-related variation of nasal and paranasal morphology (Figure 7.3). We begin by noting that the production of a "cold phenotype" or a "warm phenotype" does not necessarily imply mechanisms that would be opposites. Note that variation is expected in the phenotypic response to the temperature-sensitive genes and associated pathways. Some of these phenotypic

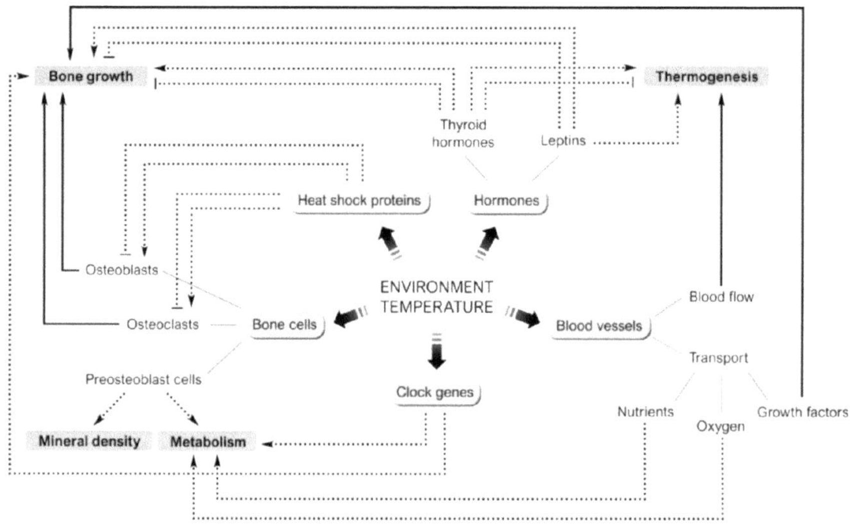

FIGURE 7.3 Diagram summarizing the major physiological pathways through which environmental temperature could influence nasal and paranasal morphology. Solid lines show direct temperature influences on bone growth and metabolism. Dashed lines indicate indirect ways through which temperature can alter bone cells. Indirect pathways include the endocrine system (e.g., thyroid hormones and leptin play a role in thermogenesis and bone growth), the circulatory system (temperature variation can induce a vasoconstriction or a vasodilatation, thus affecting blood flow and the transport of nutrients, oxygen, and hormones involved in bone metabolism), and other temperature-sensitive proteins and genes (e.g., HSP and clock genes both influence bone growth and metabolism).

responses might be continuous and proportional to the temperature variation, while others might be expressed only when a threshold is reached. Additionally, temperature variation at certain developmental stages can have different phenotypic outcomes varying both in pattern and intensity (e.g., Hall 2015).

Some researchers have demonstrated that temperature can have a direct effect on bone formation and growth by altering the activity of bone-forming osteoblasts and bone-resorbing osteoclasts. An in vitro study using cells derived from rat trabecular bone (Patel et al. 2012) has shown that, after 14–16 days of culture, the activity of calvarial osteoblasts was reduced by 75% in mild hypothermia and by 95% in severe hypothermia. A reciprocal effect of hypothermia was also observed on osteoclastogenesis, that showed a 1.5- to 2-fold stimulation, thus increasing bone resorption.

Temperature was also shown to affect preosteoblast activity in vitro, including mesenchymal stem cell differentiation (Shui and Scutt 2001; Chung and Rylander 2012). When subject to heat stress, preosteoblast cells can promote expression of osteocalcin and osteopontin, two proteins involved in bone mineral density and metabolic regulation. However, the combination of heating and osteoinductive growth factors leads to the expression of heat shock proteins (HSPs), osteoprotegerin, and vascular endothelial growth factor (VEGF). Osteoprotegerin is involved in bone resorption through its role as a decoy receptor in the RANKL/RANK pathway (Aubin and Bonnelye 2000), and VEGF stimulates angiogenesis and controls bone formation (Dai and Rabie 2007). HSPs are multifunctional proteins that can be induced by heat stress or cold stress (Rylander 2005; Barna et al. 2012; Patil and Paul 2014; Hang et al. 2018). HSP70, for example, increases the proliferation and differentiation of osteoprogenitor cells, which are bone marrow-derived stromal cells (Shui and Scutt 2001), and regulates both the resorption activity of osteoclasts via the RANKL/RANK pathway and the bone production activity within osteoblasts by activating the ERK and Wnt/β-catenin pathway (Hang et al. 2018). The temperature of the cell culture also influences HSP27, which is particularly involved in the regulation of bone cell physiology through upregulation of TGF-β (Hatakeyama et al. 2002), estrogen (Cooper and Uoshima 1994), endothelin-1 (Tokuda et al. 2003), and prostaglandins (Kozawa and Tokuda 2002). These results point to HSP27 as a potentially important factor in the modulation of cellular events in bone and cartilage.

Clock genes are also involved in cell proliferation in cartilage. Clock genes are transcriptional activators that play a central role in the regulation of circadian rhythms, the 24-hour cycles in physiology that precisely regulate organ function (Reppert and Weaver 2001). The expression of clock genes is directly dependent on temperature, and their mutation can lead to an altered regulation of bone volume or deficiencies in long bone growth because of their control of chondrocyte differentiation (Gossan et al. 2013; Marks 2018; Steindal and Whitmore 2019). Circadian clocks might also have a potentially important role in bone growth and maintenance and the production of morphology (Takarada et al. 2012).

Temperature can also affect global human DNA methylation and RNA editing (Garrett and Rosenthal 2012; Shi et al. 2020), as well as telomere length (e.g., Romano et al. 2013). Genes with methylation status have been shown to be affected by temperature changes. For example, low temperature induces the hypermethylation of ZKSCAN4, expressed by an increase of blood pressure (Xu et al. 2020), an indirect pathway that could affect bone growth. Another study addressed the contribution of cold-inducible RNA-binding protein (CIRP) to tissue remodeling in chronic rhinosinusitis (Shi et al. 2020). This cold shock protein is a chaperone that is upregulated under mild hypothermia and facilitates mRNA translation. CIRP seems to be involved in cold-induced suppression of cell proliferation, but its precise role is still poorly understood.

Finally, temperature can alter the endocrine system, which has a fundamental role in homeostasis. By affecting extracellular matrix proteins, cold stress can modify the diffusion rates of endocrine and paracrine growth regulators affecting the diffusion and transport of hormones and ultimately skeletal morphology. Among the endocrine hormones that could affect the skeleton (Massaro and Rogers 2004), thyroid hormones and leptin are probably the best studied.

It has been shown that thyroid hormones play a very important role in the regulation of homeostasis (Iwen et al. 2018). These hormones modify the transcription rate of uncoupling protein 1, localized in brown adipose tissue and involved in nonshivering thermogenesis (see Chapter 6 by Devlin; Enerbäck et al. 1997; Golozoubova et al. 2001), increase metabolic rate, and can directly influence the sodium/potassium and calcium pumps in skeletal muscle (Silva 2006), as well as the vasoconstriction and vasodilatation of blood vessels (Warner et al. 2013). Thyroid hormone levels also influence intramembranous and endochondral ossification and, consequently, craniofacial development (Bassett and Williams 2016). Indeed, skeletal hypothyroidism is expressed in a delayed ossification of the skull, which can cause defects such as wider and/or persistent cranial sutures and fontanelles. Conversely, skeletal hyperthyroidism causes advanced ossification that can manifest as malformations including craniosynostosis. Interestingly, the phenotypes produced by an altered thyroid status display similarities with loss-of-function or gain-of-function mutations affecting FGF (fibroblast growth factor), IGF (insulin-like growth factor) and WNT (Wingless-related integration site) signaling pathways, which are key pathways in craniofacial development. Since thyroid hormone T3 induces FGF, FGFR, IGF1, and IGF1R expression and enhances MAPK signaling in chondrocytes and osteoblasts but also enhances WNT signaling and RUNX2 (runt related transcription factor 2) expression in chondrocytes and inhibits WNT signaling in osteoblasts, it is possible that these signaling pathways interact in the regulation of craniofacial development together with thyroid hormones (see for review Leitch et al. 2020).

Leptin, a pleiotropic adipocyte derived hormone, is also known to increase with cold exposure and plays an important role in cold acclimatization and thermogenesis (Korhonen et al. 2008; Zhao 2011; Robbins et al. 2018). For example, studies have identified leptin target neurons that are involved in the sympathetic control of brown adipose tissue (e.g., Cannon and Nedergaard 2004). Brown adipose tissue is prevalent in newborns and hibernating mammals but is also metabolically active in human adults for nonshivering thermogenesis (Shum et al. 1991; Nedergaard et al. 2007; Saito et al. 2009; Devlin 2015; Oreskovich et al. 2019). Leptin is also involved in the regulation of bone growth (Kishida et al. 2005) and can regulate angiogenesis (new vessel formation) (Rezai-Zadeh and Münzberg 2013), which would permanently modify the blood flow and directly affect intramembranous ossification (Percival and Richtsmeier 2013).

7.4 IMPLICATIONS FOR HUMAN CRANIOFACIAL EVOLUTION

To improve our understanding of climate-related patterns of craniofacial morphological variation in human evolution and the developmental processes underlying this variation, several paths could be explored.

First, to refine our understanding of the phenotypic expression of climate-related craniofacial variation, additional quantitative studies are needed to evaluate morphological differences between populations living in regions with recorded differences in temperature and/or humidity. Medical images of modern individuals living in different climates would enable measures of bone and nasal mucosa morphology that are crucial to obtain valid estimates used in the evaluation of the covariation between these two anatomical structures (Heuzé 2019). Studying the covariation between nasal cavity and nasal airway will help with the interpretation of the results obtained on dry skulls in terms of respiratory energetics and air conditioning.

Second, studies of the influence of ambient temperature on growth and development of rodents (e.g., Rae et al. 2006; Serrat 2014) have shown that experimental studies of animals could be used to better understand the effects of temperature on the morphology of nasal structures. The use of laboratory animals can help sort genetic causes from other variables, like temperature, that potentially affect morphology. Experiments on inbred mice could enable quantitative comparison of the volume and morphology of upper airway structures between groups of animals exposed to either cold or warm environments. These studies could help determine the extent to which morphological variation can be explained by a genetically driven adaptation and/or by a physiological response to environmental factors, that is, by phenotypic plasticity. Animal models would also allow exploration of the cellular processes involved in the morphological response to temperature. An in vitro study on hyperthermia effects on the proliferation of bone and cartilage cells (Flour et al. 1992) showed that chondrocytes might be thermoresistant and osteogenic cells thermosensitive. To our knowledge, such results still need to be tested in vivo. Future studies based on animal models could investigate the effects of temperature (both hypo- and hyperthermia) on the differentiation of osteochondroprogenic cells into osteoblasts and chondrocytes (Hall 2015) and on the proliferation of bone and cartilage cells. Laboratory mice could serve as a useful model system for this purpose, which would then help us discuss the effects of temperature on nasal morphology for other mammals such as primates.

Finally, the use of computational fluid dynamics offers great opportunities to achieve a better understanding of respiratory energetics. The complex structures of the nasal airway tend to restrict in vivo studies of nasal airflow, but computational fluid dynamics enables a valid and accurate numerical simulation of airflow patterns within the nasal cavity (e.g., Inthavong et al. 2007; Chen et al. 2010; Keck and Lindemann 2010; de Gabory et al. 2020). Thoughtful use of this technique applied on modern samples of healthy or pathological individuals could contribute greatly to the discussion about how thermoregulation, respiratory energetics, and climate interact to produce differential phenotypes in humans.

Integrating these three approaches would allow a more precise definition of the anatomical and functional units of the craniofacial skeleton showing climate-related patterns of morphological variation, which would in turn expand our knowledge of developmental processes that are sensitive to temperature, providing potential explanations at the cellular, organ, and organismal levels of this observed morphological variation.

7.5 CONCLUSION

Previous research has shown that human morphological variation can correspond with differences in climate. We provided a review of the temperature-related morphological variation of nasal and paranasal structures and a discussion of genetic, cellular, and systemic temperature-sensitive pathways that might have a role in the production of morphological variation of the nasal cavity. We observe that temperature both directly and indirectly affect bone formation, either by altering the activity of preosteoblast cells, bone-forming osteoblasts, and bone-resorbing osteoclasts or by affecting proteins or hormones (e.g., heat shock proteins, clock genes, thyroid hormones, leptin) involved in the activity of bone cells. Beyond providing a review on ecogeographic patterns of morphological variation in upper airway and cellular processes that potentially influence this morphology, our purpose in this chapter is to highlight the need for studies integrating these two areas of research. Such research could ultimately improve our understanding of the role of climate in the evolution of craniofacial morphology.

7.6 ACKNOWLEDGMENTS

We thank Joan T. Richtsmeier and M. Kathleen Pitirri for inviting us to participate in this book and for their helpful suggestions. We are also grateful for the insightful reviews and constructive comments of Scott Maddux and one anonymous reviewer, which significantly improved the overall quality of this manuscript.

7.7 REFERENCES

Adhikari, K., M. Fuentes-Guajardo, M. Quinto-Sánchez, M. Mendoza Revilla, et al. 2016. A genome-wide association scan implicates DCHS2, RUNX2, GLI3, PAX1 and EDAR in human facial variation. *Nat Commun* 7:1–11.

Albert, A. M., K. Ricanek, and P. Patterson. 2007. A review of the literature on the aging adult skull and face: implications for forensic science research and applications. *Foren Sci Int* 172:1–9.

Al Dayeh, A. A., K. L. Rafferty, M. Egbert, and S. W. Herring. 2013. Realtime monitoring of the growth of the nasal septal cartilage and the nasofrontal suture. *Am J Orthodont Dentofacial Orthoped* 143:773–83.

Alexander, C. M., I. Kasza, C.-L. E. Yen, S. B. Reeder, et al. 2015. Dermal white adipose tissue: a new component of the thermogenic response. *J Lipid Res* 56:2061–9.

Alhajeri, B. H., and S. J. Steppan. 2015. Association between climate and body size in rodents: a phylogenetic test of Bergmann's rule. *Mamm Biol* 81:219–25.

Allen, J. A. 1877. The influence of physical conditions in the genesis of species. *Radical Review* 1:108–40.

Ashton, K. G., M. C. Tracy, and A. de Queiroz. 2000. Is Bergmann's rule valid for mammals? *Am Nat* 156:390–415.

Aubin, J. E., and E. Bonnelye. 2000. Osteoprotegerin and its ligand: a new paradigm for regulation of osteoclastogenesis and bone resorption. *Osteoporos Int* 11:905–13.

Barna, J., A. Princz, M. Kosztelnik, B. Hargitai, et al. 2012. Heat shock factor-1 intertwines insulin/IGF-1, TGF-β and cGMP signaling to control development and aging. *BMC Dev Biol* 12:1–12.

Bassett, J. H. D., and G. R. Williams. 2016. Role of thyroid hormones in skeletal development and bone maintenance. *Endocrine Rev* 37:135–87.

Bastir, M. 2019. Big choanae, larger face: scaling patterns between cranial airways in modern humans and African apes and their significance in Middle and Late Pleistocene hominin facial evolution. *Bull Mém Soc d'Anthropol Paris* 31:5–13.

Bastir, M., P. Godoy, and A. Rosas. 2011. Common features of sexual dimorphism in the cranial airways of different human populations. *Am J Phys Anthropol* 146:414–22.

Bastir, M., I. Megía, N. Torres-Tamayo, D. García-Martínez, et al. 2020. Three-dimensional analysis of sexual dimorphism in the soft tissue morphology of the upper airways in a human population. *Am J Phys Anthropol* 171(1):65–75.

Bastir, M., and A. Rosas. 2013. Cranial airways and the integration between the inner and outer facial skeleton in humans: facial modularity and integration. *Am J Phys Anthropol* 152:287–93.

Bastir, M., and A. Rosas. 2016. Cranial base topology and basic trends in the facial evolution of *Homo*. *J Hum Evol* 91:26–35.

Bergmann, C. 1847. Über die Verhältnisse der Wärmeökonomie der Thiere zu ihrer Größe. *Göttinger Studien* 3:595–708.

Blackburn, T. M., and B. A. Hawkins. 2004. Bergmann's rule and the mammal fauna of northern North America. *Ecography* 277:715–24.

Bourke, J. M., W. M. Porter, R. C. Ridgely, T. R. Lyson, et al. 2014. Breathing life into dinosaurs: tackling challenges of soft-tissue restoration and nasal airflow in extinct species. *Anat Rec (Hoboken)* 297(11):2148–86.

Brookes, M., and W. J. Revell. 1998. *Blood Supply of Bone: Scientific Aspects*. New York: Springer.

Brown, J. S., B. P. Kotler, and W. P. Porter. 2017. How foraging allometries and resource dynamics could explain Bergmann's rule and the body-size diet relationship in mammals. *Oikos* 126:224–30.

Buck, L. T., C. B. Stringer, A. M. MacLarnon, and T. C. Rae. 2019. Variation in paranasal pneumatisation between Mid-Late Pleistocene hominins. *Bull Mém Soc d'Anthropol Paris* 31:14–33.

Burgos, M. A., E. Sanmiguel-Rojas, C. del Pino, M. A. Sevilla-García, et al. 2017. New CFD tools to evaluate nasal airflow. *Eur Arch Oto-Rhino-Laryngol* 274:3121–8.

Butaric, L. N. 2015. Differential scaling patterns in maxillary sinus volume and nasal cavity breadth among modern humans. *Anat Rec* 298:1710–21.

Butaric, L. N., and R. P. Klocke. 2018. Nasal variation in relation to high-altitude adaptations among Tibetans and Andeans. *Am J Hum Biol* 30:e23104.

Butaric, L. N., and S. D. Maddux. 2016. Morphological covariation between the maxillary sinus and midfacial skeleton among sub-Saharan and circumpolar modern humans. *Am J Phys Anthropol* 160(3):483–97.

Butaric, L. N., R. C. McCarthy, and D. C. Broadfield. 2010. A preliminary 3D computed tomography study of the human maxillary sinus and nasal cavity. *Am J Phys Anthropol* 143:426–36.

Cannon, B., and J. Nedergaard. 2004. Brown adipose tissue: function and physiological significance. *Physiol Rev* 84:277–359.

Caputa, M. 2004. Selective brain cooling: a multiple regulatory mechanism. *J Thermal Biol* 29:691–702.

Carey, J. W., and A. T. Steegmann. 1981. Human nasal protrusion, latitude, and climate. *Am J Phys Anthropol* 56:313–19.

Cauna, N. 1982. Blood and nerve supply of the nasal lining. In *The Nose: Upper Airway Physiology and the Atmospheric Environment*, ed. D. F. Proctor and I. Andersen, 45–69. New York: Elsevier Biomedical Press.

Chae, Y. S., G. Aguilar, E. J. Lavernia, and B. J. F. Wong. 2003. Characterization of temperature dependent mechanical behavior of cartilage. *Lasers Surg Med* 32:271–8.

Charles, C. M. 1930. The cavum nasi of the American Negro. *Am J Phys Anthropol* 14:177–253.

Chen, X. B., H. P. Lee, V. F. H. Chong, and D. Y. Wang. 2010. Numerical simulation of the effects of inferior turbinate surgery on nasal airway heating capacity. *Am J Rhinol Allergy* 24:118–22.

Chung, E., and M. N. Rylander. 2012. Response of preosteoblasts to thermal stress conditioning and osteoinductive growth factors. *Cell Stress Chaperones* 17:203–14.

Churchill, S. E., L. L. Shackelford, J. N. Georgi, and M. Y. Black. 2004. Morphological variation and airflow dynamics in the human nose. *Am J Hum Biol* 16:625–38.

Claes, P., J. Roosenboom, J. D. White, T. Swigut, et al. 2018. Genome-wide mapping of global-to-local genetic effects on human facial shape. *Nat Genet* 50:414–23.

Clauss, M., M. T. Dittmann, D. W. H. Müller, C. Meloro, C., et al. 2013. Bergmann's rule in mammals: a cross-species interspecific pattern. *Oikos* 122:1465–72.

Cohen, M. M., and R. E. MacLean. 2000. *Craniosynostosis: Diagnosis, Evaluation, and Management*, 2nd ed. New York: Oxford University Press.

Cole, P. 1982a. Modification of inspired air. In *The Nose: Upper Airway physiology and the Atmospheric Environment*, ed. D. F. Proctor and I. Andersen, 351–75. New York: Elsevier Biomedical Press.

Cole, P. 1982b. Upper respiratory airflow. In *The Nose: Upper Airway Physiology and the Atmospheric Environment*, ed. D. F. Proctor and I. Andersen, 163–90. New York: Elsevier Biomedical Press.

Collins, J. C., T. C. Pilkington, and K. Schmidt-Nielsen. 1971. A model of respiratory heat transfer in a small mammal. *Biophys J* 11:886–914.

Coon, C. S. 1962 *The Origin of Races*. New York: Knopf.

Cooper, L. F., and K. Uoshima. 1994. Differential estrogenic regulation of small M(r) heat shock protein expression in osteoblasts. *J Biol Chem* 269:7869–73.

Cottle, M. H. 1955. The structure and function of the nasal vestibule. *A.M.A. Arch Otolaryngol* 62:173–81.

Croen, K. D. 1993. Evidence for antiviral effect of nitric oxide. Inhibition of herpes simplex virus type 1 replication. *J Clin Invest* 91:2446–52.

Crognier, E. 1981a. Climate and anthropometric variations in Europe and the Mediterranean area. *Ann Hum Biol* 8:99–107.

Crognier, E. 1981b. The influence of climate on the physical diversity of European and Mediterranean populations. *J Hum Evol* 10:611–14.

Dai, J., and A. B. Rabie. 2007. VEGF: an essential mediator of both angiogenesis and endochondral ossification. *J Dent Res* 86:937–50.

Davies, A. 1932. A re-survey of the morphology of the nose in relation to climate. *J Roy Anthropol Instit G Brit Ireland* 62:337.

de Azevedo, S., M. F. González, C. Cintas, V. Ramallo, et al. 2017. Nasal airflow simulations suggest convergent adaptation in Neanderthals and modern humans. *PNAS* 114:12442–7.

de Gabory, L., M. Kérimian, Y. Baux, N. Boisson, et al. 2020. Computational fluid dynamics simulation to compare large volume irrigation and continuous spraying during nasal irrigation. *International Forum of Allergy & Rhinology* 10:41–8.

Devlin, M. J. 2015. The "skinny" on brown fat, obesity and bone. *Am J Phys Anthropol* 156:98–115.

Doorly, D. J., D. J. Taylor, A. M. Gambaruto, R. C. Schroter, et al. 2008. Nasal architecture: form and flow. *Phil Trans Roy Soc A* 366:3225–46.

Eccles, R. 1982. Neurological and pharmacological considerations. In *The Nose: Upper Airway Physiology and the Atmospheric Environment*, ed. D. F. Proctor and I. Andersen, 191–214. Amsterdam: Elsevier.

Eccles, R. 1996. A role for the nasal cycle in respiratory defence. *Eur Respirat J* 9:371–6.

Eccles, R. 2002. An explanation for the seasonality of acute upper respiratory tract viral infections. *Acta Oto-Laryngolog* 122:183–91.

Egeli, E., L. Demirci, B. Yazýcý, and U. Harputluoglu. 2004. Evaluation of the inferior turbinate in patients with deviated nasal septum by using computed tomography. *The Laryngoscope* 114:113–17.

Elad, D., M. Wolf, and T. Keck. 2008. Air-conditioning in the human nasal cavity. *Respirat Physiol Neurobiol* 163:121–7.

Enerbäck, S., A. Jacobsson, E. M. Simpson, C. Guerra, et al. 1997. Mice lacking mitochondrial uncoupling protein are cold-sensitive but not obese. *Nature* 387:90–4.

Enlow, D. H. 1990. *Facial Growth*, 3rd ed. Philadelphia: Saunders WB Co.

Evteev, A. A., A. L. Cardini, I. Morozova, and P. O'Higgins. 2014. Extreme climate, rather than population history, explains mid-facial morphology of Northern Asians: mid-facial cold adaptation in Northern Asians. *Am J Phys Anthropol* 153:449–62.

Evteev, A. A., and A. N. Grosheva. 2019. Nasal cavity and maxillary sinuses form variation among modern humans of Asian descent. *Am J Phys Anthropol* 169:513–25.

Evteev, A. A., and Y. Heuzé. 2018. Impact of sampling strategies and reconstruction protocols in nasal airflow simulations in fossil hominins. *PNAS* 115:E4737–8.

Feldhamer, G. A., L. C. Drickamer, S. H. Vessey, J. F. Merritt, et al. 2007. *Mammalogy: Adaptation, Diversity, Ecology*. Baltimore: The Johns Hopkins University Press.

Finlayson, C., F. Giles Pacheco, J. Rodríguez-Vidal, D. A. Fa, et al. 2006. Late survival of Neanderthals at the southernmost extreme of Europe. *Nature* 443:850–3.

Fischer, A. W., R. I. Csikasz, R. I., G. von Essen, B. Cannon, et al. 2016. No insulating effect of obesity. *Am J Physiol-Endocrinol and Metabol* 311:E202–13.

Flour, M.-P., X. Ronot, F. Vincent, B. Benoit, et al. 1992. Differential temperature sensitivity of cultured cells from cartilaginous or bone origin. *Biol Cell* 75:83–7.

Foster, F., and M. Collard. 2013. A reassessment of Bergmann's rule in modern humans. *PLoS One* 8:e72269.

Franciscus, R. G. 1995. *Later Pleistocene Nasofacial Variation in Western Eurasia and Africa and Modern Human Origins*. Ph.D. thesis. Albuquerque: University of New Mexico.

Franciscus, R. G., and S. E. Churchill. 2002. The costal skeleton of Shanidar 3 and a reappraisal of Neandertal thoracic morphology. *J Hum Evol* 42:303–56.

Franciscus, R. G., and J. C. Long. 1991. Variation in human nasal height and breadth. *Am J Phys Anthropol* 85:419–27.

Franciscus, R. G., and E. Trinkaus. 1988. Nasal morphology and the emergence of *Homo erectus*. *Am J Phys Anthropol* 75:517–27.

Froehle, A. W. 2008. Climate variables as predictors of basal metabolic rate: new equations. *Am J Hum Biol* 20:510–29.

Froehle, A. W., T. R. Yokley, and S. E. Churchill. 2013. Energetics and the origin of modern humans. In *The Origins of Modern Humans*, ed. F. H. Smith and J. C. M. Ahern, 285–320. Hoboken: John Wiley & Sons, Inc.

Fukase, H., T. Ito, and H. Ishida. 2016. Geographic variation in nasal cavity form among three human groups from the Japanese archipelago: ecogeographic and functional implications. *Am J Hum Biol* 28:343–51.

Gannon, P. J., W. J. Doyle, E. Ganjian, S. Marquez, et al. 1997. Maxillary sinus mucosal blood flow during nasal vs tracheal respiration. *Arch Otolaryngol* 123:1336–40.

García-Martínez, D., N. Torres-Tamayo, I. Torres-Sánchez, F. García-Río, F., et al. 2018. Ribcage measurements indicate greater lung capacity in Neanderthals and Lower Pleistocene hominins compared to modern humans. *Comm Biol* 1:117.

Garrett, S. C., and J. J. C. Rosenthal. 2012. A role for A-to-I RNA editing in temperature adaptation. *Physiol* 27:362–9.

Gohli, J., and K. L. Voje. 2016. An interspecific assessment of Bergmann's rule in 22 mammalian families. *BMC Evol Biol* 16:1–12.

Golozoubova, V., E. Hohtola, A. Matthias, A. Jacobsson, et al. 2001. Only UCP1 can mediate adaptive nonshivering thermogenesis in the cold. *FASEB J* 15:2048–50.

Gossan, N., L. Zeef, J. Hensman, A. Hughes, et al. 2013. The circadian clock in murine chondrocytes regulates genes controlling key aspects of cartilage homeostasis. *Arthritis Rheum* 65:2334–45.

Hall, B. K. 2015. *Bones and Cartilage: Developmental and Evolutionary Skeletal Biology*, 2nd ed. Amsterdam: Elsevier/Academic Press.

Hall, B. K., and D. S. Precious. 2013. Cleft lip, nose, and palate: the nasal septum as the pacemaker for midfacial growth. *Oral Surg Oral Med Oral Pathol Oral Radiol* 115:442–7.

Hall, B. K., and P. E. Witten. 2018. Plasticity and variation of skeletal cells and tissues and the evolutionary development of Actinopterygian fishes. In *Evolution and Development of Fishes*, ed. Z. Johanson, C. Underwood, and M. Richter, 1st ed., 126–43. Cambridge: Cambridge University Press.

Hall, R. L. 2005. Energetics of nose and mouth breathing, body size, body composition, and nose volume in young adult males and females. *Am J Hum Biol* 17:321–30.

Hang, K., C. Ye, E. Chen, W. Zhang, et al. 2018. Role of the heat shock protein family in bone metabolism. *Cell Stress Chaperones* 23:1153–64.

Hanna, L. M., and P. W. Scherer. 1986. A theoretical model of localized heat and water vapor transport in the human respiratory tract. *J Biomech Eng* 108:19–27.

Harrison, G. A, and E. J. Clegg. 1969. Environmental factors influencing mammalian growth. In *Physiology and Pathology of Adaptation Mechanisms*, ed. E. Bajusz, 74. Oxford: Pergamon Press.

Hartman, C., N. Holton, S. Miller, T. Yokley, et al. 2016. Nasal septal deviation and facial skeletal asymmetries: nasal septal deviation. *Anat Rec* 299:295–306.

Harvati, K., and T. D. Weaver. 2006a. Human cranial anatomy and the differential preservation of population history and climate signatures. *Anat Rec* 288A:1225–33.

Harvati, K., and T. D. Weaver. 2006b. Reliability of cranial morphology in reconstructing Neanderthal phylogeny. In *Neanderthals Revisited: New Approaches and Perspectives*, ed. J.-J. Hublin, K. Harvati, and T. Harrison, 239–54. Dordrecht: Springer Netherlands.

Hasegawa, M., and E. B. Kern. 1977. The human nasal cycle. *Mayo Clin Proc* 52:28–34.

Hatakeyama, D., O. Kozawa, M. Niwa, H. Matsuno, et al. 2002. Upregulation by retinoic acid of transforming growth factor-β-stimulated heat shock protein 27 induction in osteoblasts: involvement of mitogen-activated protein kinases. *Biochim Biophys Acta* 1589:15–30.

Havenith, G. 2005. Temperature regulation, heat balance and climatic stress. In *Extreme Weather Events and Public Health Responses*, ed. W. Kirch, R. Bertollini, and B. Menne, 69–80. Berlin/Heidelberg: Springer-Verlag.

Heuzé, Y. 2019. What does nasal cavity size tell us about functional nasal airways? *Bull Mém Soc d'Anthropol Paris* 31:69–76.

Hiernaux, J., and A. Froment. 1976. The correlations between anthropobiological and climatic variables in sub-Saharan Africa: revised estimates. *Hum Biol* 48:757–67.

Hildebrandt, T., W. Heppt, U. Kertzscher, and L. Goubergrits. 2013. The concept of rhinorespiratory homeostasis—a new approach to nasal breathing. *Facial Plastic Surgery* 29:085–92.

Holden, W. E., J. P. Wilkins, M. Harris, H. A. Milczuk, et al. 1999. Temperature conditioning of nasal air: effects of vasoactive agents and involvement of nitric oxide. *J Appl Physiol* 87:1260–5.

Holliday, T. W. 1997. Body proportions in late Pleistocene Europe and modern human origins. *J Hum Evol* 32:423–48.

Holliday, T. W., and C. E. Hilton. 2010. Body proportions of circumpolar peoples as evidenced from skeletal data: Ipiutak and Tigara (Point Hope) versus Kodiak Island Inuit. *Am J Phys Anthropol* 142:287–302.

Holton, N. E., A. Alsamawi, T. R. Yokley, and A. W. Froehle. 2016. The ontogeny of nasal shape: an analysis of sexual dimorphism in a longitudinal sample. *Am J Phys Anthropol* 160:52–61.

Holton, N. E., A. Piche, and T. Yokley. 2018. Integration of the nasal complex: implications for developmental and evolutionary change in modern humans. *Am J Phys Anthropol* 166:791–802.

Holton, N. E., T. R. Yokley, and L. Butaric. 2013. The morphological interaction between the nasal cavity and maxillary sinuses in living humans. *Anat Rec* 296(3):414–26.

Holton, N. E., T. R. Yokley, and R. G. Franciscus. 2011. Climatic adaptation and Neandertal facial evolution: a comment on Rae et al. *J Hum Evol* 61:624–7.

Holton, N. E., T. R. Yokley, A. W. Froehle, and T. E. Southard. 2014. Ontogenetic scaling of the human nose in a longitudinal sample: implications for genus *Homo* facial evolution. *Am J Phys Anthropol* 153:52–60.

Hoyme, L. E. 1965. The nasal index and climate—a spurious case of natural selection in man. *Am J Phys Anthropol* 23:336–7.

Hubbe, M., T. Hanihara, and K. Harvati. 2009. Climate signatures in the morphological differentiation of worldwide modern human populations. *Anat Rec* 292:1720–33.

Ingelstedt, S. 1956. Studies on the conditioning of air in the respiratory tract. *Acta Otolaryngol Suppl* 131:1–80.

Inthavong, K., A. Chetty, Y. Shang, and J. Tu. 2018. Examining mesh independence for flow dynamics in the human nasal cavity. *Comput Biol Med* 102:40–50.

Inthavong, K., Z. F. Tian, and J. Y. Tu. 2007. CFD simulations on the heating capability in a human nasal cavity. In *Proceedings of the 16th Australasian Fluid Mechanical Conference (AFMC)*. Gold Coast, Australia.

Irmak, M. K., A. Korkmaz, and O. Erogul. 2004. Selective brain cooling seems to be a mechanism leading to human craniofacial diversity observed in different geographical regions. *Med Hypotheses* 63:974–9.

Iwen, K. A., R. Oelkrug, and G. Brabant. 2018. Effects of thyroid hormones on thermogenesis and energy partitioning. *J Mol Endocrinol* 60:R157–70.

Jain, B., I. Rubinstein, R. A. Robbins, K. L. Leise, et al. 1993. Modulation of airway epithelial cell beat frequency by nitric oxide. *Biochem Biophys Res Comm* 191:83–8.

James, F. C. 1970. Geographic size variation in birds and its relationship to climate. *Ecology* 51:365–90.

Jankowski, R. 2013. *The Evo-Devo Origin of the Nose, Anterior Skull Base and Midface.* Paris: Springer Paris.

Jaskulska, E. 2014. *Adaptation to Cold Climate in the Nasal Cavity Skeleton. A Comparison of Archaeological Crania from Different Climatic Zones.* PhD dissertation. University of Warsaw.

Joganic, J. L., and Y. Heuzé. 2019. Allometry and advancing age significantly structure craniofacial variation in adult female baboons. *J Anat* 235:217–32.

Kasza, I., D. Hernando, A. Roldán-Alzate, C. M. Alexander, C. M., et al. 2016. Thermogenic profiling using magnetic resonance imaging of dermal and other adipose tissues. *JCI Insight* 1:e87146.

Kasza, I., Y. Suh, D. Wollny, R. J. Clark, et al. 2014. Syndecan-1 is required to maintain intradermal fat and prevent cold stress. *PLoS Genet* 10:e1004514.

Katzmarzyk, P. T., and W. R. Leonard. 1998. Climatic influences on human body size and proportions: ecological adaptations and secular trends. *Am J Phys Anthropol* 106:483–503.

Keck, T., R. Leiacker, H. Riechelmann, and G. Rettinger. 2000. Temperature profile in the nasal cavity. *Laryngoscope* 110:651–4.

Keck, T., and J. Lindemann. 2010. Numerical simulation and nasal air-conditioning. *GMS Curr Topics Otorhinolaryngol—Head Neck Surg* 9.

Keir, J. 2009. Why do we have paranasal sinuses? *J Laryngol Otol* 123:4–8.

Keyhani, K., P. W. Scherer, and M. M. Mozell. 1995. Numerical simulation of airflow in the human nasal cavity. *J Biomech Engin* 117:429–41.

Kiliaridis, S., C. Engström, and B. Thilander. 1985. The relationship between masticatory function and craniofacial morphology: I. A cephalometric longitudinal analysis in the growing rat fed a soft diet. *Eur J Orthodon* 7:273–83.

Kim, D.-W., S. K. Chung, and Y. Na. 2017. Numerical study on the air conditioning characteristics of the human nasal cavity. *Comput Biol Med* 86:18–30.

Kishida, Y., M. Hirao, N., Tamai, A. Nampel, et al. 2005 Leptin regulates chondrocyte differentiation and matrix maturation during endochondral ossification. *Bone* 37:607–21.

Korhonen, T., A. M. Mustonen, P. Nieminen, and S. Saarela. 2008. Effects of cold exposure, exogenous melatonin and short-day treatment on the weight regulation and body temperature of the Siberian hamster (*Phodopus sungorus*). *Regul Pept* 149:60–6.

Kozawa, O., and H. Tokuda. 2002. Heat shock protein 27 in osteoblasts. *Nihon Yakurigaku Zasshi* 119:89–94.

Leitch, V. D., J. H. D. Bassett, and G. R. Williams. 2020. Role of thyroid hormones in craniofacial development. *Nat Rev Endocrinol* 16:147–64.

Leong, S. C., and R. Eccles. 2009. A systematic review of the nasal index and the significance of the shape and size of the nose in rhinology. *Clin Otolaryngol* 34:191–8.

Lesciotto, K. M., and J. T. Richtsmeier. 2019. Craniofacial skeletal response to encephalization: how do we know what we think we know? *Am J Phys Anthropol* 168:27–46.

Li, C., A. A. Farag, J. Leach, B. Deshpande, et al. 2017. Computational fluid dynamics and trigeminal sensory examinations of empty nose syndrome patients: computational and trigeminal studies of ENS. *The Laryngoscope* 127:E176–84.

Lieberman, D. E. 2011. *The Evolution of the Human Head.* Cambridge, MA: Belknap Press of Harvard University Press.

Lindemann, J., E. Tsakiropoulou, T. Keck, R. Leiacker, R., et al. 2009. Nasal air conditioning in relation to acoustic rhinometry values. *Am J Rhinol Allergy* 23:575–7.

Lovejoy, C. O., M. A. McCollum, P. L. Reno, and B. A. Rosenman. 2003. Developmental biology and human evolution. *Ann Rev Anthropol* 32:85–109.

Lowell, B. B., and B. M. Spiegelman. 2000. Towards a molecular understanding of adaptive thermogenesis. *Nature* 404:652–60.

Lundberg, J. O. 2008. Nitric oxide and the paranasal sinuses. *Anat Rec* 291:1479–84.

Lundberg, J. O., T. Farkas-Szallasi, E. Weitzberg, J. Rinder, et al. 1995. High nitric oxide production in human paranasal sinuses. *Nat Med* 1:370–3.

Lundberg, J. O., E. Weitzberg, J. Rinder, A. Rudehill, A., et al. 1996. Calcium-independent and steroid-resistant nitric oxide synthase activity in human paranasal sinus mucosa. *Eur Respir J* 9:1344–7.

Ma, J., J. Dong, Y. Shang, K. Inthavong, et al. 2018. Air conditioning analysis among human nasal passages with anterior anatomical variations. *Med Eng Phy* 57:19–28.

Maddux, S. D., and L. N. Butaric. 2017. Zygomaticomaxillary morphology and maxillary sinus form and function: how spatial constraints influence pneumatization patterns among modern humans. *Anat Rec (Hoboken)* 300(1):209–25.
Maddux, S. D., L. N. Butaric, T. R. Yokley, and R. G. Franciscus. 2016b. Ecogeographic variation across morphofunctional units of the human nose. *Am J Phys Anthropol* 162:103–19.
Maddux, S. D., T. R. Yokley, B. M. Svoma, and R. G. Franciscus. 2016a. Absolute humidity and the human nose: a reanalysis of climate zones and their influence on nasal form and function. *Am J Phys Anthropol* 161:309–20.
Maier, W., and I. Ruf. 2014. Morphology of the nasal capsule of Primates with special reference to *Daubentonia* and *Homo*. *Anat Rec* 297:1985–2006.
Mancinelli, R. L., and C. P. McKay. 1983. Effects of nitric oxide and nitrogen dioxide on bacterial growth. *Appl Environ Microbiol* 46:198–202.
Marks, R. 2018. Circadian clock: potential role in cartilage integrity and disruption. *Int J Orthopaed* 5:936–42.
Marks, T. N., S. D. Maddux, L. N. Butaric, and R. G. Franciscus. 2019. Climatic adaptation in human inferior nasal turbinate morphology: evidence from arctic and equatorial populations. *Am J Phys Anthropol* 169:498–512.
Márquez, S., and J. T. Laitman. 2008. Climatic effects on the nasal complex: a CT imaging, comparative anatomical, and morphometric investigation of *Macaca mulatta* and *Macaca fascicularis*. *Anat Rec* 291:1420–45.
Márquez, S., A. S. Pagano, E. Delson, W. Lawson, et al. 2014. The nasal complex of Neanderthals: an entry portal to their place in human ancestry. *Anat Rec* 297:2121–37.
Martin, J. M., A. B. Leece, S. Neubauer, S. E. Baker, et al. 2021. Drimolen cranium DNH 155 documents microevolution in an early hominin species. *Nat Ecol Evol* 5:38–45.
Massaro, E. J., and J. M. Rogers. 2004. *The Skeleton: Biochemical, Genetic, and Molecular Interactions in Development and Homeostasis*. Totowa, NJ: Humana Press.
Masuda, S. 1992. Role of the maxillary sinus as a resonant cavity. *Nippon Jibiinkoka Gakkai Kaiho* 95:71–80.
Mayr, E. 1963. *Animal Species and Evolution*. Cambridge, MA: Harvard University Press.
McNab, B. K. 1971. On the ecological significance of Bergmann's rule. *Ecology* 52:845–54.
McNab, B. K. 2010. Geographical and temporal correlations of mammalian size reconsidered: a resource rule. *Oecologia* 164:13–23.
McNab, B. K. 2012. *Extreme Measures: The Ecological Energetics of Birds and Mammals*. Chicago: University of Chicago Press.
Meiri, S., and T. Dayan. 2003. On the validity of Bergmann's rule. *J Biogeo* 30:331–51.
Meiri, S., T. Dayan, and D. Simberloff. 2004. Carnivores, biases and Bergmann's rule. *Biol J Linn Soc* 81:579–88.
Menéndez, L., V. Bernal, P. Novellino, and S. I. Perez. 2014. Effect of bite force and diet composition on craniofacial diversification of Southern South American human populations: diet and cranial variation in South America. *Am J Phys Anthropol* 155:114–27.
Mlynski, G., S. Grützenmacher, S. Plontke, and B. Mlynski. 2001. Correlation of nasal morphology and respiratory function. *Rhinology* 39:197–201.
Moore, W. J. 1981. *The Mammalian Skull*. Cambridge: Cambridge University Press.
Moss, M. L., and R. W. Young. 1960. A functional approach to craniology. *Am J Phys Anthropol* 18:281–92.
Na, Y., K. S. Chung, S.-K. Chung, and S. K. Kim. 2012. Effects of single-sided inferior turbinectomy on nasal function and airflow characteristics. *Respir Physiol Neurobiol* 180:289–97.
Naclerio, R. M., J. Pinto, P. Assanasen, and F. M. Baroody. 2007. Observations on the ability of the nose to warm and humidify inspired air. *Rhinology* 45:102–11.

Naftali, S., M. Rosenfeld, M. Wolf, and D. Elad. 2005. The air-conditioning capacity of the human nose. *Ann Biomed Engin* 33:545–53.
Nedergaard, J., T. Bengtsson, and B. Cannon. 2007. Unexpected evidence for active brown adipose tissue in adult humans. *Am J Physiol-Endocrinol Metabol* 293:E444–52.
Negus, V. E. 1952. Humidification of the air passages. *Thorax* 7:148–51.
Negus, V. E. 1960. Further observations on the air conditioning mechanism of the nose. *Ann R Coll Surg Engl* 27:171–204.
Noback, M. L., K. Harvati, and F. Spoor. 2011. Climate-related variation of the human nasal cavity. *Am J Phys Anthropol* 145:599–614.
Noback, M. L., E. Samo, C. van Leeuwen, N. Lynnerup, N., et al. 2016. Paranasal sinuses: a problematic proxy for climate adaptation in Neanderthals. *J Hum Evol* 97:176–9.
Nunes, G. T., P. L. Mancini, and L. Bugoni. 2017. When Bergmann's rule fails: evidences of environmental selection pressures shaping phenotypic diversification in a widespread seabird. *Ecography* 40:365–75.
Ochocinska, D., and J. R. E. Taylor. 2003. Bergmann's rule in shrews: geographical variation of body size in Palearctic Sorex species. *Biol J Linn Soc* 78:365–81.
O'Malley, J. F. 1923. Evolution of the nasal cavities and sinuses in relation to function. *Proc Roy Soc Med* 16:83–4.
Opperman, L. A., P. T. Gakunga, and D. S. Carlson. 2005. Genetic factors influencing morphogenesis and growth of sutures and synchondroses in the craniofacial complex. *Semin Orthod* 11:199–208.
Oreskovich, S. M., F. J. Ong, B. A. Ahmed, N. B. Konyer, N. B., et al. 2019. MRI reveals human brown adipose tissue is rapidly activated in response to cold. *J Endocrine Soc* 3:2374–84.
Paschetta, C., de Azevedo, S., Castillo, L., Martínez-Abadías, N., et al. 2010. The influence of masticatory loading on craniofacial morphology: a test case across technological transitions in the Ohio Valley: masticatory stress and technological transitions. *Am J Phys Anthropol* 141:297–314.
Patel, J. J., J. C. Utting, M. L. Key, I. R. Orriss, et al. 2012. Hypothermia inhibits osteoblast differentiation and bone formation but stimulates osteoclastogenesis. *Exp Cell Res* 318:2237–44.
Patil, S., and S. Paul. 2014. A comprehensive review on the role of various materials in the osteogenic differentiation of mesenchymal stem cells with a special focus on the association of heat shock proteins and nanoparticles. *Cells Tissues Organs* 199:81–102.
Pendolino, A. L., V. J. Lund, E. Nardello, and G. Ottaviano. 2018. The nasal cycle: a comprehensive review. *Rhinol Online* 1:67–76.
Percival, C. J., and J. T. Richtsmeier. 2013. Angiogenesis and intramembranous osteogenesis, *Dev Dyn* 242:909–22.
Pickrell, J. K., T. Berisa, J. Z. Liu, L. Ségurel, et al. 2016. Detection and interpretation of shared genetic influences on 42 human traits. *Nat Genet* 48:709–17.
Ponganis, P. J., R. P. Van Dam, D. H. Levenson, T. Knower, et al. 2003. Regional heterothermy and conservation of core temperature in emperor penguins diving under sea ice. *Compar Biochem Physiol A* 135:477–87.
Proctor, D. 1982. The upper airway. In *The Nose: Upper Airway Physiology and the Atmospheric Environment*, ed. D. F. Proctor and I. Andersen, 23–43. New York: Elsevier Biomedical Press.
Proetz, A. W. 1951. Air currents in the upper respiratory tract and their clinical importance. *Ann Otol Rhinol Laryngol* 60:439–67.
Proetz, A. W. 1953. *Applied Physiology of the Nose*. St. Louis: Annals Publishing Co.

Rae, T. C., R. A. Hill, Y. Hamada, and T. Koppe. 2003. Clinical variation of maxillary sinus volume in Japanese Macaques (*Macaca fuscata*). *Am J Primatol* 59:153–8.
Rae, T. C., T. Koppe, and C. B. Stringer. 2011. The Neanderthal face is not cold adapted. *J Hum Evol* 60:234–9.
Rae, T. C., U. S. Viðarsdóttir, N. Jeffery, and A. T. Steegmann. 2006. Developmental response to cold stress in cranial morphology of *Rattus*: implications for the interpretation of climatic adaptation in fossil hominins. *Proc Roy Soc B* 273:2605–10.
Reppert, S. M., and D. R. Weaver. 2001. Molecular analysis of mammalian circadian rhythms. *Ann Rev Physiol* 63:647–76.
Rezai-Zadeh, K., and H. Münzberg. 2013. Integration of sensory information via central thermoregulatory leptin targets. *Physiol Behav* 121:49–55.
Richtsmeier, J. T., K. Aldridge, V. B. DeLeon, J. Panchal, J., et al. 2006. Phenotypic integration of neurocranium and brain. *J Exp Zool* 306B:360–78.
Richtsmeier, J. T., and K. M. Lesciotto. 2020. From phenotype to genotype and back again. *Bull Mém Soc d'Anthropol Paris* 32:8–17.
Robbins, A., C. Tom, M. N. Cosman, C. Moursi, C., et al. 2018. Low temperature decreases bone mass in mice: implications for humans. *Am J Phys Anthropol* 167:1–12.
Rodriguez, M. A., I. L. Lopez-Sanudo, and B. A. Hawkins. 2006. The geographic distribution of mammal body size in Europe. *Global Ecol Biogeog* 15:173–81.
Romano, G. H., Y. Harari, T. Yehuda, A. Podhorzer, et al. 2013. Environmental stresses disrupt telomere length homeostasis. *PLoS Genet* 9:e1003721.
Romanovsky, A. A. 2018. The thermoregulation system and how it works. *Handb Clin Neurol* 156:3–43.
Roseman, C. C. 2004. Detecting interregionally diversifying natural selection on modern human cranial form by using matched molecular and morphometric data. *PNAS* 101:12824–9.
Roseman, C. C., and T. D. Weaver. 2004. Multivariate apportionment of global human craniometric diversity. *Am J Phys Anthropol* 125:257–63.
Ruff, C. 1994. Morphological adaptation to climate in modern and fossil hominids. *Am J Phys Anthropol* 37:65–107.
Ruff, C. 2002. Variation in human body size and shape. *Ann Rev Anthropol* 31:211–32.
Rylander, M. N. 2005. Thermally induced injury and heat-shock protein expression in cells and tissues. *Ann New York Acad Sci* 1066:222–42.
Sahin-Yilmaz, A., and R. M. Naclerio. 2011. Anatomy and physiology of the upper airway. *Proc Am Thoracic Soc* 8:31–9.
Saito, M., Y. Okamatsu-Ogura, M. Matsushita, K. Watanabe, et al. 2009. High incidence of metabolically active brown adipose tissue in healthy adult humans: effects of cold exposure and adiposity. *Diabetes* 58:1526–31.
Sardi, M. L. 2018. Craniofacial morphology and adaptation. In *The International Encyclopedia of Biological Anthropology*, ed. W. Trevathan, M. Cartmill, D. Dufour, C. Larsen, et al., 1–2. Hoboken, NJ: John Wiley & Sons.
Sardi, M. L., G. G. Joosten, C. D. Pandiani, M. M. Gould, et al. 2018. Frontal sinus ontogeny and covariation with bone structures in a modern human population. *J Morphol* 279:871–82.
Sargis, E. J., V. Milien, N. Woodman, and L. E. Olson. 2018 Rule reversal: ecogeographical patterns of body size variation in the common treeshrew (Mammalia, Scandentia). *Ecol Evol* 8:1634–45.
Schmidt-Nielsen, K., F. R. Hainsworth, and D. E. Murrish. 1970. Counter-current heat exchange in the respiratory passages: effect on water and heat balance. *Respir Physiol* 9:263–76.

Scholander, P. F. 1955. Evolution of climatic adaptation in homeotherms. *Evolution* 9:15–26.

Schroter, R. C., and N. V. Watkins. 1989. Respiratory heat exchange in mammals. *Respir Physiol* 78:357–67.

Scott, N. A., A. Strauss, J.-J. Hublin, P. Gunz, et al. 2018. Covariation of the endocranium and splanchnocranium during great ape ontogeny. *PLoS One* 13:e0208999.

Serrat, M. A. 2014. Environmental temperature impact on bone and cartilage growth. In *Comprehensive Physiology*, ed. R. Terjung, 621–55. Hoboken, NJ: John Wiley & Sons.

Serrat, M. A., D. King, and C. O. Lovejoy. 2008. Temperature regulates limb length in homeotherms by directly modulating cartilage growth. *PNAS* 105:19348–53.

Serrat, M. A., T. J. Schlierf, M. L. Efaw, F. D. Shuler, et al. 2015. Unilateral heat accelerates bone elongation and lengthens extremities of growing mice. *J Orthopaed Res* 33:692–8.

Serrat, M. A., R. M. Williams, and C. E. Farnum. 2010. Exercise mitigates the stunting effect of cold temperature on limb elongation in mice by increasing solute delivery to the growth plate. *J Applied Physiol* 109:1869–79.

Shaffer, J. R., E. Orlova, M. K. Lee, E. J. Leslie, et al. 2016. Genome-wide association study reveals multiple loci influencing normal human facial morphology. *PLOS Genet* 12:e1006149.

Shea, B. T. 1977. Eskimo craniofacial morphology, cold stress and the maxillary sinus. *Am J Phys Anthropol* 47:289–300.

Shi, L.-L., J. Ma, Y.-K. Deng, C.-L. Chen, et al. 2020. Cold-inducible RNA-binding protein contributes to tissue remodeling in chronic rhinosinusitis with nasal polyps. *Allergy* doi.org/10.1111/all.14287

Shui, C., and A. Scutt. 2001. Mild heat shock induces proliferation, alkaline phosphatase activity, and mineralization in human bone marrow stromal cells and Mg-63 cells in vitro. *J Bone Min Res* 16:731–41.

Shum, A. Y., F. Y. Liao, C. F. Chen, and J. Y. Wang. 1991. The role of interscapular brown adipose tissue in cold acclimation in the rat. *Chin J Physiol* 34:427–37.

Silva, J. E. 2006. Thermogenic mechanisms and their hormonal regulation. *Physiol Rev* 86:435–64.

Smith, T., J. Rossie, and P. Doherty. 2006. Primate olfaction: anatomy and evolution. In *Olfaction and the Brain*, ed. W. J. Brewer, D. Castle, and C. Pantelis, 135–66. Cambridge: Cambridge University Press.

Sommer, F., R. Kroger, and J. Lindemann. 2012. Numerical simulation of humidification and heating during inspiration within an adult nose. *Rhinology* 50:157–64.

Steegmann, A. T., F. J. Cerny, and T. W. Holliday. 2002. Neandertal cold adaptation: physiological and energetic factors. *Am J Hum Biol* 14:566–83.

Steegmann, A. T., and W. S. Platner. 1968. Experimental cold modification of cranio-facial morphology. *Am J Phys Anthropol* 28:17–30.

Steindal, I. A., and D. Whitmore. 2019. Circadian clocks in fish—what have we learned so far? *Biology* 8:17.

St. Hoyme, L. E., and M. Y. Işcan. 1989. Determination of sex and race: accuracy and assumptions. In *Reconstruction of life from the skeleton*, ed. M. Y. Iscan and K. A. R. Kennedy, 53–93. New York: Alan R. Liss Inc.

Swift, D., and D. Proctor. 1977. Access of air to the respiratory tract. In *Respiratory defense mechanisms*, ed. J. Brain, D. Proctor, and L. Reid, 63–91. New York: Dekker M.

Takarada, T., A. Kodama, S. Hotta, M. Mieda, et al. 2012. Clock genes influence gene expression in growth plate and endochondral ossification in mice. *J Biol Chem* 287:36081–95.

Tansey, E. A., and C. D. Johnson. 2015. Recent advances in thermoregulation. *Adv Physiol Educ* 39:139–48.

Tattersall, G. J., B. J. Sinclair, P. C. Withers, P. A. Fields, et al. 2012. Coping with thermal challenges: physiological adaptations to environmental temperatures. *Compr Physiol* 2:2151–202.

Thomson, A. 1913. The correlations of isotherms with variations in the nasal index. *Int Congr Med Lond* 17:89–90.

Thomson, A., and L. H. Buxton. 1923. Man's nasal index in relation to certain climatic conditions. *J Roy Anthropol Instit Gr Brit Ireland* 53:92.

Tillier, A.-M. 1977. La pneumatisation du massif cranio-facial chez les hommes actuels et fossiles. *Bull Mém Soc d'anthropol Paris* 4:177–89.

Tokuda, H., M. Niwa, H. Ito, Y. Oiso, Y., et al. 2003. Involvement of stress-activated protein kinase/c-Jun N-terminal kinase in endothelin-1-induced heat shock protein 27 in osteoblasts. *Eur J Endocrinol* 239–45.

Tos, M. 1982. Goblet cells and glands in the nose and paranasal sinuses. In *The Nose: Upper Airway Physiology and the Atmospheric Environment*, ed. D. F. Proctor, and I. Andersen, 99–144. New York: Elsevier Biomedical Press.

Walker, J. E. C., and R. E. Wells. 1961. Heat and water exchange in the respiratory tract. *Am J Med* 30:259–67.

Wang, K., T. S. Denney, E. E. Morrison, and V. J. Vodyanoy. 2005. Numerical simulation of air flow in the human nasal cavity. *Conf Proc IEEE Eng Med Biol Soc*: 5607–10.

Warner, A., Rahman, A., Solsjo, P., Gottschling, K., et al. 2013. Inappropriate heat dissipation ignites brown fat thermogenesis in mice with a mutant thyroid hormone receptor 1. *PNAS* 110:16241–6.

Watelet, J. B., and P. Van Cauwenberge. 1999. Applied anatomy and physiology of the nose and paranasal sinuses. *Allergy* 54:14–25.

Wealthall, R. J., and S. W. Herring. 2006. Endochondral ossification of the mouse nasal septum. *Anat Rec* 288A:1163–72.

Weaver, T. 2009. The meaning of Neandertal skeletal morphology. *PNAS* 160(38):16028–33.

Weinberg, S. M., R. Cornell, and E. J. Leslie. 2018. Craniofacial genetics: where have we been and where are we going? *PLOS Genet* 14:e1007438.

Weiner, J. S. 1954. Nose shape and climate. *Am J Phys Anthropol* 12:615–18.

Wen, J., K. Inthavong, J. Tu, and S. Wang. 2008. Numerical simulations for detailed airflow dynamics in a human nasal cavity. *Respirat Physiol Neurobiol* 161:125–35.

West-Eberhard, M. J. 2005. Developmental plasticity and the origin of species differences. *PNAS* 102:6543–9.

White, D. E., J. Bartley, and R. J. Nates. 2015. Model demonstrates functional purpose of the nasal cycle. *BioMed Engin OnLine* 14:1–11.

White, M. D. 2006. Components and mechanisms of thermal hyperpnea. *J Appl Physiol* 101:655–63.

Williams, R. B. 1998. The effects of excessive humidity. *Respir Care Clin N Am* 4:215–28.

Wolf, M., S. Naftali, R. C. Schroter, and D. Elad. 2004. Air-conditioning characteristics of the human nose. *J Laryngol Otol* 118:87–92.

Wolpoff, M. H. 1968. Climatic influence on the skeletal nasal aperture. *Am J Phys Anthropol* 29:405–23.

Woo, T. L., and G. M. Morant. 1934. A biometric study of the 'flatness' of the facial skeleton in man. *Biometrika* 26:196.

Wroe, S., W. C. H. Parr, J. A. Ledogar, J. Bourke, et al. 2018. Computer simulations show that Neanderthal facial morphology represents adaptation to cold and high energy demands, but not heavy biting. *Proc Roy Soc B* 285:20180085.

Xiong, G.-X., J.-M. Zhan, H.-Y. Jiang, J.-F. Li, et al. 2008. Computational fluid dynamics simulation of airflow in the normal nasal cavity and paranasal sinuses. *Am J Rhinol* 22:477–82.

Xiong, Z., G. Dankova, L. J. Howe, M. K. Lee, et al. 2019. Novel genetic loci affecting facial shape variation in humans. *eLife* 8:e49898.

Xu, R., S. Li, S. Guo, Q. Zhao, et al. 2020. Environmental temperature and human epigenetic modifications: a systematic review. *Environ Pollut* 259:113840.

Yokley, T. R. 2006. *The Functional and Adaptive Significance of Anatomical Variation in Recent and Fossil Human Nasal Passages*. PhD dissertation. Duke University.

Yokley, T. R. 2009. Ecogeographic variation in human nasal passages. *Am J Phys Anthropol* 138:11–22.

Zaidi, A. A., B. C. Mattern, P. Claes, B. McEcoy, et al. 2017. Investigating the case of human nose shape and climate adaptation. *PLoS Genet* 13:e1006616.

Zhao, K. 2004. Effect of anatomy on human nasal air flow and odorant transport patterns: implications for olfaction. *Chem Senses* 29:365–79.

Zhao, K., and J. Jiang. 2014. What is normal nasal airflow? A computational study of 22 healthy adults. *Int Forum Allergy Rhinol* 4:435–46.

Zhao, Z. J. 2011. Serum leptin, energy budget, and thermogenesis in striped hamsters exposed to consecutive decreases in ambient temperatures. *Physiol Biochem Zool* 84:560–72.

Zhu, J. H., H. P. Lee, K. M. Lim, S. J. Lee, et al. 2011. Evaluation and comparison of nasal airway flow patterns among three subjects from Caucasian, Chinese and Indian ethnic groups using computational fluid dynamics simulation. *Respirat Physiol Neurobiol* 175:62–9.

Zollikofer, C. P. E., and J. D. Weissmann. 2008. A morphogenetic model of cranial pneumatization based on the invasive tissue hypothesis. *Anat Rec* 291:1446–54.

Zuckerkandl, E. 1893. *Normale und pathologishe Anatomie der Nasenhöle und ihrer pneumatischen Anhänge*. Wien: W. Braumuller.

8 Evolution and Development of the Nasal Airways in Primates

The Influence of Eye Size and Position on Chondrogenesis and Ossification of the Nasal Skeleton

Timothy D. Smith and Valerie B. DeLeon

CONTENTS

8.1 Introduction 190
8.1.1 Variations in the Nasal Skeleton of Living Primates 191
8.2 Influence of Soft Tissue Growth on Development of the Nasal Region 193
8.2.1 Skeletal Tissues of the Nasal Region 193
8.2.2 Functional Matrices in Late Embryonic and Fetal Primates 193
8.2.3 Timing of Eye Growth 196
8.2.4 Perinatal Development of the Nasal Capsule 198
8.3 Tissue-Level Changes in Development of the Nasal Region 200
8.3.1 Late Embryonic and Early Fetal Stages: Formation of the Nasal Capsular Cartilage 200
8.3.2 Perinatal Bone Modeling Patterns and Fate of the Posterior Nasal Capsule 201
8.3.3 Postnatal Factors Influencing Nasal Complexity 205
8.4 Conclusions 205
8.5 References 207

8.1 INTRODUCTION

The nasal cavity is multifunctional, bearing mucosa specialized for filtering incoming air (trapping particles and pathogens), transporting mucous, warming and moistening air, preventing water loss, and chemosensation. The mammalian nasal cavity is complicated by recesses (accessory compartments) and elaborate scrolls called turbinals (Van Valkenburgh et al. 2014; Smith et al. 2015). These scrolls and/or branching projections of bone lined by nasal mucosae fill much of the nasal cavity and greatly augment surface area for both olfactory and respiratory function (Smith et al. 2007, 2019; Van Valkenburgh et al. 2014; Pang et al. 2016; Yee et al. 2016). The turbinals are highly complex in animals reliant on olfaction (e.g., carnivores) and simplified in other animals (e.g., humans). Turbinal reductions in primates, described in more detail in the following section, disproportionately reduce surface area for olfaction (Smith et al. 2014a). Olfactory reductions in primates (including reduction in turbinals, or size of brain regions) are commonly discussed as a sensory trade-off in which vision and olfaction are negatively correlated with one another, with structures associated with vision being relatively large (e.g., Barton et al. 1995). The relative size of sensory structures is also known to correlate with ecological variables (e.g., Barton et al. 1995; Kirk 2006; Nevo and Heymann 2015).

Living primates show a complex trend in nasal cavity reduction (Smith et al. 2015). The most profound reductions can be seen in haplorhine primates. As adults, haplorhine primates have at least a 50% reduction in number of ethmoturbinals compared to strepsirrhines (Kollmann and Papin 1925; Le Gros Clark 1959), and this likely yields a far greater reduction in turbinal surface area (Smith et al. 2014a). Most strepsirrhine primates, while arguably having many primitive nasal traits, have less elaborate arrays of turbinals compared to other euarchontans (Smith et al. 2015; perhaps less true of *Daubentonia*—Maier and Ruf 2014). Thus, while strepsirrhines have historically been regarded as "primitive" in nasal anatomy (e.g., Napier and Napier 1967; Cave 1973), they (like haplorhines) may have undergone some degree of simplification of internal nasal anatomy. More importantly, they bear certain traits of the eyes and orbits in common with haplorhines. In all living primates, the bony orbits and the eyes that they support are relatively forward facing (Ross 1995). Cartmill (1972) conceived of orbits as conical spaces (with the optic canals at the apices). Because in primates the "orbital cones" converge toward the midline, interorbital space is diminished by comparison with mammals possessing non-convergent or less convergent orbits (Cartmill 1972).

Ecological variables may explain the complexity of the trends we see in living primates (see previously), but the trend toward turbinal simplification may be explained by mechanistic constraints and the reduction of the interorbital space. This interpretation has historically been based on inferences from adult primate anatomy. The observation that adult primates have highly convergent orbits (Ross 1995; Ravosa et al. 2006) and that interorbital space varies according to size and position of the orbits (Cartmill 1972) is suggestive of a developmental trade-off. This idea has been the subject of much debate (e.g., Cartmill 1972; Szalay and Delson 1979; Simons and Rasmussen 1989) but invariably based on extrapolation from adult morphology. The development of orbital and nasal anatomy has only rarely been observed using subadult primate samples (e.g., Jeffery et al. 2007; Smith et al. 2013; Nett and Ravosa 2019). In this chapter, we discuss novel observations and recent published work that elucidates the entire trajectory of nasal cavity development at the tissue level from late fetal to subadult stages.

8.1.1 Variations in the Nasal Skeleton of Living Primates

Studies prior to the availability of micro-computed tomography (micro-CT) established the basic morphology of nasal cavity structures in primates. Through painstaking dissection of a wide taxonomic breadth of primates, Kollmann and Papin (1925) and Cave (1973) established that strepsirrhines (lemurs and lorises) possess a larger array of delicate scrolled projections, or turbinals, of the ethmoid bone compared to haplorhines. With the widespread availability of micro-CT, we now have the ability to efficiently examine large samples of primates. Recent studies have established that most euarchontans possess five or six turbinals of the ethmoid bone that project close to the septum (four to five ethmoturbinals and one nasoturbinal), between one and three smaller ethmoidal turbinals tucked between ethmoturbinals (i.e., interturbinals) or within paranasal spaces (frontoturbinals), and the maxilloturbinal. *Daubentonia madagascarensis* possesses an even greater suite of smaller turbinals of the ethmoid bone (four interturbinals; six frontoturbinals) (Maier and Ruf 2014; Lundeen and Kirk 2019).

All living haplorhines possess far fewer turbinals, displaying only half of the complement of ethmoturbinals and lacking any frontoturbinals or interturbinals (Kollmann and Papin 1925; Cave 1973; Lundeen and Kirk 2019). The stark difference in complexity between extant strepsirrhines and haplorhines presents a basic dichotomy between suborders: haplorhines have reduced complexity, and strepsirrhines are more similar to the ancestral state (Lundeen and Kirk 2019). However, the nasal cavity of all living primates is smaller in some respects compared to other euarchontans, for instance, in the breadth of the olfactory recess, a cul-de-sac at the posterodorsal limit of the nasal cavity which is proportionally much wider and more complex in tree shrews (Smith et al. 2015).

If living strepsirrhines have reductions in the nasal fossae compared to the sister taxa of primates (e.g., Scandentia), they are subtle changes. These include compressed paranasal spaces or narrow olfactory recesses, and each of these is readily explained by spatial encroachment by the eyes (Smith et al. 2011, 2019), which are convergent (forward facing and close-set) to a varying degree in all primates (Ross 1995). Whereas it is true that living strepsirrhines possess a similar number of turbinals as tree shrews, colugos, and many other mammals, their relatively narrow nasal passageways appear to have different airflow patterns compared to mammals such as canids. In canids, there are distinct ventral and dorsal airstreams during inspiration; the former direct air along a strictly respiratory mucosa, and the latter shunt air toward the olfactory recess (Craven et al. 2010). Only one primate has been studied using airflow simulations, *Nycticebus pygmaeus* (pygmy slow loris). In the loris, airstreams are not as distinctly segregated as in canids (Smith et al. 2019). Moreover, the frontal recess is a compressed region immediately adjacent to the orbits; though it contains three frontoturbinals, only one of them bears olfactory mucosa. Smith et al. (2019) suggest the compressed frontal recess and olfactory recess in most primates described to date is the result of a developmentally induced trade-off of a part of the nasal fossa via encroachment by growing eyes.

These observations are consistent with Cartmill's (1972) mechanistic hypothesis that converging orbital cones constrain interorbital space. However, there have been few *developmental* studies to assess how functional matrices, the soft tissue regions that the skull supports, are in direct opposition to the nasal region in a comparative sense (Figure 8.1; see Table 8.1 for stages of development of the specimens pictured in this chapter). Moreover, the microanatomical evidence for structural trade-offs has rarely been reported. In this chapter, we review studies of prenatal and postnatal growth pertinent to structural trade-offs and examine microanatomy of prenatal and

TABLE 8.1
Characteristics of Mammalian Specimens Pictured in Figures 8.1–8.5

Fig. #	Species	Characteristics	Approx. Stage	SN/Collection
1a	*Tenrec ecaudatus*	Unflexed posture; palate formed; FL well formed; ankle joint as yet not well-formed	Early fetal	I2a/Bluntschli
1b	*Microcebus murinus*	Palate formed; some turbinals as yet in mesenchymal state	Early fetal	M54/Bluntschli
1c	*T. ecaudatus*	Unflexed posture; palate formed; FL well formed; ankle joint as yet not well-formed	Early fetal	I20/Bluntschli
1d	*M. myoxinus*	CRL = 2× stage 23 embryo[1]	Mid or late fetal	M24/Bluntschli
1e	*Rousettus leschenaultii*	Palate formed; some turbinals as yet in mesenchymal state	Early fetal	KPB/TDS
1f	*Nycticebus coucang*	Flexed posture; palate fully formed	Early fetal	N36/Hubrecht
2a,c	*Tarsius bancanus*	Humerus chondrified with beginning of bone collar formation; secondary palatal shelves completely separated by the tongue	Late embryo	T633/Hubrecht
2b,d	*T. bancanus*	Fully formed palate; nasal capsule chondrified with scant matrix production as yet	Early fetal	T72/Hubrecht
2e	*T. bancanus*	CRL = ~2× late embryo[1]; but less than ~½ newborn[2,3]	Mid fetal	T981/Hubrecht
3a	*Saimiri sciureus*	Much of nasal capsule in condensed mesenchymal state	Late embryo	P22/Bluntschli
3b	*S. boliviensis*	Known age[4]	Neonate	Ss2662/TDS
3c	*S. boliviensis*	Known age[4]	Neonate	Ss2680L/TDS
4c	*E. flavifrons*	Known age[4]	Neonate (left)/ adult (right)	DLC6778, DLC6187/TDS
5a	*Tupaia belangeri*	Known age[4]	Neonate	Tupaia1a/TDS
5b	*Otolemur crassicaudatus*	Known age[4]	Late fetal	DLC2810/TDS
5c-e	*Lemur catta*	Known age[4]	Late fetal	DLC6888/TDS

Sources: 1, Smith and Rossie (2008); 2, Smith et al. (2003); 3, Lucket and Maier (1982); 4, Smith et al. (2015).

Abbreviations: CRL, crown-rump length; DLC, Duke Lemur Center; FL, forelimb; HL, hindlimb; SN, specimen number.

postnatal primates in order to identify specific tissue-level mechanisms by which adjacent functional matrices influence skull form.

8.2 INFLUENCE OF SOFT TISSUE GROWTH ON DEVELOPMENT OF THE NASAL REGION

8.2.1 Skeletal Tissues of the Nasal Region

An important concept to formation of the nasal skeleton is that primitive connective tissue (i.e., mesenchyme) forms more than one skeletal tissue; in the cranium, these tissues may remain cartilage, undergo endochondral ossification, or form bone without cartilaginous precursor (Hall 2015). Both cartilage and bone are composed of specialized connective tissue cells and the matrix that these cells produce. The matrices of many connective tissue types comprise a mix of fibrous and non-fibrous elements (the latter is termed *ground substance* and is viscous and hydrophilic to a variable degree). After differentiation, the key cellular activity that distinguishes the tissue type is the matrix production. Cartilage cells of hyaline cartilage (chondroblasts and chondrocytes) produce a highly viscous matrix that becomes compression resistant once mature. Osteoblasts, the cells that produce bone matrix, secrete collagen and orchestrate mineralization of the matrix (Hall 2015).

In the nasal region, there is a time period, mostly restricted prenatally, during which the entirety of the nasal skeleton exists as hyaline cartilage (Maier 1993a). This template for the nasal skeleton is called the nasal capsule (Maier, 1980). Beginning late prenatally, portions of the nasal capsule begin to ossify (endochondrally) or resorb (Maier and Ruf 2014; Smith et al. 2012, 2021b). In the adult, the skeletal architecture of the nasal region is formed by a variety of skeletal tissues, including parts of the nasal capsule that remain cartilage (e.g., septal cartilage), an endochondral bone (the ethmoid), and other bones of the face which ossify membranously (without a cartilaginous precursor) and in some regions form the boundaries of the nasal cavity. We emphasize that the ensuing discussion follows on a lengthy trajectory of nasal development, beginning with its precartilage (mesenchymal) and cartilage framework and extending to formation of bone elements of nasal architecture.

8.2.2 Functional Matrices in Late Embryonic and Fetal Primates

Functional matrices such as the eye (and supporting tissues), the nasal mucosae, and the brain are supported by *skeletal elements*. The latter are not usually individual bones but are instead shared osseous borders between adjacent functional matrices (Moss and Young 1960; Moss and Greenberg 1967). Examples include the soft tissues of the anterior cranial fossa and orbit, for which the orbital portion of the frontal bone is a shared skeletal element. In humans and other primates, multiple bones articulate as a shared skeletal element separating the soft tissue contents of the orbit and nasal cavity.

Growth of the brain and other soft tissue organs has primacy over bone growth in a directional sense. That is to say, growing bones will yield as soft tissues enlarge or expand (O'Connor 2006; Smith et al. 2014b). During growth, selective osteoblastic and osteoclastic activity (bone deposition and resorption, respectively) causes bones

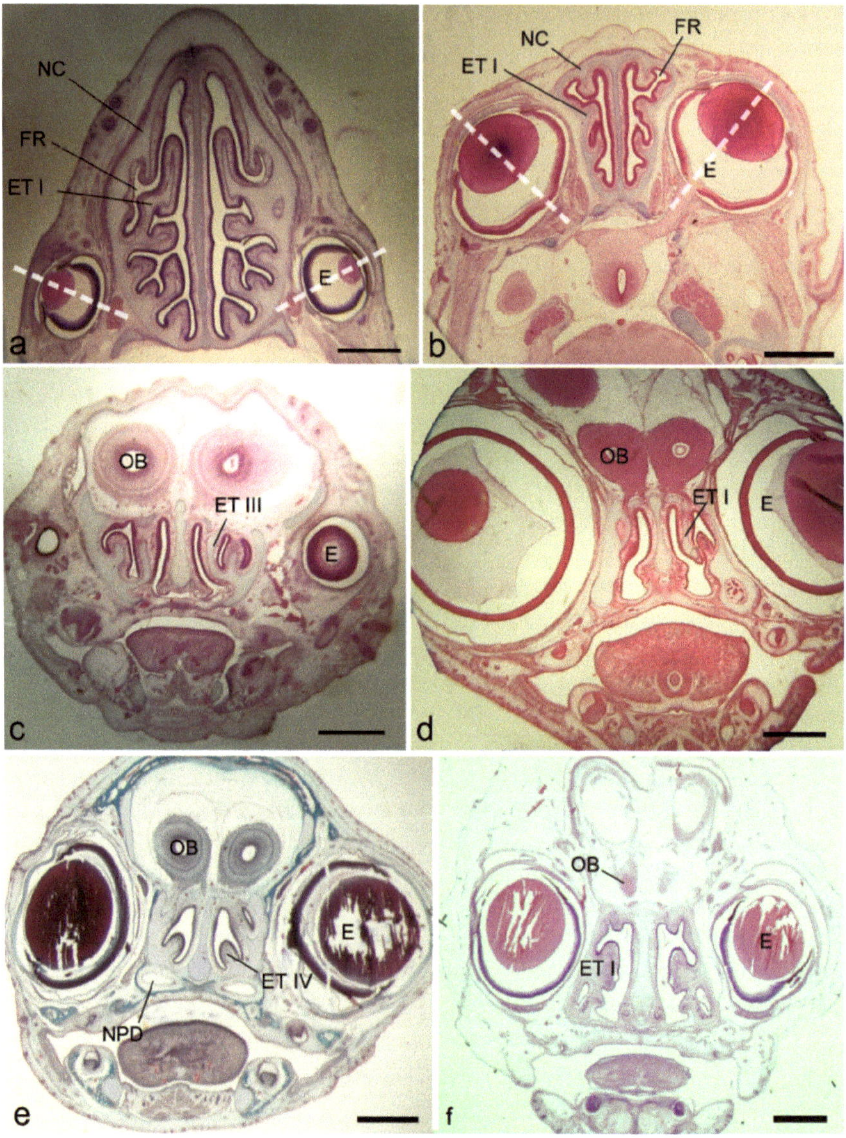

FIGURE 8.1 The cartilaginous nasal capsule (NC) in late embryonic mammals. Top row: Horizontal sections through the crania of 23.5-mm crown-rump length (CRL) tenrec (*Tenrec ecaudatus*—a) and 16-mm CRL mouse lemur (*Microcebus murinus*—b) late embryos. Note in the former, the eyes are adjacent only to the more posterior ethmoturbinals, while in the mouse lemur, the proportionally larger and more convergent eyes are directly apposed to the entire ethmoturbinal region of the nasal capsule and also the frontal recess (FR). The differing degree of convergence is qualitatively compared by superimposing a dashed line along the midline (axial plane) of each eye. Middle row: Coronal sections through the largest circumference of at least one eye in a 23-mm CRL tenrec embryo (*T. ecaudatus*—c) and a 31-mm CRL mid-to-late fetal mouse lemur (*M. myoxinus*—d). Note the broader interorbital

to "drift" in space as soft tissues grow (Enlow and Hans 1996). This process is an example of bone modeling, a depositional and resorptive process in which bone cell activity (i.e., of osteoclasts and osetoblasts) are considered uncoupled (Barak 2020). In contrast, in bone remodeling, resorption and deposition of bone is orchestrated in a sequential and coupled manner to *replace* existing bone (Barak 2020). There is a key difference, whether in reference to accommodation to biomechanical demands or growth and development: bone modeling causes bone tissue to shift in space, while remodeling does not (Smith et al. 2014b; Barak 2020).

Bone drift involves continual turnover, or the generation of new bone tissue: whole bones and parts of bone shift spatially, both to accommodate and support growing soft tissues and to orchestrate overall organismal growth (e.g., Robling and Stout 1999; Verna et al. 1999). In the midfacial region, as the brain, eyes, teeth, and nasal mucosae grow, they effectively "compete" for space. The fastest-growing elements or spatial movement of growing elements (e.g., shifting tooth position) dictates overall patterns of bone deposition and resorption during growth (Smith et al. 2014b).

Later in the chapter, we will discuss modeling patterns further. In this section, we begin our discussion of growing soft tissues by focusing on early stages of skeletogenesis, here meaning differentiation and morphogenesis of cartilage or bone. In relation to the nasal capsule, the eyes of primates are much more rostrally positioned relative to other mammals (Figures 8.1a, b). The eyes of non-primates tend to "bracket" the nasal capsule at the level of the ethmoturbinals (Figures 8.1a, c, e). In contrast, the eyes of primates also border more rostral parts, such as the paranasal recesses (Figures 8.1b, d, f). As a result, the convergent eyes of developing primates have narrowed the interorbital dimensions across a broad extent of the nasal capsule. It is both the relative size and convergence of the eyes that present a "packaging" problem within the growing midface (Figure 8.2b). In prenatal stages, the eyes border a large anteroposterior extent of the nasal capsule. This is also true in developing tarsiers, primates that eventually form a small snout that has been hypothesized to be compressed by extremely enlarged eyes (Cartmill 1972). Although the eyes of prenatal tarsiers are not yet as disproportionately large as in adults (Cummings et al. 2012), they still present a packaging problem for the neighboring nasal capsule (Figure 8.2). In part, this is an allometric phenomenon. Eyes do not scale to body size across primates; they are proportionately larger in the smallest species (Schultz 1940; Kirk 2006), although at what stage of development this becomes true has not been fully explored. A comparison of prenatal *Tarsius bancanus* at different sizes reveals two phenomena

---

FIGURE 8.1 (Continued)
breadth in the former, with eyes adjacent to posterior ethmoturbinals (ET III, third ethmoturbinal). In the latter, the interorbital space is much narrower, and the eyes' largest diameter is adjacent to ethmoturbinal I (ET I). Bottom row: coronal sections of two large-eyed mammals at early fetal stages, including a fruit bat (*Rousettus leschenaultii*,—e) and a primate (*Nycticebus coucang*—f). Note the more posterior level at which the maximal eye diameter apposes the nasal capsule in the bat (e, the fourth ethmoturbinal, ET IV) compared to the loris, in which maximal eye diameter is adjacent to more anterior ethmoturbinals (f—ET 1, first ethmoturbinal). NPD, nasopharyngeal duct; OB, olfactory bulb. Scale bars = 1 mm. Stains: (a–d, f) hematoxylin and eosin (cell nuclei, blue or black; cytoplasm, pink or red; fetal cartilage, pale or unstained matrix); (e) Gomori trichrome (bone, dark green or red; collagen, green; cartilage, pale), with hematoxylin counterstain (cell nuclei, blue or black).

of eye development that impact the developing nasal capsule. The smaller specimen in Figure 8.2a (13-mm crown-rump length, or CRL) has embryonic characteristics of a late embryo (Table 8.1). Notably, the eyes are not as fully convergent as older specimens. By comparison, in a 20-mm CRL early fetal *T. bancanus* (Figure 8.2b), the eyes are both proportionately larger and more convergent toward the midline. Figures 8.2a and 8.2b, both at a relatively dorsal position near the "roof" of the nasal cavity, reveal how differentially faster embryonic growth of the lateral face (including the orbital region) leaves the slower growing nasal capsule in a proportionately narrow interorbital space. Figures 8.2c and 8.2d show the same two specimens at a more inferior position within the nasal cavity. In the late embryo, the posterior margin of the lateral wall (paries nasi) remains mesenchymal, while more anteriorly chondrification has commenced (Figure 8.2c). In the fetus, the lateral wall is fully chondrified, but matrix production is not yet apparent (Figure 8.2d). Thus, the differential growth that shifted the eyes to a more convergent position in haplorhines occurs when the nasal capsular cartilage is in a primordial state. This provides a context for further discussion in Section 8.3.2.

Convergence of orbital cones has been most frequently discussed based on adult morphology (e.g., Cartmill 1972; Ross 1995; Ravosa et al. 2006), with the process of developmental competition for space being implied. A comparison of late embryos is instructive for context. In two small-bodied primates, such as *Microcebus* (Figure 8.1b) and *Tarsius* (Figure 8.2b), the degree of eye convergence is different, being greater in the tarsier. Based on the timing of embryonic orbital convergence, we can speculate that the bilateral orbital functional matrices impose constraint on the nasal capsule.

Given the primordial state of the nasal capsular cartilages, differentially faster growth of the lateral face (including eyes), which leads to embryonic shifting of eyes toward the midline, may limit the capacity for matrix production and chondrocyte proliferation in the nasal capsule. A proximate basis for interdependence among cranial regions has rarely been discussed at the tissue or cellular level. However, it is known that both mesenchymal stem cells and chondroblasts or chondrocytes are capable of responding to differing mechanical stimuli (Li et al. 2001; Thorpe et al. 2008). In particular, dynamic (variant) loading results in increased cartilage volume, while static compressive loading has little or no effect (Elder et al. 2000; Li et al. 2001). At the cellular level, dynamic and particularly cyclic loads promote increased synthesis of components of the matrix by cartilage cells (e.g., Li et al. 2001; Lee et al. 2003). In contrast, static loading is known to decrease matrix production by cartilage cells, both in developing and mature cartilage (Li et al. 2001; Lee et al. 2003), whereas static loads can inhibit matrix production, and may even promote cartilage degeneration (Lee et al. 2003; and see Responte et al. 2012). *Because the cranium is a self-contained space comprising multiple partially dependent regions (Lieberman 2011), we propose that growing eyes may inhibit nasal capsular growth by placing developing cartilage in a state of static compression.* In particular, this may be the case when growing eyes are recessed deeply within the growing face (as opposed to more ectopically positioned eyes).

8.2.3 Timing of Eye Growth

Following embryonic eye convergence, an additional factor that may limit interorbital dimensions is volumetric growth of the eyes. Available data indicate that eyes

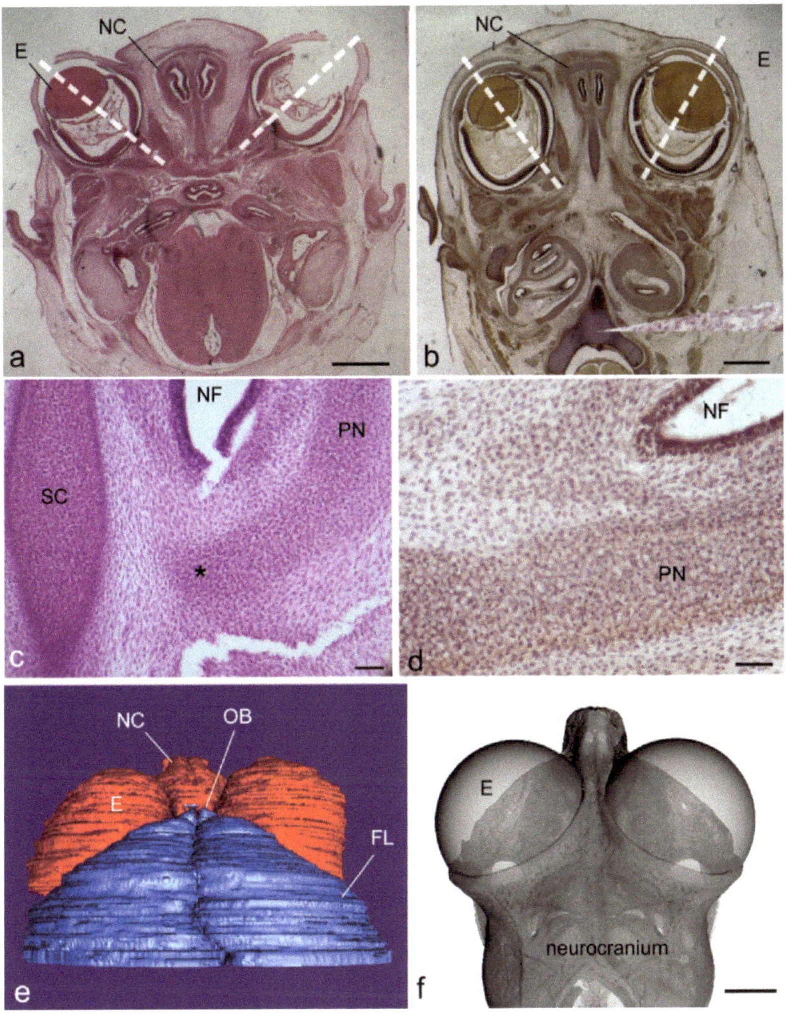

FIGURE 8.2 Top row: A 13-mm crown-rump length (CRL) embryonic tarsier (*Tarsius bancanus*, a) and 20 mm CRL early fetal tarsier (b). Note the orientation of the eyes (E), wider set in the embryo (dashed line approximates axial midline of eyes). Both are horizontal planes at approximately the greatest eye circumference. Middle row: same two specimens, shown at a more ventral cross-sectional level. The posterior end of the lateral wall, or paries nasi (PN), is shown at higher magnification. In the embryo (c), it is still mesenchymal posteriorly (note dense cellularity), while in the fetus (d), it is entirely in an early phase of chondrification (note the enlarged chondrocytes but not dispersed by matrix). Bottom row: A 25-mm CRL fetal tarsier, reconstructed in Amira from segmented histological sections (e) and adult tarsier (f) (both *T. bancanus*), shown in dorsal view. The fetal specimen represents a reconstruction of soft tissues alone. Note the eyes are partially tucked ventral to the brain (FL = frontal lobe). In contrast, the eyes of the adult are more ectopic, and project anterior to the neurocranium. NF, nasal fossa; NC, nasal capsule; OB, olfactory bulb; SC, septal cartilage. Scale bars: (a, b) 1 mm; (c, d,) 50 μm; (F) 5 mm. Stains: (a, c) Hematoxylin and eosin (cell nuclei, blue or black; cytoplasm, pink or red; fetal cartilage, pale or unstained matrix); (b, d) unspecified, but includes hematoxylin (cell nuclei, blue or black).

are proportionally more progressed to adult size at birth in anthropoids compared to strepsirrhines and tarsiers, as indicated by the ratio of newborn to adult axial eye diameter (Cummings et al. 2012; Smith et al. 2020). In strepsirrhines, newborn eyes are 41 to 62% of adult axial eye diameter. There are fewer data available for anthropoids, but the percentage range for known anthropoids overlaps minimally and reaches a considerably higher maximum (61–70%; Smith et al. 2020). This basic contrast suggests anthropoids have a greater degree of interorbital constraint based on precocious eye growth, and this rapid growth may be especially constraining in smaller species (see subsequently).

Although differential timing of eye growth has an influence, large relative eye size alone is insufficient to explain the extent of reduction of nasal anatomy in primates. Eyes of all small-bodied primates are proportionally large, including strepsirrhines (Kirk 2006). This is true of adult primates, but it is also readily apparent prenatally in cross-sections of fetal mouse lemurs and lorises (Figure 8.1). However, in both of these small-bodied, nocturnal primates, the developing eyes are partially ectopic (Figures 8.1c, f); that is, they extrude beyond the confines of the primordial orbit.

We hypothesize that the degree to which eyes are ectopic *during* nasal capsular development is a critical factor mitigating the degree of interorbital constraint, perhaps as important as the degree of convergence. Conversely, eyes that are more deeply recessed (confined more deeply within the bony orbit) during development may impose a relatively greater constraint, depending on eye size. Notably, the eyes of tarsiers are deeply recessed in the early fetus shown in Figure 8.2b. A comparison of the degree to which the eyes protrude in later fetal (Figure 8.2e) compared to adult stages (Figure 8.2f) shows that across age, the centroid or anterior surface of the eye becomes much more anteriorly displaced from the brain, and thus the eyes become far more ectopic and separated from shared skeletal elements. (For a more quantitative description of eye protrusion see the "ectopy index" in Rosenberger et al. 2016.) This may explain why tarsiers have nasal cavity reduction despite relatively slow prenatal eye growth (Cummings et al. 2012) and a high degree of eye protrusion in adults (Rosenberger et al. 2016). The most profound nasal cavity reductions (e.g., absence of posterior ethmoturbinals) relate to late embryonic and fetal constraint on the nasal capsule, especially in the context of deeply recessed eyes.

8.2.4 Perinatal Development of the Nasal Capsule

The extent to which the precociously large eyes of newborn anthropoids may impact the developing nasal skeleton (including both cartilaginous and osseous elements) likely depends, in part, on body size. Eye size increases with negative allometry in primates (i.e., small primates have larger eyes relative to body size than do large primates) (Kirk 2006). Eyes of smaller anthropoids should be occupying proportionally more space within the head than in larger anthropoids. For example, consider two catarrhines of disparate body sizes, the Rhesus macaque (*Macaca mulatta*) and humans. At birth, the eyes of these species are similar in size, relative to adults (~68–70%) (Cummings et al. 2012). But *M. mulatta* is of far smaller body size than humans. Using previously published data, Cartmill (1980) showed that in *M. mulatta*, eye volume expressed as a percentage of body mass is two-fold greater compared to humans. Thus, we may expect eyes present a bigger "packaging" problem for smaller primates as they develop than they do for larger primates.

We see this most clearly in small-bodied New World monkeys, perhaps most especially in *Saimiri* spp. As adults *Saimiri* form an interorbital fenestra, an "open" communication between orbits (open in skeletonized heads; in life, a membrane fills this opening), but the osseous fenestra is not yet present at birth (Maier 1983). Although the septal cartilage separates the orbits in newborn *Saimiri*, the nasal fossae are

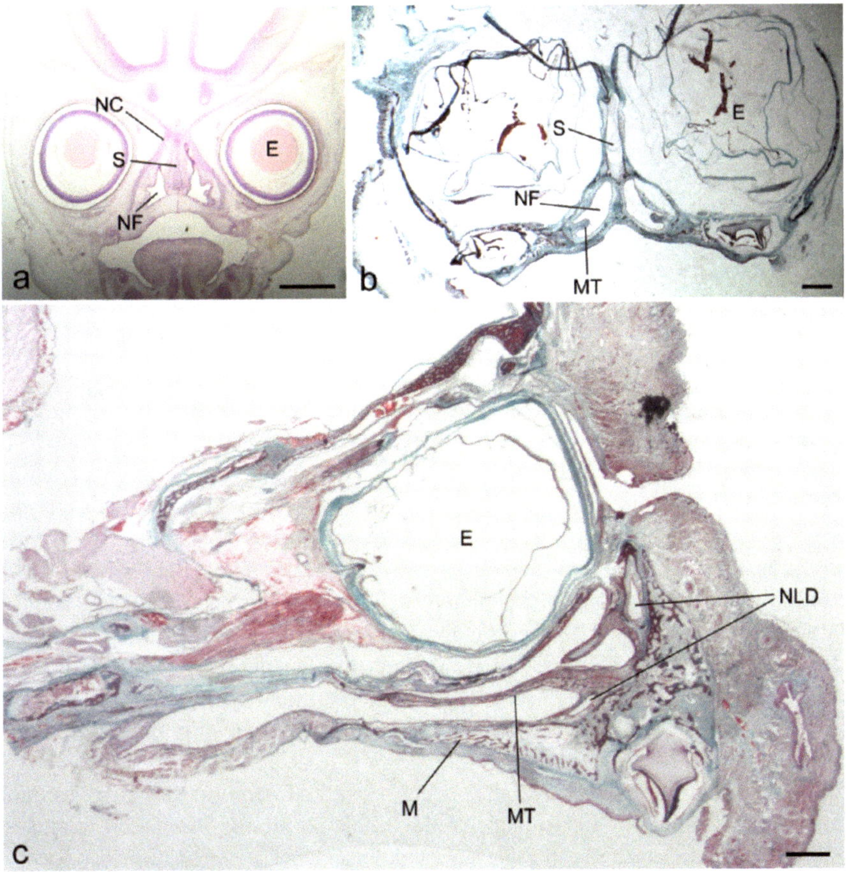

FIGURE 8.3 Orbitonasal development in *Saimiri*. (a) Coronal section of a 25-mm crown-rump length early fetal squirrel monkey (*Saimiri sciureus*) shown near the greatest eye (E) circumference. (b) Neonatal *Saimiri boliviensis* also shown at greatest eye circumference, also shown in coronal section. Note the far greater degree of interorbital encroachment at this level in the newborn. (c) Sagittal section of a neonatal *S. boliviensis* in sagittal section. Note the eye is positioned over the most posterior part of the nasal fossa. M, maxilla; MT, maxilloturbinal; NF, nasal fossa; NLD, nasolacrimal duct; S, septal cartilage. Scales bars: 1 mm. Stains: (a) hematoxylin and eosin (cell nuclei, blue or black; cytoplasm, pink or red; fetal cartilage, pale or unstained matrix); (b, c) Gomori trichrome (bone, dark green or red; collagen, green; cartilage, pale), with hematoxylin counterstain (cell nuclei, blue or black).

prescribed to narrow wedges at orbital levels, as in adults. Based on a comparison to fetal stage *Saimiri*, this restriction of the nasal fossa at orbital levels progressively develops as eye growth accelerates; constraint at levels of maximal eye diameter may become greater across fetal development (Figures 8.3a, b). At birth, large eye size in *Saimiri* likely explains the resorptive activity in the bones making up the orbital floor, such that this bone drifts inferiorly (Smith et al. 2011); this is the opposite of what is more broadly observed in primates—depositional activity in the floor of the orbit (Enlow and Hans 1996; Smith et al. 2011). In *Saimiri*, osteoclastic activity on the orbital floor results in downward drift of the orbital surface of the maxilla. In turn, this leads to narrowing and inferior displacement of the nasal airway at all levels that correspond to a relatively large orbital circumference (Figure 8.3c); pneumatization of the maxillary bone is also suppressed (Rossie 2006; Smith et al. 2011). Thus, growing eyes appear to dictate modeling patterns, directly explaining reduced interorbital space, especially in species with the largest relative eye size (Smith et al. 2014b).

8.3 TISSUE-LEVEL CHANGES IN DEVELOPMENT OF THE NASAL REGION

8.3.1 Late Embryonic and Early Fetal Stages: Formation of the Nasal Capsular Cartilage

Throughout prenatal and postnatal development, there are cell- and tissue-level mechanisms that promote greater complexity of the internal nasal skeleton. The earliest such mechanism, documented in all classes of vertebrates by Dieulafé (1906), is "fissuration" of the lateral walls of the nasal fossa. This is a process by which clefts penetrate from the surface contour of the developing lateral nasal wall, making the lateral wall appear to possess multiple bulges. This occurs even prior to the first indications of skeletogenesis, in which mesenchymal cells become more densely distributed (the mesenchymal condensation) in regions of incipient cranial cartilages. In this phase of tissue maturation, mesenchymal cells congregate more closely and form gap junctions, communicating more frequently (see Hall 2015, for more details). Mesenchymal condensations subsequently undergo chondrification.

The formation of mesenchymal condensations has been discussed by multiple authors (e.g., Zeller 1987; Hall and Miyake 2000; Smith and Rossie 2008; and see de Beer 1937). Importantly, the mesenchymal condensations for turbinals form as distinct structures after condensations destined to become the septum and the lateral walls have formed. Subsequently, chondrification is not synchronized across the nasal capsule, that is, some turbinals achieve a chondrified state before others (Smith and Rossie 2008). Additional smaller turbinals form from mesenchymal condensations on the inner surface of the fully chondrified capsule. Thus, the ethmoturbinal region develops in an iterative manner spanning late embryonic and early fetal stages of development (Smith and Rossie 2008; Smith et al. 2021a). More recently, similar findings by Kaucka et al. (2017) on mice suggest this is broadly true of mammals. With this context in mind, the posterior nasal capsule of at least some haplorhines is highly compressed by convergent eyes at a timeframe in which the chondrogenesis of the nasal capsule has only been initiated (Figures 8.2c, d). Turbinal development may be inhibited by space limitations.

Once chondrified, turbinals possess the capacity for interstitial and appositional growth. As fetal development proceeds, some portions of turbinals undergo endochondral ossification. Many aspects of endochondral ossification, including chondrocyte hypertrophy, are observed; presumptive chondroclasts are also observed even before turbinals ossify (Smith et al. 2012). Thus, we may expect that these processes shape turbinals as they proceed to an ossified state. Recently, Smith et al. (2021a) identified late fetal/perinatal evidence for appositional chondroblastic activity that presages tertiary lamellae of turbinals in fruit bats. Thus, during prenatal development, the turbinals form as primordia and then are shaped into the lamellar and possibly scrolled structures by the specialized activity of cartilage and bone cells. As discussed in the following, the process by which turbinals become more complex spans a late prenatal and early postnatal timeframe.

8.3.2 Perinatal Bone Modeling Patterns and Fate of the Posterior Nasal Capsule

Drawing a causal relationship between eye size, position, or growth and the extent of nasal reduction is challenging. Bone modeling patterns offer one line of evidence, since bone cell activity (e.g., Howship's lacunae of osteoclasts) leaves behind indications of the influence of soft tissues on the skeletal elements that support them. In humans, late prenatal and early postnatal orbital modeling patterns assume the form of an enlarging and anteriorly advancing cone. Bone cell activity is largely depositional, except superiorly where the "roof" of the orbit ascends due to osteoclast activity (Enlow and Hans 1996). An important caveat is that humans are large-bodied primates, and the interorbital region is comparatively less constrained (especially as head size increases) compared to smaller bodied primates.

Late fetal orbital modeling patterns in small-bodied primates reflect a larger packaging dilemma, due to the large relative eye size (see previously) and forward-facing orbits. In perinatal primates, there is a predominantly resorptive activity where bone supports the orbital contents superomedially (e.g., orbital surface of the frontal bone) (Smith et al. 2014b). The coronal sections shown in Figure 8.4 transect the first permanent molar in three species of primates (fetal) and one non-primate (*Tupaia belangeri*, a perinatal tree shrew). In all primates, this coronal plane transects a level through the axial line of the eye quite near its largest circumference, whereas most of the volume of the nonconvergent eyes of *Tupaia* lies posterior to this cross-sectional level. Relative eye size and the degree of convergence are great in any primate compared to *Tupaia*, and this appears to be true prenatally and at birth, as in adults (Figure 8.4, Cummings et al. 2012). This commonality explains the modeling patterns (denoted by arrows) that were recorded in these species by Smith et al. (2014b): the superomedial orbit (formed by the frontal bone) drifts medially due to osteoclastic activity on the orbital surface and osteoblastic activity on the opposing side. As a result, the interorbital region widens at a slower pace than the orbital region. In *Tupaia*, the reverse is true. Smith et al. (2014b) hypothesized that olfactory

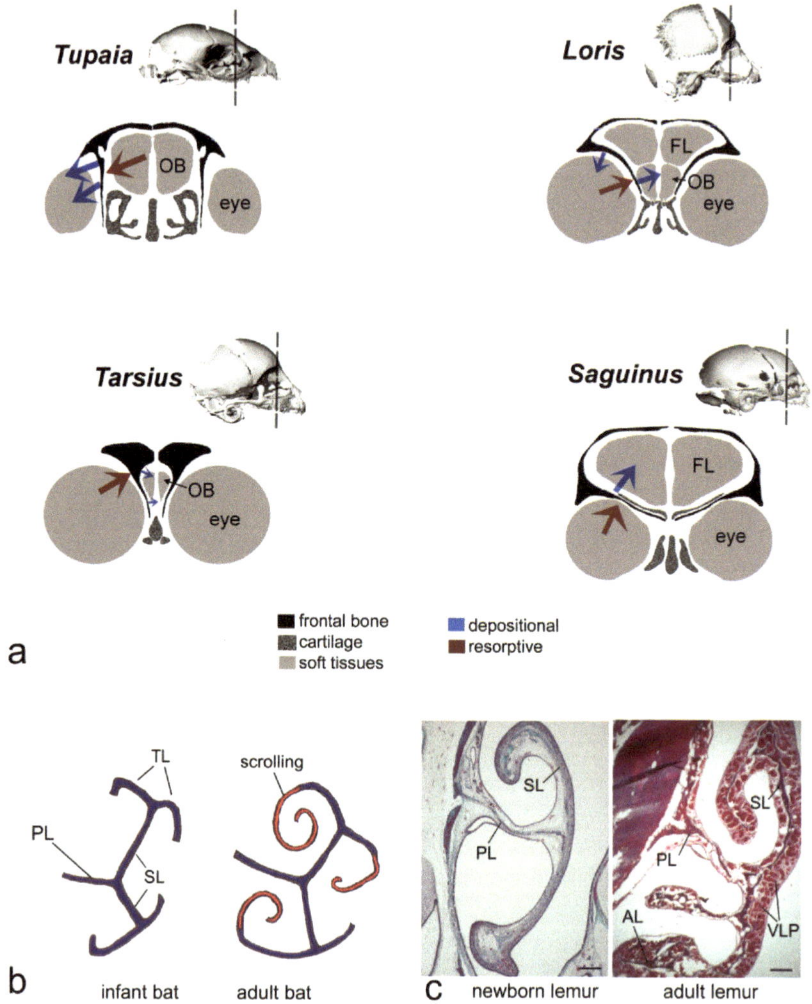

FIGURE 8.4 (a) Schematic views of non-primate (*Tupaia*, newborn) and three primate species (late fetal). The color-coded arrows indicate resorption and deposition patterns described in Smith et al. (2014b) at the coronal plane at M1 (first permanent molar). At M1, the nasal capsule of *Tupaia* is wide; the eyes are relatively non-convergent. Note the outward drift of orbital bone at this level could be explained by brain growth (OB = olfactory bulb). Note that in all primate species, modeling patterns suggest the eye has primacy over brain growth. (b) Schematic of the maxilloturbinal in a bat (*Rousettus leschenaulti*) in infancy and in the adult. Adults possess greater inward scrolling of tertiary lamellae (TL) compared to infants. Smith et al. (2021a) interpret this as postnatal appositional growth. (c) Similarly, adult lemurs possess accessory lamellae compared to newborns and also develop a more vascular lamina propria (VLP).

Abbreviations: FL, frontal lobe; PL, primary (root) lamellae; SL, secondary lamellae. Scale bars: 0.5 mm. Stains: (c) Gomori trichrome (bone, dark green or red; collagen, green; cartilage, pale), with hematoxylin counterstain (cell nuclei, blue or black).

bulb growth at this cross-sectional level outpaces growth of more lateral soft tissues in *Tupaia*, resulting in outward drift of the superomedial orbit.

Other structures influence remodeling patterns of the nasal skeleton, notably the developing dentition. Developing teeth may impact the more inferior paranasal spaces, but the resulting modeling patterns depend on complex factors, such as relative tooth size and position (Smith et al. 2011). Once again, the significance of teeth depends, in part, on head size. At birth, the alveolar crypts of posterior postcanine teeth are generally shifting laterally, the result of modeling fields that enlarge the face as a whole (Smith et al. 2011). This is also associated with enlargement of the maxillary sinus, which broadens laterally as the palate enlarges (Smith et al. 2011). But in *Saimiri*, small monkeys with relatively large posterior maxillary teeth, the medial wall of the alveolar crypt are largely resorptive at birth (Smith et al. 2011). The combination of relatively large teeth and eyes are thus thought to inhibit secondary pneumatization of the maxillary recess (Rossie 2006; Smith et al. 2011).

The constraint of large eyes on the nasal cavity may be mitigated by several factors. First, more dorsal positioning of the eyes relative to the midface (e.g., lorises) creates more space for the posterior nasal cavity (Cartmill 1972). Due to negative allometry of eyes in primates (Kirk 2006), larger-bodied species have greater interorbital space. And finally, as described previously for fetal primates, primate eyes may protrude from the orbit (i.e., ectopic eyes) across fetal and postnatal development. Despite these avenues by which strepsirrhines may preserve room for the ethmoturbinal recess, the posterior space may be considered less complex than in mammals with less convergent eyes (e.g., *Tupaia*). Smith and colleagues (Smith and Rossie, 2008; Smith et al. 2019) have described the olfactory recess of adult *Microcebus murinus* as relatively small and described the diminutive frontal recess in adult *Nycticebus pygmaeus*. Each of these conditions may be hypothesized to result from postnatal constraint due to relatively large, adjacent eyes. In particular, Smith et al. (2019) observed the adjacency of eyes to the dorsally positioned frontal recess and suggested the small frontal recess may bear a reduced role in olfaction due to the reduced surface area for olfactory mucosa and altered airflow patterns.

The cartilaginous nasal capsule, like the remainder of the chondrocranium, has a complex fate in postembryonic phases of development, with many portions that ossify but others that are replaced (or substituted) by dermal bone and still other portions that are resorbed (Kawasaki and Richtsmeier 2017; Pitirri et al. 2020). However, the nasal capsule has a unique trajectory during development, and portions of its future osseous elements (e.g., the ethmoid) may be notably slow to ossify compared to the remainder of the chondrocranium, though there is much taxonomic variation (Smith et al. 2021b). Unlike bone, cartilage is an ideal connective tissue for resisting compression (Hall 2015; Jones et al. 2020). We suggest the fully chondrified nasal capsule of mammals provides a buffer in which internal evaginations of the lateral nasal wall may harness interstitial growth properties of cartilage to enhance turbinal complexity.

The nasal capsule remains cartilaginous for a variable time in mammals. In primates, the posterior nasal capsule has a relatively early breakdown. The posterior end of the nasal capsule is a thimble-shaped cartilaginous "cap," called the posterior nasal cupula (PNC). The PNC surrounds the end of the most dorsal part of the nasal cavity, which is diminutive in haplorhines (at best a small recess) and forms a larger

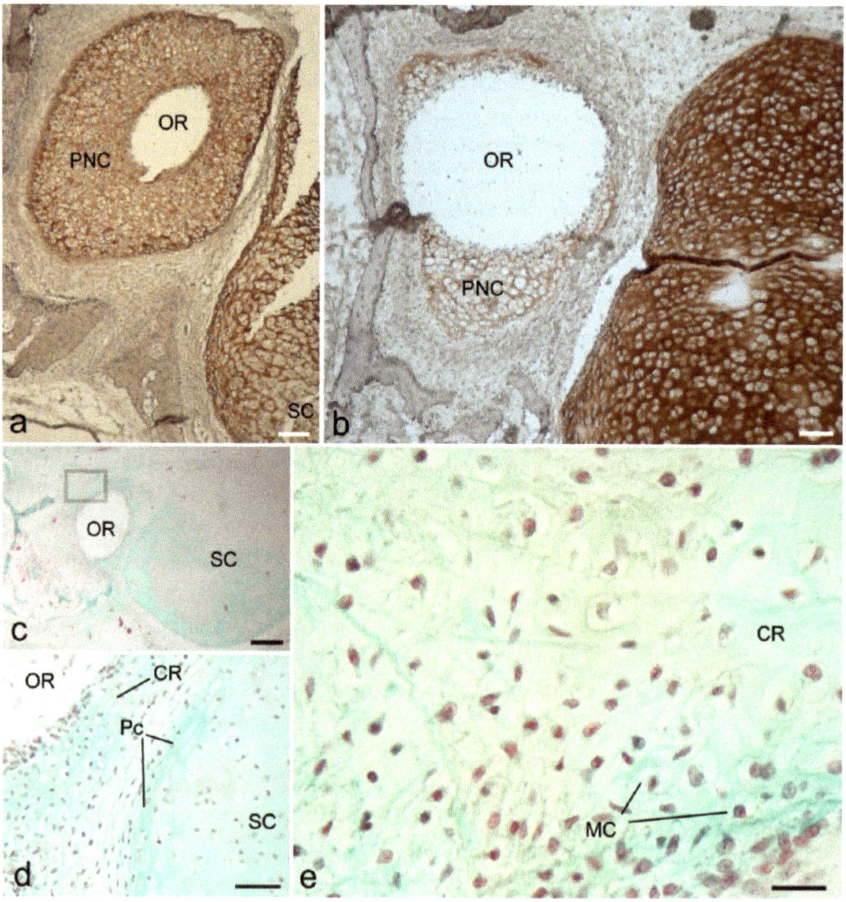

FIGURE 8.5 The posterior nasal cupula (PNC) in a newborn tree shrew (*Tupaia belangeri*—a) and bushbaby (*Otolemur crassicaudatus*—b). Relative immunoreactivity of the PNC and septal cartilage (SC) to collagen II antibodies is shown; note the attenuated expression in the PNC of the primate. (c–e) A late fetal lemur (*Lemur catta*) is shown in coronal sections near the posterior end of the PNC. d) Note that only cartilaginous remnants (CRs) of the PNC are found in the lamina propria of the olfactory recess (OR). These remnants lack a fibrous perichondrium, which is emphasized by trichrome stain in 5d (see Pc). (e) An enlarged view of the boxed region in 5c reveals that near the margins, the remnants exhibit marginal chondrocytes (MC) which are in contact with loose connective tissue. Scale bars: (a, b) 100 μm; (c) 250 μm; (d) 50 μm; (e) 20 μm. Stains: (c–e) Gomori trichrome (bone, dark green or red; collagen, green; cartilage, pale green), with hematoxylin counterstain (cell nuclei, blue or black).

recess in strepsirrhines, lined mostly with olfactory mucosa (Maier 1993a, 1993b; Smith et al. 2019). The PNC has been described to remain cartilaginous in many mammals at birth, including species of scandentians (Figure 8.5a), dermopterans, afrosoricids, chiropterans (Smith et al. 2017), and at least some rodents (Ruf 2020). However, few primates possess an intact cartilaginous PNC at birth, and it has only been observed in one strepsirrhine (*Varecia* spp.). Late fetal specimens of several species have a more complete PNC, but the cartilage exhibits uneven or weak reactivity to type II collagen antibodies, the main collagenous fiber of hyaline cartilages (Figure 8.5b; Smith et al. 2021b). This characteristic, as well as a diminished amount of matrix, and some chondroclast activity suggest that the PNC simply ceases to manufacture matrix; this is likely related to the loss of the supporting perichondrium, which occurs at the same time (Figure 8.5c–e). We suggest that due to early loss of the PNC, the consolidation of the basicranial and facial skeletons happens ontogenetically earlier in primates than other mammals. We further hypothesize that early loss of capsular cartilage at the posterior end limits chondral mechanisms for nasal complexity, such as interstitial expansion or endochondral ossification.

8.3.3 Postnatal Factors Influencing Nasal Complexity

A final mechanism for turbinal growth is available after turbinals are fully ossified. The newly deposited bone, in this case, radiates outward from the ends of existing turbinal lamellae. Such bone has been termed "appositional bone" or *Zuwachsknochen* (Starck 1967; Maier 1983, 1987), and this bone accounts for increased complexity that occurs postnatally. Turbinals are enhanced by appositional bone in numerous mammals, including primates (Smith et al. 2016), bats (Smith et al. 2021a), tree shrews (Zeller 1987), canids (Wagner and Ruf 2020), and likely many others. The addition of accessory parts, including scrolls, is inferred by comparing the turbinal after it initially ossifies (e.g., during infancy) with adults (Figures 8.4b–c).

Smith et al. (2016) considered these added, accessory lamellae to form epigenetically. The postnatal timing, as well as the degree to which respiratory surfaces of turbinals appear to be preferentially expanded by these newly formed lamellae, is notable and suggests that appositional bone is either added to enhance the air-conditioning properties of respiratory mucosa, to optimize airflow patterns for respiration and olfaction, or both. Experimental evidence supports the second hypothesis, since turbinal morphology of growing mice is altered following naris occlusion (Coppola et al. 2014).

8.4 CONCLUSIONS

The mammalian nasal fossa has fascinated morphologists for centuries (e.g., Paulli 1900; Negus 1958; Van Valkenburgh et al. 2011; DeLeon and Smith, 2014; Werneburg 2020). Paulli (1900) scrutinized the complexity of the largest of the world's terrestrial mammals in skeletonized crania, while the smallest of mammals were most often studied by microscopy (e.g., Adams 1972; Bhatnagar and Kallen 1975, LaRochelle and Baron 1989; Kollmann and Papin 1925). A most minute sensory organ of the nasal cavity, the vomeronasal organ, has similarly fascinated anatomists and histologists for centuries (e.g., Jacobson 1811; Gratiolet 1845; Bhatnagar and Smith 2007). The recent widespread availability of micro-computed tomography has yielded a new wave of highly detailed information on mammalian nasal anatomy. Such work confirms that living haplorhine primates share in common an extreme reduction in nasal complexity (e.g., Lundeen and Kirk 2019), but the developmental mechanisms that influence this reduction have mostly been the subject of inferences drawn from adult morphology (e.g., le Gros Clark 1959; Cartmill 1972).

In this chapter, we examined microanatomical evidence to glean details concerning nasal development, with a particular eye toward connective tissue maturation. Mechanisms that promote nasal cavity complexity span a lengthy duration of prenatal and postnatal time, beginning at the phase of embryonic skeletogenesis and culminating in postnatal appositional growth and pneumatic skull cavitation. The earliest mechanisms involve the formation of mesenchymal condensations that presage elements of the nasal capsule. In this stage, the size of adjacent functional matrices might impose spatial limitations on the rate of proliferation within mesenchymal condensations. Some authors have noted the extreme size of eyes in avian embryos, notably proportionately larger than in mammals, including primates (Abramyan et al. 2015). Because of this, eye size is suggested to affect other skull regions (Kawabe et al. 2013), yet experimental eye ablation in birds suggest the effect does not manifest itself during embryogenesis (Jomaa et al. 2020). In this regard, we point to another fundamental difference among vertebrates. The large eyes of sauropsid embryos (birds and reptiles) are largely ectopic, and this may diminish their constraining influence. In contrast, the eyes of mammals are more deeply "recessed" within the head during embryogenesis, even in primate species that eventually develop ectopic eyes (e.g., tarsiers: Figures 8.2a, b). As such, the eyes of mammals may have a greater impact on regional skeletogenesis, especially near directly adjacent functional matrices. If primates are deficient in nasal capsular components due to relative eye size and position, it appears not to be manifested at the stage of mesenchymal condensation but perhaps at stages of chondrogenesis.

Because nasal complexity unfolds during later prenatal and postnatal stages in mammals, integration of skeletogenesis in the orbit and neighboring regions involves multiple time points and developmental processes that may be interrupted. For example, turbinal formation is an iterative process that involves the generation of new mesenchymal condensations across prenatal ontogeny (Smith and Rossie 2008; Smith et al. 2021a; Kaucka et al. 2017), and thus prolonged spatial constraint may limit turbinal complexity even if the total number of turbinals remains constant, which is more or less true for strepsirrhines (Lundeen and Kirk 2019). It has been proposed that artificial selection may impose the same influence, limiting space for turbinals in domestic dogs with short faces (Wagner and Ruf 2021).

Although in vivo work is scant, evidence using cultured tissues indicates that chondrocyte differentiation and matrix production are generally inhibited by static compression (Lee et al. 2003; and see Responte et al. 2012). Thus, the basis for reduction by prenatal spatial constraint should be evidenced by cellular processes, including the rates of proliferation of chondrocytes of the nasal capsule as well as measurement of matrix production (e.g., type II collagen and ground substance components). Other evidence may be sought in terms of the extent of chondral modeling (e.g., chondroclast activity) or chondrocyte apoptosis. A comparative study of mammal embryos with varying eye size may be most informative, given the rarity of primate specimens available for study.

As the nasal skeleton ossifies, the direct adjacency of the eyes to some parts of the nasal fossa but not others can diminish size of specific parts of the nasal chamber by altering bone modeling patterns (Smith et al. 2019). This hypothesis has undergone some investigation using late fetal stage primates and other mammals. Smith et al. (2014b) show that the degree of eye convergence in tree shrews (*Tupaia* spp.) compared to primates is associated with markedly different patterns of osteoclastic activity along the walls of the bone orbit, and, further, the size of the olfactory bulb is a competing factor that actually may promote interorbital breadth. Thus, bone modeling patterns may affect nasal airway dimensions because of an influence on interorbital breadth. We hypothesize that midfacial modeling patterns may be influenced by constraint due to eye size in just such a manner; the functional implications may become clear as we accumulate more comparative and experimental data on nasal airflow patterns (e.g., Craven et al. 2010; Eiting et al. 2014; Smith et al. 2019). Overall, this demands a better understanding of cranial integration in primates, and we propose ontogenetic skeletal samples of primates offer ample opportunities to test predictions based on relative eye size and position.

8.5 REFERENCES

Abramyan, J., B. Thivichon-Prince, and J. M. Richman. 2015. Diversity in primary palate ontogeny of amniotes revealed with 3D imaging. *J Anat* 226:420–33.

Adams, D. R. 1972. Olfactory and non-olfactory epithelia in the nasal cavity of the mouse, *Peromyscus*. *Am J Anat* 133:37–50.

Barak, M. M. 2020. Bone modeling or bone remodeling: that is the question. *Am J Phys Anthropol* 172:153–5.

Barton, R. A., A. Purvis, and P. H. Harvey. 1995. Evolutionary radiation of visual and olfactory brain systems in primates, bats and insectivores. *Philos Trans R Soc Lond B Biol Sci* 348:381–92.

Bhatnagar, K. P., and F. C. Kallen. 1975. Quantitative observations on the nasal epithelia and olfactory innervation in bats. *Acta Anat* 91:272–82.

Bhatnagar, K. P., and T. D. Smith. 2007. Light microscopic and ultrastructural observations on the vomeronasal organ of *Anoura* (Chiroptera: Phyllostomidae). *Anat Rec* 290:1341–54.

Cartmill, M. 1972. Arboreal adaptations and the origin of the order primates. In *The functional and evolutionary biology of primates*, ed. R. Tuttle, 97–122. Chicago: Aldine.

Cartmill, M. 1980. Morphology, function, and evolution of the anthropoid postorbital septum. In *Evolutionary biology of the New World Monkeys and Continental Drift*, ed. R. L. Ciochon and A. Chiarelli, 243–74. New York: Plenum Press.

Cave, A. J. E. 1973. The primate nasal fossa. *Biol J Linn Soc, Lond* 5:377–87.

Coppola, D. M., B. A. Craven, J. Seeger, and E. Weiler. 2014. The effects of naris occlusion on mouse nasal turbinate development. *J Exp Biol* 217:2044–52.

Craven, B. A., E. G. Paterson, and G. S. Settles. 2010. The fluid dynamics of canine olfaction: unique nasal airflow patterns as an explanation of macrosmia. *J Roy Soc Interface* 7:933–43.

Cummings, J. R., M. N. Muchlinski, E. C. Kirk, S. J. Rehorek, et al. 2012. Eye size at birth in prosimian primates: life history correlates and growth patterns. *PLoS One* 7:e36097.

de Beer, G. R. 1937. *The Development of the Vertebrate Skull*. Chicago: Chicago University Press.

DeLeon, V. B., and T. D. Smith. 2014. Mapping the nasal airways: using histology to enhance CT-based three-dimensional reconstruction in *Nycticebus*. *Anat Rec* 297:2113–20.

Dieulafé, L. 1906. Morphology and embryology of the nasal fossae of vertebrates. *Ann Otol, Rhinol, Laryngol* 15:1–584.

Eiting, T. P., T. D. Smith, J. B. Perot, and E. R. Dumont. 2014. The role of the olfactory recess in olfactory airflow. *J Exp Biol* 217:1799–803.

Elder, S. H., J. H. Kimura, L. J. Soslowsky, M. Lavagnino, et al. 2000. Effect of compressive loading on chondrocyte differentiation in cultures of chick limb bud cells. *J Orthop Res* 18:78–86.

Enlow, D., and M. Hans. 1996. *Essentials of Facial Growth*. Philadelphia: W.B. Saunders.

Gratiolet, L. P. 1845. *Researches on Jacobson's organ*. Thése pour le Doctoral Médecine. Paris.

Hall, B. K. 2015. *Bones and Cartilage*, 2nd ed. New York: Academic Press.

Hall, B. K., and T. Miyake. 2000. All for one and one for all: condensations and the initiation of skeletal development. *Bioessays* 22:138–47.

Jacobson, L. 1811. *Description anatomique d'un organe observe´ dans le mammiferes*. Paris: Ann Mus Nat d'Hist Nat.

Jeffery, N., K. Davies, W. Köckenberger, and S. Williams. 2007. Craniofacial growth in fetal *Tarsius bancanus*: brains, eyes and nasal septa. *J Anat* 210:703–22.

Jones, M. E. H., F. Gröning, R. M. Aspden, H. Dutel, et al. 2020. The biomechanical role of the chondrocranium and the material properties of cartilage. *Vertebr Zool* 70:699–715. doi:10.26049/VZ70-4-2020-10.

Jomaa, J., J. Martinez-Vargas, S. Essaili, N. Hainder, et al. 2020. Disconnect between the developing eye and craniofacial prominences in the avian embryo. *Mech Dev* 161:103596.

Kaucka, M., T. Zikmund, M. Tesarova, D. Gyllborg, et al. 2017. Oriented clonal cell dynamics enables accurate growth and shaping of vertebrate cartilage. *eLife* 6:e25902

Kawabe, S., T. Shimokawa, H. Miki, S. Matsuda, et al. 2013. Variation in avian brain shape: relationship with size and orbital shape. *J Anat* 223:495–508.

Kawasaki, K., and J. T. Richtsmeier. 2017. Association of the chondrocranium and dermatocranium in early skull formation. In *Building Bones: Bone Formation and Development in Anthropology* ed. C. J. Percival and J. T. Richtsmeier, 52–78. Cambridge: Cambridge University Press.

Kirk, E. C. 2006. Effects of activity pattern on eye size and orbital aperture size in primates. *J Hum Evol* 51:159–70.

Kollmann, M., and L. Papin. 1925. Etudes sur lémuriens. Anatomie compareé des fosses nasales et de leurs annexes. *Arch Morphol Gén Expér* 22:1–60.

Larochelle, L., and G. Baron. 1989. Comparative morphology and morphometry of the nasal fossae of four species of North American shrews (Soricinae). *Am J Anat* 186:306–14.

Lee, C. R., A. J. Grodzinsky, and M. Spector. 2003. Biosynthetic response of passaged chondrocytes in a type II collagen scaffold to mechanical compression. *J Biomed Mat Res A* 64:560–9.

Le Gros Clark, W. E. 1959. *The Antecedents of Man: An Introduction to the Evolution of the Primates*. Edinburgh: Edinburgh University Press.

Li, K. W., A. K. Williamson, A. S. Wang, and R. L. Sah. 2001. Growth responses of cartilage to static and dynamic compression. *Clin Orthop Relat Res* 391:S34–48.

Lieberman, D. E. 2011. *The Evolution of the Human Head*. Cambridge, MA: Harvard University Press.

Luckett, W. P., and W. Maier. 1982. Development of deciduous and permanent dentition in *Tarsius* and its phylogenetic significance. *Folia Primatol* 37:1–36.

Lundeen, I. K., and E. C. Kirk. 2019. Internal nasal morphology of the Eocene primate *Rooneyia viejaensis* and extant Euarchonta: using μCT scan data to understand and infer patterns of nasal fossa evolution in primates. *J Hum Evol* 132:137–73.

Maier, W. 1980. Nasal structures in old and new world primates. In *Evolutionary Biology of the New World Monkeys and Continental Drift*, ed. R. L. Ciochon and A. B. Chiarelli, 219–41. New York: Plenum Press.

Maier, W. 1983. Morphology of the interorbital region of *Saimiri sciureus*. *Folia Primatol* 42:277–303.

Maier, W. 1987. The ontogenetic development of the orbitotemporal region in the skull of *Monodelphis domestica* (Didelphidae, Marsupialia), and the problem of the mammalian alisphenoid. In *Morphogenesis of the Mammalian Skull (Mammalia depicta 13)*, ed. H. J. Kuhn, and U. Zeller, 71–90. Hamburg: Parey.

Maier, W. 1993a. Cranial morphology of the therian common ancestor, as suggested by the adaptations of neonate marsupials. In *Mammal Phylogeny: Mesozoic Differentiation, Multituberculates, Monotremes, Early Therians, and Marsupials*, ed. F. S. Szalay, M. J. Novacek, and M. C. McKenna, 165–81. New York: Springer-Verlag.

Maier, W. 1993b. Zur evolutiven und funktionellen Morphologie des Gesichtsschädels der Primaten. *Z Morphol Anthropol* 79:279–99.

Maier, W., and Ruf I. 2014. Morphology of the nasal capsule of primates—with special reference to *Daubentonia* and *Homo*. *Anat Rec* 297:2018–30.

Moss, M. L., and S. Greenberg. 1967. Functional cranial analysis of the human maxillary bone. *Angle Orthodontist* 37:151–64.

Moss, M. L., and R. W. Young. 1960. A functional approach to craniology. *Am J Phys Anthropol* 18:281–92.

Napier, J. P., and P. Napier. 1967. *A Handbook of Living Primates*. London: Academic Press.

Negus, V. 1958. *The Comparative Anatomy and Physiology of the Nose and Paranasal Sinuses*. Livingston: Edinburgh and London.

Nett, E. M., and M. J. Ravosa. 2019. Ontogeny of orbit orientation in primates. *Anat Rec* 302:2093–104.

Nevo, O., and E. W. Heymann. 2015. Led by the nose: olfaction in primate feeding ecology. *Evol Anthropol* 24:137–48.

O'Connor, P. M. 2006. Postcranial pneumaticity: an evaluation of soft-tissue influences on the postcranial skeleton and the reconstruction of pulmonary anatomy in archosaurs. *J Morphol* 267:1199–226.

Pang, B., K. K. Yee, F. W. Lischka, N. E. Rawson, N. E., et al. 2016. The influence of nasal airflow on respiratory and olfactory epithelial distribution in felids. *J Exp Biol* 219:1866–74.

Paulli, S. 1900. Über die pneumaticität des Schädels bei den Säugethieren. Eine morphologische Studie III. Über die morphologie des Siebbeins und die der Pneumaticität bei den Insectivoren, Hyracoideen, Chiropteren, Carnivoren, Pinipedien, Edentaten, Rodentiern, Prosimiern und Primaten. *Morph Jb* 28:483–564.

Pitirri, M. K., K. Kawasaki, and J. T. Richtsmeier. 2020. It takes two: building the vertebrate skull from chondrocranium and dermatocranium. *Vertebr Zool* 70:587–600.

Ravosa, M. J., D. G. Savakova, K. R. Johnson, and W. L. Hylander. 2006. Primate origins and the function of the circumorbital region: what load got to do with it? In *Primate Origins: Adaptations and Evolution*, ed. M. J. Ravosa and M. Dagosto, 285–28. New York: Springer.

Responte, D. J., J. K. Lee, J. C. Hu, and K. A. Athanasiou. 2012. Biomechanics-driven chondrogenesis: from embryo to adult. *FASEB J* 26(9):3614-24. doi:10.1096/fj.12-207241.

Robling, A. G., and S. D. Stout. 1999. Morphology of the drifting osteon. *Cells Tissues Organs* 164:192–204.

Rosenberger, A. L., T. D. Smith, V. B. DeLeon, A. M. Burrows, et al. 2016. Eye size and set in small-bodied fossil primates: a three-dimensional method. *Anat Rec* 299:1671–89.

Ross, C. F. 1995. Allometric and functional influences on primate orbit orientation and the origins of the Anthropoidea. *J Hum Evol* 29:201–27.

Rossie, J. B. 2006. Ontogeny and homology of the paranasal sinuses in Platyrrhini (Mammalia: Primates). *J Morphol* 267:1–40.

Ruf, I. 2020. Ontogenetic transformations of the ethmoidal region in Muroidea (Rodentia, Mammalia): new insights from perinatal stages. *Vertebr Zool* 70:383–415.

Schultz, A. H. 1940. The size of the orbit and of the eye in primates. *Am J Phys Anthropol* 26:389–408.

Simons, E. L., and D. T. Rasmussen. 1989. Cranial morphology of *Aegyptopithecus* and *Tarsius* and the question of the tarsier-anthropoidean clade. *Am J Phys Anthropol* 79:1–23.

Smith, T. D., K. P. Bhatnagar, J. B. Rossie, B. A. Docherty, et al. 2007. Scaling of the first ethmoturbinal in nocturnal strepsirrhines: olfactory and respiratory surfaces. *Anat Rec* 290:215–37.

Smith, T. D., B. A. Craven, S. M. Engel, and C. J. Bonar, et al. 2019. Nasal airflow in the pygmy slow loris (*Nycticebus pygmaeus*) based on a combined histological, computed tomographic and computational fluid dynamics methodology. *J Exp Biol* 222. doi:10.1242/jeb.207605.

Smith, T. D., A. Curtis, K. P. Bhatnagar, and S. E. Santana. 2021a. Fissures, folds and scrolls: the ontogenetic basis for complexity of the nasal cavity in a fruit bat (*Rousettus leschenaultii*). *Anat Rec* 304:883–900. doi:10.1002/ar.24488.

Smith, T. D., V. B. DeLeon, and A. L. Rosenberger. 2013. At birth, tarsiers lack a postorbital bar or septum. *Anat Rec* 296:365–77.

Smith T. D., V. D. DeLeon, C. J. Vinyard, and J. W. Young. 2020. *Skeletal Anatomy of the Newborn Primate*. Cambridge: Cambridge University Press.

Smith, T. D., T. P. Eiting, and K. P. Bhatnagar. 2015. Anatomy of the nasal passages in mammals. In *Handbook of Olfaction and Gustation*, 3rd ed, ed. R. L. Doty, 37–62. New York: Wiley.

Smith, T. D., T. P. Eiting, C. J. Bonar, and B. A. Craven. 2014a. Nasal morphometry in marmosets: loss and redistribution of olfactory surface area. *Anat Rec* 297:2093–104.

Smith, T. D., E. S. Kentzel, J. M. Cunningham, A. E. Bruening, et al. 2014b. Mapping bone cell distributions to assess ontogenetic origin of primate midfacial form. *Am J Phys Anthropol* 154:424–35.

Smith, T. D., M. C. Martell, J. B. Rossie, and C. J. Bonar, et al. 2016. Ontogeny and microanatomy of the nasal turbinals in lemuriformes. *Anat Rec* 299:1492–510.

Smith, T. D., M. J. McMahon, M. E. Millen, C. Llera, et al. 2017. Growth and development at the sphenoethmoidal junction in perinatal primates. *Anat Rec* 300:2115-37.

Smith, T. D., and J. B. Rossie. 2008. The nasal fossa of mouse and dwarf lemurs (Primates, Cheirogaleidae). *Anat Rec* 291:895–915.

Smith, T. D., J. B. Rossie, G. M. Cooper, E. L. Durham, et al. 2012. Microanatomical variation of the nasal capsular cartilage in newborn primates. *Anat Rec* 295:950–60.

Smith, T. D., J. B. Rossie, G. M. Cooper, R. M. Schmieg, et al. 2011. Comparative micro CT and histological study of maxillary pneumatization in four species of New World monkeys: the perinatal period. *Am J Phys Anthropol* 144:392–410.

Smith, T. D., M. I. Siegel, and Bhatnagar, K. P. 2003. Observations on the vomeronasal organ of prenatal *Tarsius bancanus borneanus* with implications for ancestral characteristics. *J Anat* 203:473–81.

Smith, T. D., A. Ufelle, J. J. Cray, S. B. Rehorek, et al. 2021b. Inward collapse of the nasal cavity: perinatal consolidation of the midface and cranial base in primates. *Anat Rec* 304:939-57. doi:10.1002/ar.24537.

Starck, D. 1967. Le crâne des mammifères. In *Traité de Zoologie, XVI (1)*, ed. P.-P. Grassé, 405–549. Paris: Masson.

Szalay, F. S., and E. Delson. 1979. *Evolutionary History of the Primates*. New York: Academic Press.

Thorpe, S. D., C. T. Buckley, T. Vinardell, F. J. O'Briem, et al. 2008. Dynamic compression can inhibit chondrogensis of mesenchymal stem cells. *Biochem Biophys Res Comm* 377:458–62.

Van Valkenburgh, B., A. Curtis, J. X. Samuels, D. Bird, et al. 2011. Aquatic adaptations in the nose of carnivorans: evidence from the turbinates. *J Anat* 218:298–310.

Van Valkenburgh, B., T. D. Smith, and B. A. Craven. 2014. Tour of a labyrinth: exploring the vertebrate nose. *Anat Rec* 297:1975–84.

Verna C., D. Zaffe, and G. Siciliani. 1999. Histomorphometric study of bone reactions during orthodontic tooth movement in rats. *Bone* 24:371–9.

Wagner F., and Ruf I. 2021. 'Forever young'—postnatal growth inhibition of the turbinal skeleton in brachycephanic dog breeds (*Canis lupus familiaris*, Canidae, Carnivora). *Anat Rec* 304:154-89. doi:10.1002/ar.24422.

Werneburg, I. 2020. Editorial to the special issue (virtual Issue) 2019/2020 – recent advances in chondrocranium research. *Vertebr Zool* 69/70:I-III. doi:10.26049/VZ-69-70-Special-Issue.

Yee, K. K., B. A. Craven, C. J. Wysocki, and B. Van Valkenburgh. 2016. Comparative morphology and histology of the nasal fossa in four mammals: gray squirrel, bobcat, coyote, and white-tailed deer. *Anat Rec* 299:840–52.

Zeller, U. 1987. Morphogenesis of the mammalian skull with special reference to *Tupaia*. In *Morphogenesis of the Mammalian Skull*, ed. H. J. Kuhn and U. Zeller, 17–50. New York: Springer.

9 Stem Cells in Primate Evolution

Emily L. Durham and M. Kathleen Pitirri

CONTENTS

9.1 Introduction 213
9.2 Stem Cells: A Definition 214
9.2.1 Stem Cell Classifications 214
9.2.2 Stem Cell Location 215
9.2.3 Stem Cell Phenotype 216
9.3 Comparative Stem Cell Biology in Primates 218
9.3.1 Conservation in Primate Stem Cell Biology 219
9.3.2 Variation in Primate Stem Cell Biology 220
9.4 Stem Cells and the Evolution of Primate Characteristics 222
9.4.1 Primate Life History: Long, Slow Development and Delayed Aging 223
9.4.2 Stem Cells Facilitate Increased Length of Development in Primates 224
9.4.3 Stem Cells Contribute to Delayed Aging in Primates 225
9.5 Summary 227
9.6 Acknowledgments 228
9.7 References 228

9.1 INTRODUCTION

As a consequence of its origins in comparative anatomy, biological anthropology has a long history of exploring morphological and behavioral similarities and differences among primate species as fundamental to the study of primate and human evolution. Throughout the history of our field, this approach has focused on the phenotypic homology of structures and traits, which continues to prevail in modern biological anthropology (Lieberman 1999, 2000; Baab et al. 2012). One significant aspect of homology that has received relatively little attention in biological anthropology is the importance of shared cellular mechanisms throughout Primates. In addition to similar structures in similar spaces and relative positions, closely related species also share fundamental cell types and morphogenetic processes, such as differentiation, proliferation, migration, and interactions (Atchley and Hall 1991; Hall 2007, 2012; Pfefferle and Wray 2013; Wunderlich et al. 2014; Glinsky and Barakat 2019; Li et al. 2019). As suggested by Hall (2007, 2012), it is modification to these fundamental processes that catalyze the evolution of structures—and it is those traits that are of interest to biological anthropologists. Stem cells, because of their role in initiating development and continued function throughout life, offer a particularly informative line of inquiry. Though primates are incredibly diverse, primates have homology in stem cell types, niches, and processes. Given their role in development, stem cells possess the capacity to produce variation (Yang et al. 2012; Skelly et al. 2020), which is critical for natural selection and can directly influence the evolution of a species. Exploring differences and similarities in the function of stem cells across an order as diverse and complex as Primates can enhance our understanding of the evolution of characteristics that are specific to this group. Here, we provide a review for biological anthropologists that defines stem cells, highlights their important functions in development and maintenance throughout life, and outlines what is currently known about stem cells in different primate species. We then use an important topic in human and primate evolution, the evolution of slow life history characteristics, as an example of how studying stem cells can be applied in biological anthropology to understand the mechanisms underlying significant events in human and primate evolution.

9.2 STEM CELLS: A DEFINITION

To understand the role of stem cells in the evolution of humans and nonhuman primates, it is necessary to accurately define this special type of cell. Typically, all cells undergo a process, differentiation, by which they mature to take on a more distinct morphology and function. Stem cells are characteristically undifferentiated or only partially differentiated. In their earliest uncommitted state, stem cells have the ability to replace all cell types within an organism while maintaining an adequate population of stem cells (Li and Xie 2005; Oh et al. 2014; Zakrzewski et al. 2019). The ability of a stem cell to give rise to more stem cells is termed *self-renewal*. The characteristic of self-renewal coupled with the state of being undifferentiated are characteristics specific to stem cells (Li and Xie 2005; Morrison and Spradling 2008; Hartenstein 2013; Zakrzewski et al. 2019). Like all other cell types, stem cells are primarily characterized by their function and secondarily identified by their location and phenotype.

9.2.1 Stem Cell Classifications

Stem cells are classified according to a hierarchy based on their specific capacity to differentiate (Figure 9.1). *Totipotent* stem cells can differentiate into any cell type, giving them the ability to create an entire organism, including extra-embryonic tissues. These cells are completely unrestricted (Li and Xie 2005; Suwińska 2012; Baker and Pera 2018). *Pluripotent* stem cells can differentiate into all the different cell types that contribute to an individual. In contrast to totipotent cells, pluripotent

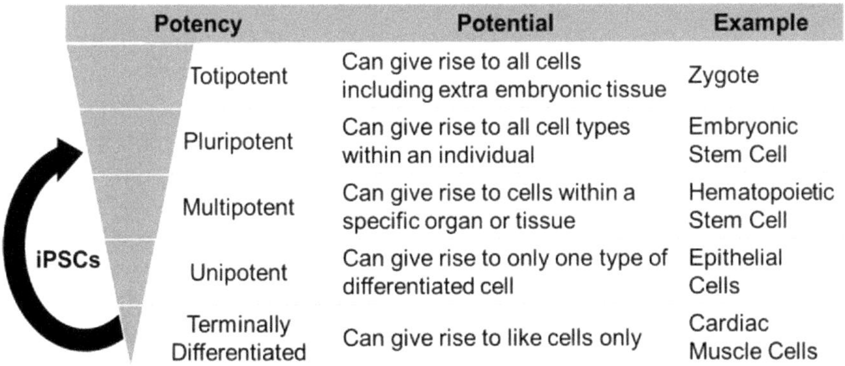

FIGURE 9.1 Stem cell hierarchy. Potency of stem cells decreases from top to bottom. Induced pluripotent stem cells (iPSCs) are made with terminally differentiated cells that are reprogrammed to a pluripotent embryonic stem cell like state. Once reprogrammed, iPSCs can be directed to differentiate into any cell type. (From Tewary et al. 2018.)

cells cannot create the extra-embryonic tissues necessary for development (Li and Xie 2005; Zakrzewski et al. 2019). *Multipotent* stem cells are a step down the stem cell potency hierarchy because they are further restricted in their differentiation potential. Each tissue or organ is composed of a variety of specialized cells, each with specific functions and abilities. Multipotent stem cells can differentiate into all cells specific to a tissue or organ mostly during initial development or as a response to injury or damage caused by disease (Slack 2000; Voog and Jones 2010; Zakrzewski et al. 2019). Unipotent stem cells produce only one type of differentiated cell and are largely responsible for steady state stem cell renewal (Gu and Sarvetnick 1993; Rosenberg et al. 1996; Alison et al. 1998; Slack 2000). Most cells that make up an organism are terminally differentiated, meaning they cannot give rise to other cell types and thus are at the bottom of the hierarchy. Terminally differentiated cells can proliferate to make more cells of the same type and are responsible for the general functions of life (Figure 9.1) (Li and Xie 2005; Lo and Parham 2009; Tewary et al. 2018).

9.2.2 Stem Cell Location

All stem cells reside within a microenvironment (niche) that supports their characteristic undifferentiated state and self-renewal while also facilitating their differentiation (Li and Xie 2005; Jones and Wagers 2008; Morrison and Spradling 2008; Voog and Jones 2010). Often stem cells are identified by their niche, an anatomically distinct location (the bulge region of the hair follicle or the basal layer of the epidermis) that supports the lifelong self-renewal of a specific stem cell type (Schofield 1978; Hoggatt et al. 2016). For example, hematopoietic stem cells reside within the bone marrow and vasculature and are responsible for forming and maintaining blood cell populations (Lemischka et al. 1986; Hoggatt et al. 2016). Stem cell niches can regulate the behavior of stem cells, coordinating activity in a dynamic fashion in response to short-term metabolic flux or other whole-organism changes, including injury and aging (Xie and Spradling 2000; Jones and Wagers 2008; Voog and Jones 2010). Though there is evidence that hematopoietic stem cells may help to create and regulate their own niche by regulating vascular leakage and responding to damage, these efforts are in coordination with adjacent terminally differentiated cells (Hoggatt et al. 2016). Stem cell niches thus provide an environment in which stem cells and adjacent differentiated cells can communicate and support one another towards proper development and maintenance of an organism (Hsu et al. 2011; Sato et al. 2011; Zhou et al. 2015; Hoggatt et al. 2016).

Development of an organism is in large part determined by the timing of differentiation and proliferation of stem cells, which in turn is dependent upon their niche and can be specific to a particular species (Barry et al. 2017; Baker and Pera 2018). For example, the rate of differentiation of human derived or isolated stem cells reflect a 9-month gestation requiring several months to achieve optimal differentiation and/or research utility, while similarly derived or isolated mouse stem cells differentiate much more quickly (Ebert et al. 2009; Shi et al. 2012; Espuny-Camacho et al. 2013; Barry et al. 2017). Stem cell niche is so important to the proper function of stem cells that if removed from their natural environment, stem cells tend to decrease in potential and become irregular in self-renewal and proliferative activities (Li and Xie 2005; Voog and Jones 2010). This indicates that stem cell niche is vital to facilitating the proper signals between stem cells and adjacent differentiated cells and that, without their niche, stem cells might act in unpredictable ways and can even fail to thrive (Hoggatt et al. 2016).

There is, however, an advantage to taking stem cells from their niche. In an artificial environment (cell culture system), adult multipotent stem cells can be given signals to differentiate more widely or along specific differentiation trajectories. For example, hematopoietic stem cells can be signaled through additives in cell culture media to become adipose, bone, or cartilage cells rather than being constrained to differentiating into blood-related cell types (Lemischka et al. 1986; Morrison and Spradling 2008; Voog and Jones 2010). The ability of stem cells taken from their niche to differentiate into multiple cell types makes them ideal for research (Schofield 1978; Tropepe et al. 2000; Pfefferle and Wray 2013; Cosgrove et al. 2014; Wunderlich et al. 2014; Tewary et al. 2018; Li et al. 2019). Stem cell plasticity allows for cells to be collected from abundantly available tissue, such as blood or fat, and then differentiated to other cell types, such as bone or cartilage, in order to answer questions about these less abundant yet vital tissues. Further, the ability to derive stem cells allows for non-destructive investigations of endangered and protected species. These advantages outweigh the potential confounding factors caused by removing a cell from its niche and thus removing the vital signals from adjacent cells (Medvedev et al. 2010).

9.2.3 Stem Cell Phenotype

In addition to being identified, supported, and at the same time constrained by their location/niche, stem cells, like all other cells, are covered with a coat of specialized proteins (Zhao et al. 2012; Lv et al. 2014; Maleki et al. 2014). These cell surface proteins can help to identify individual stem cells and define a stem cell population (Durham et al. 2019; Chen et al. 2020). For example, stem cells from the bone marrow niche express cell surface marker CD44 but do not have markers associated with hematopoietic stem cells (CD45 and CD34) on their surface (Hartenstein 2013; Lv et al. 2014; Maleki et al. 2014). When attempting to identify a specific stem cell population using cell surface marker phenotype, it is important to remember that protein markers can be transient and that populations of cells can overlap, expressing some but not all of the same markers (Lv et al. 2014; Maleki et al. 2014; Doro et al. 2017; Durham et al. 2019; Chen et al. 2020). For example, calvarial stem cells have been defined by different investigations as being positive for the protein Gli1 (Gli1+) (Zhao et al. 2015), Axin2+ (Maruyama et al. 2016), and Prx1+ (Lu et al. 2011). These markers describe stem cells within the calvarial suture niche. However, it remains unclear if stem cells in the calvarial suture express one, two, or all of these markers. Using an approach that utilizes multiple surface markers may more accurately define stem cell populations (Durham et al. 2019; Holmes et al. 2020).

Stem cells are often labeled as "embryonic" or "adult". Embryonic stem cells are self-renewing, limitlessly differentiating stem cells that are derived from the inner cell mass of the blastocyst during embryonic formation. Embryonic stem cells are pluripotent, being responsible for the development of a complete organism (Li and Xie 2005; Hartenstein 2013; Baker and Pera 2018; Zakrzewski et al. 2019). As development progresses, and organogenesis begins, embryonic stem cells form germ line stem cells that are responsible for reproduction later in life and somatic stem cells that are responsible for creating all the other tissues and organs that make up an individual (Li and Xie 2005; Zakrzewski et al. 2019).

Once embryonic stem cells begin to differentiate, a subset of them become adult stem cells. Adult stem cells are slightly more limited and specialized and include both germ line and somatic stem cells (Li and Xie 2005; Morrison and Spradling 2008; Voog and Jones 2010; Hartenstein 2013; Li et al. 2017, 2019; Zakrzewski et al. 2019). Adult stems cells are characterized as cells in an undifferentiated state that have the ability to self-renew, are located throughout an organism, and have the ability to differentiate into several terminally differentiated cell types (Li and Neaves 2006; Takahashi and Yamanaka 2006; Pfefferle and Wray 2013; Cosgrove et al. 2014; Maleki et al. 2014; Zhao et al. 2015; Zakrzewski et al. 2019).

Unlike embryonic stem cells, adult stem cells generate lineage specific cells emanating from a single germ layer of an embryo: mesoderm, ectoderm, or endoderm (Li and Xie 2005; Takahashi and Yamanaka 2006). Adult stem cells are involved in regeneration of damaged tissues and replenishment of dying cells in tissues that continually renew, such as blood, bone, gametes, and epithelia (Maleki et al. 2014; Zakrzewski et al. 2019). For example, hematopoietic stem cells are responsible for replenishing red and white blood cells as they wear out or are needed (Lemischka et al. 1986). Adult stem cells are also slow cycling, meaning that they only divide or proliferate when they are needed rather than on a set schedule (Voog and Jones 2010; Hartenstein 2013). Some adult stem cells lie dormant and are activated by certain life cycle stages or following injury (Morrison and Spradling 2008).

Although terminally differentiated cells cannot naturally differentiate into other cell types or regain their stem identity, induced pluripotent stem cells are an exception to this rule. Takahashi and Yamanaka (2006) identified four transcription factors necessary to program (to induce) adult terminally differentiated murine and human cells to have stem cell characteristics. To produce iPSCs, differentiated cells are transfected with Oct3/4, Sox2, c-Myc, and Klf4, four transcription factors that mimic protein expression in embryonic cells. By this process, fully mature terminally differentiated cells are transformed into iPSCs possessing increased potency relative to adult stem cells and placing them closer to embryonic stem cells within the stem cell hierarchy (Figure 9.1) (Takahashi and Yamanaka 2006; Wunderlich et al. 2014; Rowe and Daley 2019; Zakrzewski et al. 2019). In addition to expressing the same proteins and genes as embryonic stem cells, iPSCs have similar morphology, proliferative ability, and only a slightly restricted ability to differentiate (Takahashi and Yamanaka 2006; Rowe and Daley 2019). Once iPSCs are differentiated into multipotent adult stem cells or any other specific cell type, they take on the characteristic protein coat of cell surface markers associated with that cell type.

Of the multiple types of stem cells discussed previously, iPSCs are the most accessible for research involving primates, where sampling is difficult due to ethical concerns or a lack of resources when studying protected species (Lo and Parham 2009; Koplin 2019). iPSCs can be created from primary cells collected from abundant and accessible tissues, such as blood and skin, allowing stem cells and, after differentiation, other cell types, to be compared between and among primate species (Rowe and Daley 2019; Grogan and Perry 2020). Importantly, there are potential pitfalls with this methodology, as iPSCs can retain characteristics (e.g., through epigenetic "memory") of their primary cell type or have artifacts that accumulate during the de-differentiation procedure (Medvedev et al. 2010). However, the ability of iPSCs to give rise to multiple tissues and closely mimic the physiology and function of cells in vivo without the need for multiple individuals makes them a logical choice for experimental approaches aimed at evaluating hypotheses about the origins, trajectories, and evolution of specific traits in primates (Wunderlich et al. 2014; Rowe and Daley 2019).

9.3 COMPARATIVE STEM CELL BIOLOGY IN PRIMATES

Comparative stem cell biology studies the degree of relatedness among individuals within a defined population, in terms of genetic constitution and developmental potential of cell populations (Hartenstein 2013). Related to this, the field of functional genomics explores the functions and interactions of genes and proteins (Grogan and Perry 2020). Combining knowledge from comparative stem cell biology and functional genomics in primate species is essential for understanding traits that are unique to Primates. Stem cells in general, due to their role in initiating development, and iPSCs in particular, are useful for investigating the molecular basis for trait differences and similarities between species. Genomic data from adult stem cells isolated from multiple species of primates can provide vital insight into primate specific traits because stem cell genes and function are conserved across the order. Stem cells of all types (embryonic, adult, induced) offer a unique opportunity to investigate the cellular mechanisms that drive primate specific traits and their evolution (Pfefferle and Wray 2013).

9.3.1 Conservation in Primate Stem Cell Biology

Nucleic acids and proteins compared from multiple primate species indicate relatively small genetic inter-species distances, meaning that primates in general have relatively similar DNA sequences and proteins (Sibley and Ahlquist 1984; Goodman et al. 1994; Frazer et al. 2003; Perelman et al. 2011; Li et al. 2019; Grogan and Perry 2020). An assessment of iPSCs from a small sample of *Homo sapiens*, *Pan paniscus*, *Gorilla gorilla*, and *Macaca fascicularis* revealed similarities in gene regulatory networks responsible for development and the fundamental function of stem cells (Wunderlich et al. 2014). The lack of variation in gene expression related to stem cell maintenance among the few catarrhine species that have been examined suggests or agrees with a hypothesis of conservation of stem cell programming (Wunderlich et al. 2014). Catarrhines share many identical proteins in both stem and terminally differentiated cells. Proteins drive cell function, which explains the similarities in physiology across primate species (Daadi et al. 2014; Martínez and Conde-Valverde 2020). These data suggest that stem cells could be similar in both genetic constitution and function across Primates. However, this assertion needs to be evaluated in a more diverse taxonomic sample that includes strepsirrhines, tarsiers, platyrrhines, and an increased number of cercopithecine species.

Enhancer sequences, which augment the transcription of an associated gene, are highly conserved between humans and their closest evolutionary relatives, *Pan troglodytes* and *P. paniscus* (Glinsky and Barakat 2019; Li et al. 2019). When comparing *H. sapiens* with their closest living relatives, more than 90% of enhancer sequences associated with stem cells are conserved, suggesting close evolutionary ties. Expanding this comparison to include more distantly related *M. fascicularis* indicates that only 75% of enhancer sequences involved in stem cell function are conserved (Glinsky and Barakat 2019). These similar sequences are expressed in embryonic and adult stem cells and are part of the regulatory network that determines gene expression controlling development (Davidson 2010; Glinsky and Barakat 2019). Primates that possess similarities in anatomy, physiology, biochemistry, and development share similar genetic sequences governing stem cell function and activity (Daadi et al. 2014). Similar genes and proteins coupled with similar expression patterns in stem cells may be responsible, at least in part, for the phenotypic similarities among closely related primate species and perhaps for the maintenance of those traits that define the Order. A highly conserved regulatory network for stem cell function can be found in species with close phylogenetic relationships. Similarities in gene networks and transcriptional regulation across Primates could be responsible for primate-specific traits.

While highly conserved regions of the genome imply important functions (Grogan and Perry 2020), embryonic stem cells carry out the important function of initiating and controlling development, while adult stem cells maintain proper function of an organism by replenishing injured or malfunctioning cells. Assessment of gene sequences within embryonic stem cells from *M. fascicularis*, African great apes, and humans indicates conservation of the stem cell gene regulatory network since the origin of the common ancestor of all hominines (African great apes and humans) (Glinsky and Barakat 2019) ~15.1 to 12 million years ago (Moorjani et al. 2016). Thus, our current knowledge of primate stem cells indicates that across species stem cells generally serve the same purpose and perform their functions in a similar fashion, particularly in closely related Primates (e.g., hominines).

Stem cell niche is critical to stem cell function, which is controlled by cues from the cells' surroundings, as well as genetic programming (Schofield 1978; Li and Xie 2005; Voog and Jones 2010). In hominines and *M. fascicularis*, expression patterns for iPSCs cluster according to the type of somatic cell used to create them and then according to species (Wunderlich et al. 2014), following the tendency of cell types to be more tightly tied to their environment than their species (Brawand et al. 2011). Muscle-related stem cells from closely related species are more similar to one another than they are to stem cells from other niches, such as bone marrow within the same species (Morrison and Spradling 2008; Voog and Jones 2010). In Primates, who share similar stem cell niches as well as genetic constitutions, adult stem cells in particular function almost identically across species. For example, the circulatory systems of anthropoids are so similar that hematopoietic cells encountering atherosclerosis function in the same manner regardless of species (King et al. 1988; Daadi et al. 2014; Li et al. 2019; Rowe and Daley 2019).

9.3.2 Variation in Primate Stem Cell Biology

Likenesses among stem cells from different hominine species indicate that the last common ancestor of all hominines likely shared specific stem cell traits. Despite these similarities, human iPSCs have an expanded regulatory network as compared to stem cells from the closest living relatives of humans, suggesting that human and *Pan* stem cells have continued to evolve and change since diverging from each other (Glinsky and Barakat 2019). Interestingly, iPSCs from nonhuman primates are also similar to iPSCs from other mammals in terms of stem-related gene expression and stem cell function (Niu et al. 2017; Li et al. 2019). This pattern of similarity indicates that stem cells retain gene expression and function from a common ancestor while also continuing to add order-, species-, and individual-specific characteristics (Figure 9.2) (Pfefferle and Wray 2013; Daadi et al. 2014; Wunderlich et al. 2014; Niu et al. 2017; Li et al. 2019). Though there are similarities in adult stem cell function between closely related primate species, most genes show significant expression differences among *H. sapiens*, *P. paniscus*, *G. gorilla*, and *M. fascicularis*, potentially accounting for or contributing to species-specific traits (Wunderlich et al. 2014).

Genetic differences between species have a strong impact on embryonic and adult stem cell gene expression despite similar cellular characteristics (Wunderlich et al. 2014). Transcriptional regulation differs between mice and Primates. Stem cells from both Primates and mice use the same genes to express classic stem cell characteristics of potency, self-renewal, and lack of differentiation but use different gene networks and/or timing of expression during the processes involved in development and tissue maintenance. Despite similar gene expression, the epiblast and hypoblast of primate embryos form a bilaminar disc, while a cup shape is formed in mouse embryos. In addition to different shapes, the timing of gene expression differs between species with the development of an embryo taking 60 days in humans, 46 days in macaques, and 16 days in mice (Theiler 1972; O'Rahilly 1979; Bourne 2014; Li et al. 2019). The same genes expressed to different degrees (e.g., more copies of

FIGURE 9.2 Comparing stem cells between species. The core stem cell related genes and functions are the same between closely (primates and humans, blue-green area of overlap) and more distantly related mammals (humans and rodents, arrow, yellow-blue-green area of overlap). Similarities in stem cells (overlap) increases according to relatedness between species. Some expression patterns and stem cell functions are species specific (areas without overlap).

the same gene) or for different amounts of time (e.g., less time in mice, more time in humans) drive differences among species (like the ones identified previously), while embryonic, adult, and induced stem cells in general remain similar.

Conserved sequences make up most of the functional enhancers in human stem cells, meaning that the genes associated with stemness (molecular processes underlying the core stem cell properties) are conserved evolutionarily and that important differences in stem cells and the genes that regulate them are responsible for human and primate specific traits (King and Wilson 1975; Wunderlich et al. 2014; Glinsky and Barakat 2019). A closer investigation indicates some specific and important changes to embryonic and adult stem cells throughout the evolution of Primates. For example, changes to the genetic control of cell pluripotency occurred during primate evolution, allowing for increased regulation of cell differentiation and pacing of development as compared to other mammals (e.g., rodents) (Barry et al. 2017). In turn, these changes to embryonic and adult stem cell regulation may have allowed for an increase in the number of unique, terminally differentiated cells to arise from a particular stem cell (Glinsky and Barakat 2019). These changes in the control of cell pluripotency in primate evolution could, in part, lead to specialized phenotypes that typify Primates as an order, such as relatively larger brains (Jerison 1973; Martin 1990; Allman 1999) and eyes (King et al. 1988; Ross and Martin 2007; Ross and Kirk 2007) compared to other mammals.

When iPSCs from multiple cercopithecine species are compared, the similarities among stem cells tends to be concentrated on cellular functions and protein coding regions as opposed to other DNA regions (King and Wilson 1975; Wunderlich et al. 2014; Grogan and Perry 2020). This indicates that the obvious phenotypic diversity across Primates is potentially driven by small changes in other regions of DNA and/or by changes to the timing and/or patterns of gene and protein expression in stem cells. Species-specific changes to the pace of development (acceleration or deceleration) and/or changes to gene regulatory networks (when, if, and for how long specific genes are expressed) can potentially explain much of the observed phenotypic diversity throughout Primates (Davidson 2010; Wunderlich et al. 2014; Li et al. 2019). For example, though humans and chimpanzees are very closely related and share very similar embryonic and adult stem cells, stem cells from humans express more genes related to cell cycling over a longer period of development as compared to other great apes (Waterson et al. 2005; Pfefferle and Wray 2013). More cell cycling in humans relative to other great apes could result in an overall increase in cell number and contribute to the relatively increased encephalization that is characteristic of humans (Pfefferle and Wray 2013; Bakken et al. 2016; Lesciotto and Richtsmeier 2019).

Differences between human and other catarrhine (*G. gorilla*, *P. paniscus*, *P. troglodytes*, *M. fascicularis*) adult stem cells and iPSCs may in some cases be related to species-specific factors (Wunderlich et al. 2014). For example, hepatic stem cells may be unique in humans owing to an increased necessity for metabolic control and detoxification due to variation in the human diet (Brawand et al. 2011; Wunderlich et al. 2014; Martínez and Conde-Valverde 2020). Similarly, adipose stem cells from *P. troglodytes* have more immunity-related gene expression, most likely resulting from species-specific responses to environmental factors (Pfefferle and Wray 2013).

9.4 STEM CELLS AND THE EVOLUTION OF PRIMATE CHARACTERISTICS

Stem cells are responsible for the production of all cells and organs in an organism; therefore, the characteristics that are unique to Primates can be attributed, at least partly, to stem cells. As initiators and controllers of development, embryonic and adult stem cells play a role throughout primate life, including the timing of developmental events and aging (Rossi et al. 2005; Oh et al. 2014; Barry et al. 2017; Li et al. 2019; Martínez and Conde-Valverde 2020). The following sections focus on how stem cells might contribute to the analysis of primate life history characteristics that have been the focus of anthropological research. The goal of our discussion is to provide examples of how investigating stem cells can potentially identify mechanisms driving the evolution of specific traits across and within primate groups. Further, we hope that our examples may be a catalyst for future hypothesis development and testing employing stem cells to inform us about important aspects of primate evolution.

9.4.1 Primate Life History: Long, Slow Development and Delayed Aging

In mammals, life history is best summarized by key variables, such as litter size, gestation length, weaning age, interbirth interval, and maximum longevity (Charnov 1991). Primates are unique among mammals, having the slowest life histories (Harvey and Clutton-Brock 1985; Harvey et al. 1987; Marsh 1988; Promislow and Harvey 1990; Ross 1998). Even when controlling for body size, primates have significantly longer gestation periods, larger neonates, low reproductive rates, slower postnatal growth, and later ages of maturity than other mammals (Case 1978; Western 1979; Kirkwood 1985; Martin and MacLarnon 1985; Wootton 1987; Martin and MacLarnon 1988; Ross 1988; Watts 1990; Charnov 1991). While Primates are among the longest-lived mammals, they do not exceed expectations for life span when accounting for body size and metabolism (Austad and Fischer 1992; Austad 1997; Speakman 2005). As a result, compared to other mammals, Primates exhibit a unique "slow" life history that is marked by longer periods of development and growth and a relatively delayed onset of aging.

Though this pattern of elongated growth and development and delayed aging in Primates stands out among mammals, there is also variation in this life history trajectory among Primates that demonstrates a phylogenetic distribution (Harvey and Clutton-Brock 1985; Kamilar and Cooper 2013). In general, anthropoids have longer periods of development and life span than strepsirrhines. When comparing hominoids with all other Primates, humans and great apes have remarkably extended periods of development and life span (Austad and Fischer 1992; Kamilar and Cooper 2013), with humans at the most extreme end of this trend (Kamilar and Cooper 2013). Since a slow life history is a hallmark of Primates and especially hominoids, the evolutionary significance of variation in life history between Primates and other mammals, as well as variation among Primates, is of particular interest to anthropologists. In order to gain a better understanding of the evolution of slow life history within this order, research has mainly focused on reconstructing the length of development and growth, as well as longevity in fossil primates (Kelley and Smith 2003; Nargolwalla et al. 2005; Schwartz et al. 2005; Schwartz 2012). The majority of this research has used biological markers to determine the length of dental development (Kelley and Smith 2003; Nargolwalla et al. 2005; Schwartz et al. 2005; Smith et al. 2007; Schwartz 2012; Dean and Cole 2013; López-Torres et al. 2015), age at death estimations (Thompson and Trinkaus 1981; Dean et al. 1986, 1993; Caspari and Lee 2004; Dean and Liversidge 2015; Le Cabec et al. 2017), and pelvic shape and size to infer neonate brain size (Simpson et al. 2008; Weaver and Hublin 2009; Gruss and Schmitt 2015). While these approaches are extremely informative, they are also limited by availability of appropriate specimens and do not allow us to query or understand the mechanisms driving life history variation in Primates. In other words, current approaches allow us to identify *when* differences in primate life history occurred throughout evolution, but they do not allow us to understand *how* these differences occur.

9.4.2 Stem Cells Facilitate Increased Length of Development in Primates

The increase in developmental time in Primates relative to other mammals is driven, at least partially, by embryonic stem cells. Stem cells, both adult and embryonic, involved in development are autonomous and species specific and maintain developmental timing even when moved to an artificial environment. Embryonic, adult, and induced stem cells from catarrhines differentiate slowly as compared to stem cells from other mammals (Pfefferle and Wray 2013; Wunderlich et al. 2014; Barry et al. 2017). Likewise, it takes human embryonic stem cells 20 days to differentiate in culture, while mouse stem cells take only 6 days under the same conditions (Barry et al. 2017). Though niche is important to stem cell function, the autonomous function of stem cells, including the time required to differentiate, is largely maintained outside of a specific niche, making assessments using in vitro systems reasonably representative of cell functions in vivo (Li and Xie 2005; Morrison and Spradling 2008; Voog and Jones 2010; Barry et al. 2017). In vitro systems could be used to understand the role of stem cells in determining differences in developmental timing between Primates and non-primates and among Primate species that demonstrate different life spans, such as long-lived humans and chimpanzees and shorter-lived primates like marmosets (Barrickman et al. 2008; Daadi et al. 2014).

Though development is generally slower in Primates compared to other mammals, there are variations in developmental timing within Primates (Hrvoj-Mihic et al. 2014). For example, humans develop more slowly than chimpanzees and rhesus macaques. Human gestation is 280 days compared to 220 days for chimpanzees and 165 days in rhesus macaques (Holly Smith et al. 1994; De Magalhaes and Costa 2009; Yuan et al. 2011). These differences may be driven by variation in gene expression within stem cells. Stem cells are known to be autonomous and species specific, particularly in their least differentiated form (Barry et al. 2017). Potential candidate genes that deserve further exploration include genes in the Wnt, FGF, and TGFβ families, as well as PAX6, SOX1, OLIG2, and BMP4 (Yamaguchi and Rossant 1995; Hogan 1996; Sun et al. 1999; Yamaguchi 2001; Keller 2005; Barry et al. 2017). Exploring gene expression across species can be difficult, especially when the timing of expression is different depending on the species. Dynamic time warping is a technique that is used to identify patterns within a time series. Though dynamic time warping is most often used for speech recognition to allow for human interactions with machines (Amin and Mahmood 2008), it has recently been used to compare gene and/or protein expression patterns across species during a specific life stage such as infancy, adolescence, or gestation (Barry et al. 2017). This technique can identify similar progressions of expression that occur at different speeds between species. Rather than comparing the precise developmental timing of expression of a specific gene or protein between groups, dynamic time warping compares the duration and order of gene expression during a defined timespan. If applied to embryonic stem cells from humans, chimpanzees, and rhesus macaques, dynamic time warping

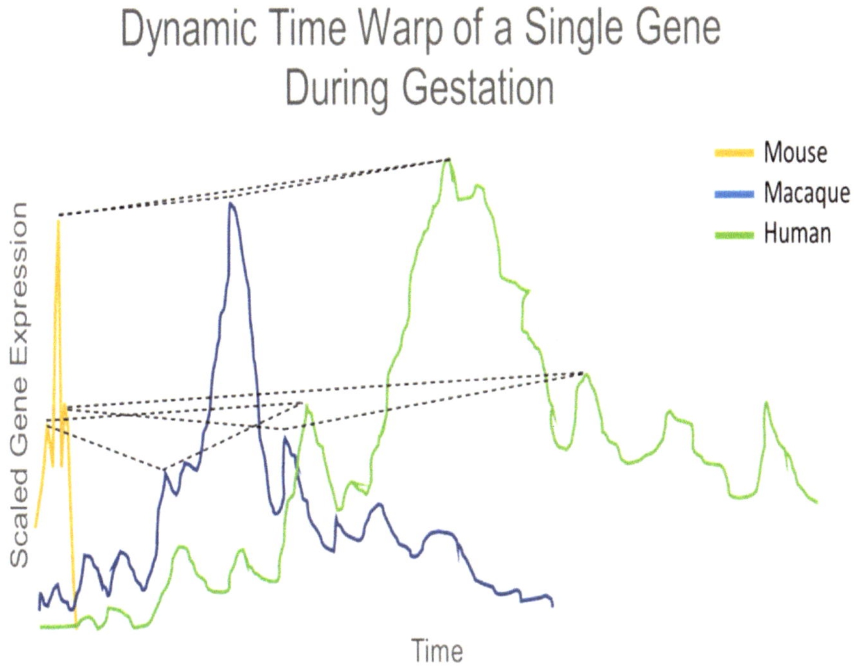

FIGURE 9.3 Dynamic time warp comparing expression between species during gestation. This gene is accelerated in mouse (yellow) compared to macaque (blue) and human (green) using dynamic time warping analysis by identifying and warping similarly patterned regions (connected by dotted lines). This technique could be applied to all genes or a target panel to identify genes expressed earlier in one species (mouse) compared to another (macaque, human). Dynamic time warping can be applied to iPSCs in vitro given the same culture conditions and could identify gene patterns such as delayed or prolonged expression associated with species specific traits. (Modified from Barry et al. 2017.)

could identify genes driving species-specific phenotypes and traits (Figure 9.3) (Yuan et al. 2011; Zhao et al. 2014; Barry et al. 2017).

9.4.3 Stem Cells Contribute to Delayed Aging in Primates

The genetic network responsible for cell cycle regulation in stem cells is highly conserved, particularly in closely related species (Pfefferle and Wray 2013; Hrvoj-Mihic et al. 2014; Barry et al. 2017). Thus, stem cells from Primates proliferate and mature at a slower pace relative to other mammals, which likely contributes to a delayed onset of aging and longer periods of development in Primates as compared to other mammals. Regulation of cell cycle and maturation rate are maintained both when stem cells are isolated directly from an individual and when they are induced (Rossi et al. 2005; Pfefferle and Wray 2013; Barry et al. 2017). The maintenance of cell cycle pace is regulated at numerous levels: broadly at the level of the species and specifically at the cellular level within an individual stem cell. Additionally, stem cells may proliferate (cycle) or mature at a different pace depending upon extracellular factors including to some degree niche (Fehrer and Lepperdinger 2005; Rossi et al. 2005; Shepherd et al. 2007; Barry et al. 2017). Primate stem cells therefore receive multiple levels of signals to extend development and delay aging, including at the organismal, system, and cellular level, allowing for and contributing to the slow life history trajectory that characterizes Primates. Therefore, iPSCs from multiple primate species can be compared to investigate hypotheses relating to the cellular mechanisms of aging.

The autonomy of stem cells from Primates is not reserved to developmental timing (Rossi et al. 2005; Shepherd et al. 2007). Like all mammals, Primates of all shapes and sizes experience age-related changes in anatomy, physiology, function, and behavior (King and Wilson 1975; Goncharova and Lapin 2002; Peters 2002; Herbig et al. 2006; Morrison and Spradling 2008; Cosgrove et al. 2014; Mattison and Vaughan 2017). These age-related changes are at least partially driven by changes to stem cells. As they age, the ability of adult stem cells to respond to injury is dampened, proliferation becomes dysregulated, and adult stem cells experience a decline or change in function (Fehrer and Lepperdinger 2005; Rossi et al. 2005; Zakrzewski et al. 2019). Degenerative changes in adult stem cells and their niches leads to a progressive decline in function and regenerative capacity that can be correlated with age-related disease mechanisms (Shepherd et al. 2007; Oh et al. 2014). In Primates, these degenerative changes to adult stem cells begin later in life, accounting for the delay of aging that is characteristic of this order.

For example, Primates, which are known for unique visual system features such as forward-facing eyes with a wide bionocular visual field, high visual acuity (3D stereoscopic vision), and expansion/proliferation of functional areas in their brains devoted to processing visual information (see chapter by Marchetto and Semindeferi in this volume) (Barton 1998; Kirk 2004; Ross and Kirk 2007), experience degradation of vision with age (Spear et al. 1994; Barton 1998; Barrickman et al. 2008). As Primates age, adult stem cells within the retina reduce their self-renewal, differentiation, and proliferation, leading to a reduction of specialized vision characteristics (Spear et al. 1994; Tropepe et al. 2000; Ross and Kirk 2007). Embryonic and adult stem cells are involved with creating highly cellularized avascular regions within the eye that are unique to Primates and specialized for visual acuity (King et al. 1988; Tropepe et al. 2000; Provis 2001; Kirk 2004; Ross and Kirk 2007). Adult stem cells in Primates must maintain a specific schedule to produce enough progenitor cells to facilitate the unique characteristics of Primates while also enduring normal degradation with age. Retinal stem cells, the adult stem cells resident to the visual system, are not known for their regenerative ability; thus, visual accommodation decreases with age in Primates due to a lack of adult stem cell activity within that niche (Tropepe et al. 2000; Ramsden et al. 2013; Chao et al. 2017). Degenerative diseases of the visual system are caused by cells within the eye dying and not being replaced; however, when aged retinal stem cells are replaced with embryonic stem cells, restoration of vision is possible (Spear et al. 1994; Ramsden et al. 2013; Chao et al. 2017). In Primates, long-lived adult stem cells, including those in the visual system, have been cultivated over evolution to support longer life span and use of specialized complex organs (Spear et al. 1994; Shepherd et al. 2007). Since hominoids have the most delayed onset of aging compared to other Primates, hominoids could have longer-lived or slower-cycling adult stem cells to facilitate this delay.

In optimal conditions, Primates essentially outlive the capacity for their adult stem cells to respond to mutation or injury. Importantly, primate adult stem cells have evolved to be longer lived, delaying the degenerative processes of aging at least for a while (Shepherd et al. 2007). Though generally associated with a decline in function, it is also possible that the evolution of delayed onset of aging in Primates has provided the appropriate variation to select for longer-lived adult stem cells or adult stem cells with heightened self-renewal capacity (Oh et al. 2014). These longer-lived adult stem cells could be selected for success at maintaining homeostasis and function of an organism throughout a lifespan. Embryonic and adult stem cells maintain the intricate timing of growth and latent periods that are necessary for the complexity of primate specific traits. The timing of growth with latent periods may be specific to each primate species, allowing stem cells to drive specific phenotypes and result in observed increases in complexity in this order (Hrvoj-Mihic et al. 2014). As initiators and controllers of development, embryonic and adult stem cells play a role throughout primate life including the timing of developmental events and aging, allowing stem cells to at least partially determine primate specific traits across evolution (Oh et al. 2014; Barry et al. 2017; Li et al. 2019; Martínez and Conde-Valverde 2020).

9.5 SUMMARY

From the very beginning of stem cell research, the potential of these cells to change how we think about growth and development and relatedness has been evident. In the last 30 years, the field of stem cell research has circumvented ethical and technical issues to produce life-changing treatments and solve evolutionary mysteries (Lo and Parham 2009; Daadi et al. 2014; Hrvoj-Mihic et al. 2014; Li et al. 2019; Zakrzewski et al. 2019). The application of functional genomics techniques to adult stem cells, and especially iPSCs, holds tremendous potential for exploring the underpinnings of phenotypic variation and testing hypotheses about the origins and trajectories of specific traits in Primates and other species. iPSCs allow for sampling of different cell types across different cellular developmental time points from a vast array of species, including endangered species (Hrvoj-Mihic et al. 2014; Glinsky and Barakat 2019; Grogan and Perry 2020). Applying techniques such as dynamic time warping to expression patterns from iPSCs from multiple species can shed light on the mechanisms responsible for primate traits. Human and nonhuman primate evolution can be explored using stem cells that can be expanded and employed in culture indefinitely to study specific biological processes otherwise difficult to interrogate in vivo. iPSCs derived from several primate species could be compared across multiple layers, genetic, molecular, cellular, and functional, as a means of investigating hypotheses related to the evolution of primate specific traits (Hrvoj-Mihic et al. 2014). The ultimate potential of stem cells in revealing the basis for primate characteristics is limited only by the inquisitive minds of scientists as they incorporate stem cells into their interrogations of life, relatedness, and evolution.

9.6 ACKNOWLEDGMENTS

We would like to thank Joan T. Richtsmeier for inviting us to contribute to this volume. We are also grateful to three anonymous reviewers whose comments and suggestions greatly improved this chapter. This work was funded in part by NICHD grant P01HD078233 and NIDCR grant R01DE027677.

9.7 REFERENCES

Alison, M., M. Golding, V. Emons, T. V. Anilkumar, and C. Sarraf. 1998. Stem cells of the liver. *Medical Electron Microscopy* 31:53–60. doi:10.1007/BF01557781.

Allman, J. M. 1999. *Evolving Brains*. New York: Scientific American Library.

Amin, T. B., and I. Mahmood. 2008, 29–30 Nov. Speech recognition using dynamic time warping. *2008 2nd International Conference on Advances in Space Technologies*, 74–79.

Atchley, W. R., and B. K. Hall. 1991. A model for development and evolution of complex morphological structures. *Biological Reviews* 66:101–57.

Austad, S. N. 1997. Comparative aging and life histories in mammals. *Experimental Gerontology* 32:23–38.

Austad, S. N., and K. E. Fischer. 1992. Primate longevity: its place in the mammalian scheme. *American Journal of Primatology* 28:251–61.

Baab, K. L., K. P. McNulty, and F. J. Rohlf. 2012. The shape of human evolution: a geometric morphometrics perspective. *Evolutionary Anthropology* 21:151–65.

Baker, C. L., and M. F. Pera. 2018. Capturing totipotent stem cells. *Cell Stem Cell* 22:25–34. doi:10.1016/j.stem.2017.12.011.

Bakken, T. E., J. A. Miller, S.-L. Ding, S. M. Sunkin, K. A. Smith, L. Ng, A. Szafer, et al. 2016. A comprehensive transcriptional map of primate brain development. *Nature* 535:367–75. doi:10.1038/nature18637.

Barrickman, N. L., M. L. Bastian, K. Isler, and C. P. van Schaik. 2008. Life history costs and benefits of encephalization: a comparative test using data from long-term studies of primates in the wild. *Journal of Human Evolution* 54:568–90. doi:10.1016/j.jhevol.2007.08.012.

Barry, C., M. T. Schmitz, P. Jiang, M. P. Schwartz, B. M. Duffin, S. Swanson, R. Bacher, et al. 2017. Species-specific developmental timing is maintained by pluripotent stem cells ex utero. *Dev Biol* 423:101–10. doi:10.1016/j.ydbio.2017.02.002.

Barton, R. A. 1998. Visual specialization and brain evolution in primates. *Proc Biol Sci* 265:1933–7. doi:10.1098/rspb.1998.0523.

Bourne, G. 2014. *The Rhesus Monkey: Volume II: Management, Reproduction, and Pathology*. New York: Elsevier Science.

Brawand, D., M. Soumillon, A. Necsulea, P. Julien, G. Csárdi, P. Harrigan, M. Weier, et al. 2011. The evolution of gene expression levels in mammalian organs. *Nature* 478:343–8. doi:10.1038/nature10532.

Case, T. J. 1978. On the evolution and adaptive significance of postnatal growth rates in the terrestrial vertebrates. *The Quarterly Review of Biology* 53:243–82.

Caspari, R., and S.-H. Lee. 2004. Older age becomes common late in human evolution. *PNAS* 101:10895–900.

Chao, J. R., A. Lamba, T. R. Klesert, A. La Torre, A. Hoshino, R. J. Taylor, A. Jayabalu, et al. 2017. Transplantation of human embryonic stem cell-derived retinal cells into the subretinal space of a non-human primate. *Translational Vision Science & Technology* 6:4. doi:10.1167/tvst.6.3.4.

Charnov, E. L. 1991. Evolution of life history variation among female mammals. *PNAS* 88:1134–7.

Chen, G., H. Xu, Y. Yao, T. Xu, M. Yuan, X. Zhang, Z. Lv, and M. Wu. 2020. BMP signaling in the development and regeneration of cranium bones and maintenance of calvarial stem cells. *Frontiers in Cell and Developmental Biology* 8:135. doi:10.3389/fcell.2020.00135.

Cosgrove, B. D., P. M. Gilbert, E. Porpiglia, F. Mourkioti, S. P. Lee, S. Y. Corbel, M. E. Llewellyn, et al. 2014. Rejuvenation of the muscle stem cell population restores strength to injured aged muscles. *Nature Medicine* 20:255–64. doi:10.1038/nm.3464.

Daadi, M. M., T. Barberi, Q. Shi, and R. E. Lanford. 2014. Nonhuman primate models in translational regenerative medicine. *Stem Cells Dev* 23 Suppl 1:83–7. doi:10.1089/scd.2014.0374.

Davidson, E. H. 2010. *The Regulatory Genome: Gene Regulatory Networks in Development and Evolution*. Boston: Elsevier.

Dean, M. C., A. D. Beynon, J. F. Thackeray, and G. A. Macho. 1993. Histological reconstruction of dental development and age at death of a juvenile *Paranthropus robustus* specimen, SK 63, from Swartkrans, South Africa. *American Journal of Physical Anthropology* 91:401–19.

Dean, M. C., and T. J. Cole. 2013. Human life history evolution explains dissociation between the timing of tooth eruption and peak rates of root growth. *PLoS One* 8:e54534.

Dean, M. C., and H. M. Liversidge. 2015. Age estimation in fossil hominins: comparing dental development in early *Homo* with modern humans. *Annals of Human Biology* 42:415–29.

Dean, M. C., C. B. Stringer, and T. G. Bromage. 1986. Age at death of the Neanderthal child from Devil's Tower, Gibraltar and the implications for studies of general growth and development in Neanderthals. *American Journal of Physical Anthropology* 70:301–9.

De Magalhaes, J. P., and J. Costa. 2009. A database of vertebrate longevity records and their relation to other life-history traits. *Journal of Evolutionary Biology* 22:1770–4.

Doro, D. H., A. E. Grigoriadis, and K. J. Liu. 2017. Calvarial suture-derived stem cells and their contribution to cranial bone repair. *Frontiers in Physiology* 8:956. doi:10.3389/fphys.2017.00956.

Durham, E., R. N. Howie, N. Larson, A. LaRue, and J. Cray. 2019. Pharmacological exposures may precipitate craniosynostosis through targeted stem cell depletion. *Stem Cell Research* 40:101528. doi:10.1016/j.scr.2019.101528.

Ebert, A. D., J. Yu, F. F. Rose, Jr., V. B. Mattis, C. L. Lorson, J. A. Thomson, and C. N. Svendsen. 2009. Induced pluripotent stem cells from a spinal muscular atrophy patient. *Nature* 457:277–80. doi:10.1038/nature07677.

Espuny-Camacho, I., K. A. Michelsen, D. Gall, D. Linaro, A. Hasche, J. Bonnefont, C. Bali, et al. 2013. Pyramidal neurons derived from human pluripotent stem cells integrate efficiently into mouse brain circuits in vivo. *Neuron* 77:440–56. doi:10.1016/j.neuron.2012.12.011.

Fehrer, C., and G. Lepperdinger. 2005. Mesenchymal stem cell aging. *Experimental Gerontology* 40:926–30. doi:10.1016/j.exger.2005.07.006.

Frazer, K. A., X. Chen, D. A. Hinds, P. V. K. Pant, N. Patil, and D. R. Cox. 2003. Genomic DNA insertions and deletions occur frequently between humans and nonhuman primates. *Genome Research* 13:341–6.

Glinsky, G., and T. S. Barakat. 2019. The evolution of Great Apes has shaped the functional enhancers' landscape in human embryonic stem cells. *Stem Cell Res* 37:101456. doi:10.1016/j.scr.2019.101456.

Goncharova, N. D., and B. A. Lapin. 2002. Effects of aging on hypothalamic–pituitary–adrenal system function in non-human primates. *Mechanisms of Ageing and Development* 123: 1191–201.

Goodman, M., W. J. Baileym, K. Hayasaka, M. J. Stanhope, J. Slightom, and J. Czelusniak. 1994. Molecular evidence on primate phylogeny from DNA sequences. *American Journal of Physical Anthropology* 94:3–24.

Grogan, K. E., and G. H. Perry. 2020. Studying human and nonhuman primate evolutionary biology with powerful in vitro and in vivo functional genomics tools. *Evolutionary Anthropology*. doi:10.1002/evan.21825.

Gruss, L. T., and D. Schmitt. 2015. The evolution of the human pelvis: changing adaptations to bipedalism, obstetrics and thermoregulation. *Philosophical Transactions of the Royal Society B: Biological Sciences* 370:20140063.

Gu, D., and N. Sarvetnick. 1993. Epithelial cell proliferation and islet neogenesis in IFN-g transgenic mice. *Development* 118:33.

Hall, B. K. 2007. Homoplasy and homology: dichotomy or continuum? *Journal of Human Evolution* 52:473–9.

Hall, B. K. 2012. *Evolutionary Developmental Biology*. Netherlands: Springer Science.

Hartenstein, V. 2013. Stem cells in the context of evolution and development. *Development Genes and Evolution* 223:1–3. doi:10.1007/s00427-012-0430-8.

Harvey, P. H., R. D. Martin, and T. H. Clutton-Brock. 1987. Life histories in comparative perspective. In *Primate Societies*, ed. B. B. Smuts, D. L. Cheney, R. M. Seyfarth, et al., pp. 181–96. Chicago: University of Chicago Press.

Harvey, P. H., and T. H. Clutton-Brock. 1985. Life history variation in primates. *Evolution* 39:559–81.

Herbig, U., M. Ferreira, L. Condel, D. Carey, and J. M. Sedivy. 2006. Cellular senescence in aging primates. *Science* 311:1257.

Hogan, B. L. 1996. Bone morphogenetic proteins: multifunctional regulators of vertebrate development. *Genes & Dev.* 10:1580–94.

Hoggatt, J., Y. Kfoury, and D. T. Scadden. 2016. Hematopoietic stem cell niche in health and disease. *Annual Review of Pathology: Mechanisms of Disease* 11:555–81. doi:10.1146/annurev-pathol-012615-044414.

Holly Smith, B., T. L. Crummett, and K. L. Brandt. 1994. Ages of eruption of primate teeth: a compendium for aging individuals and comparing life histories. *American Journal of Physical Anthropology* 37:177–231.

Holmes, G., A. Gonzalez-Reiche, N. Lu, X. Zhou, J. Rivera, D. Kriti, R. Sebra, A. Williams, et al. 2020. Integrated transcriptome and network analysis reveals spatiotemporal dynamics of calvarial suturogenesis. *Cell Reports* 32:107871.

Hrvoj-Mihic, B., M. C. Marchetto, F. H. Gage, K. Semendeferi, and A. R. Muotri. 2014. Novel tools, classic techniques: evolutionary studies using primate pluripotent stem cells. *Biol Psychiatry* 75:929–35. doi:10.1016/j.biopsych.2013.08.007.

Hsu, Y.-C., H. A. Pasolli, and E. Fuchs. 2011. Dynamics between stem cells, niche, and progeny in the hair follicle. *Cell* 144:92–105. doi:10.1016/j.cell.2010.11.049.

Jerison, H. J. 1973. *Evolution of the Brain and Intelligence*. New York: Academic Press.

Jones, D. L., and A. J. Wagers. 2008. No place like home: anatomy and function of the stem cell niche. *Nature Reviews Molecular Cell Biology* 9:11–21. doi:10.1038/nrm2319.

Kamilar, J. M., and N. Cooper. 2013. Phylogenetic signal in primate behaviour, ecology and life history. *Philosophical Transactions of the Royal Society B: Biological Sciences* 368:20120341. doi:10.1098/rstb.2012.0341.

Keller, G. 2005. Embryonic stem cell differentiation: emergence of a new era in biology and medicine. *Genes Dev.* 19:1129–55. doi:10.1101/gad.1303605.

Kelley, J., and T. M. Smith. 2003. Age at first molar emergence in early Miocene *Afropithecus turkanensis* and life-history evolution in the Hominoidea. *Journal of Human Evolution* 44:307–29.

King, F. A., C. J. Yarbrough, D. C. Anderson, T. P. Gordon, and K. G. Gould. 1988. Primates. *Science* 240:1475–82.

King, M.-C., and A. C. Wilson. 1975. Evolution at two levels in humans and chimpanzees. *Science* 188:107–16.

Kirk, E. C. 2004. Comparative morphology of the eye in primates. *The Anatomical Record Part A* 281A:1095–103. doi:10.1002/ar.a.20115.

Kirkwood, J. K. 1985. Patterns of growth in primates. *Journal of Zoology* 205:123–36.

Koplin, J. J. 2019. Human-animal chimeras: the moral insignificance of uniquely human capacities. *Hastings Center Report* 49:23–32. doi:10.1002/hast.1051.

Le Cabec, A., M. C. Dean, and D. R. Begun. 2017. Dental development and age at death of the holotype of *Anapithecus hernyaki* (RUD 9) using synchrotron virtual histology. *Journal of Human Evolution* 108:161–75.

Lemischka, I. R., D. H. Raulet, and R. C. Mulligan. 1986. Developmental potential and dynamic behavior of hematopoietic stem cells. *Cell* 45:917–27. doi:10.1016/0092-8674(86)90566-0.

Lesciotto, K. M., and J. T. Richtsmeier. 2019. Craniofacial skeletal response to encephalization: how do we know what we think we know? *American Journal of Physical Anthropology* 168:27–46. doi:10.1002/ajpa.23766.

Li, L., and W. B. Neaves. 2006. Normal stem cells and cancer stem cells: the niche matters. *Cancer Res* 66:4553–7. doi:10.1158/0008-5472.can-05-3986.

Li, L., and T. Xie. 2005. Stem cell niche: structure and function. *Annu Rev Cell Dev Biol* 21:605–31. doi:10.1146/annurev.cellbio.21.012704.131525.

Li, T., Z. Ai, and W. Ji. 2019. Primate stem cells: bridge the translation from basic research to clinic application. *Sci China Life Sci* 62:12–21. doi:10.1007/s11427-018-9334-2.

Li, Y., K. Watanabe, M. Fujioka, and K. Ogawa. 2017. Characterization of slow-cycling cells in the mouse cochlear lateral wall. *PLoS One* 12:e0179293. doi:10.1371/journal.pone.0179293.

Lieberman, D. E. 1999. Homology and hominid phylogeny: problems and potential solutions. *Evolutionary Anthropology* 7:142–51.

Lieberman, D. E. 2000. Ontogeny, homology, and phylogeny. In *Development, Growth and Evolution: Implications for the Study of the Hominid Skeleton*, ed. P. O'Higgins and M. J. Cohn, 85–122. San Diego: Elsevier.

Lo, B., and L. Parham. 2009. Ethical issues in stem cell research. *Endocrine Reviews* 30:204–13. doi:10.1210/er.2008-0031.

López-Torres, S., M. A. Schillaci, and M. T. Silcox. 2015. Life history of the most complete fossil primate skeleton: exploring growth models for Darwinius. *Royal Society Open Science* 2:150340.

Lu, X., G. R. Beck, Jr., L. C. Gilbert, C. E. Camalier, N. W. Bateman, B. L. Hood, T. P. Conrads, et al. 2011. Identification of the homeobox protein Prx1 (MHox, Prrx-1) as a regulator of osterix expression and mediator of tumor necrosis factor α action in osteoblast differentiation. *J Bone Miner Res* 26:209–19. doi:10.1002/jbmr.203.

Lv, F.-J., R. S. Tuan, K. M. C. Cheung, and V. Y. L. Leung. 2014. Concise review: the surface markers and identity of human mesenchymal stem cells. *Stem Cells* 32:1408–19. doi:10.1002/stem.1681.

Maleki, M., F. Ghanbarvand, M. Reza Behvarz, M. Ejtemaei, and E. Ghadirkhomi. 2014. Comparison of mesenchymal stem cell markers in multiple human adult stem cells. *Int J Stem Cells* 7:118–26. doi:10.15283/ijsc.2014.7.2.118.

Marsh, C. 1988. Primate societies. *Journal of Tropical Ecology* 4:317–18.
Martin, R. D. 1990. *Primate Origins and Evolution*. Princeton: Chapman and Hall.
Martin, R. D., and A. M. MacLarnon. 1985. Gestation period, neonatal size and maternal investment in placental mammals. *Nature* 313:220–3.
Martin, R. D., and A. M. MacLarnon. 1988. Comparative quantitative studies of growth and reproduction. *Symp Zool Soc Lond*. 60:39–80.
Martínez, I., and M. Conde-Valverde. 2020. Mapping the ancestry of primates. *eLife* 9:e55429. doi:10.7554/eLife.55429.
Maruyama, T., J. Jeong, T.-J. Sheu, and W. Hsu. 2016. Stem cells of the suture mesenchyme in craniofacial bone development, repair and regeneration. *Nature Communications* 7:10526. doi:10.1038/ncomms10526.
Mattison, J. A., and K. L. Vaughan. 2017. An overview of nonhuman primates in aging research. *Experimental Gerontology* 94:41–5.
Medvedev, S. P., A. I. Shevchenko, and S. M. Zakian. 2010. Induced pluripotent stem cells: problems and advantages when applying them in regenerative medicine. *Acta Naturae* 2:18–28.
Moorjani, P., C. E. G. Amorim, P. F. Arndt, and M. Przeworski. 2016. Variation in the molecular clock of primates. *PNAS* 113:10607–12.
Morrison, S. J., and A. C. Spradling. 2008. Stem cells and niches: mechanisms that promote stem cell maintenance throughout life. *Cell* 132:598–611. doi:10.1016/j.cell.2008.01.038.
Nargolwalla, M. C., D. R. Begun, M. C. Dean, D. J. Reid, and L. Kordos. 2005. Dental development and life history in *Anapithecus hernyaki*. *Journal of Human Evolution* 49:99–121.
Niu, Y., T. Li, and W. Ji. 2017. Paving the road for biomedicine: genome editing and stem cells in primates. *National Science Review* 4:543–9. doi:10.1093/nsr/nwx094.
Oh, J., Y. D. Lee, and A. J. Wagers. 2014. Stem cell aging: mechanisms, regulators and therapeutic opportunities. *Nature Medicine* 20:870–80. doi:10.1038/nm.3651.
O'Rahilly, R. 1979. Early human development and the chief sources of information on staged human embryos. *Eur J Obstet Gynecol Reprod Biol* 9:273–80. doi:10.1016/0028-2243(79)90068-6.
Perelman, P., W. E. Johnson, C. Roos, H. N. Seuánez, J. E. Horvath, M. A. M. Moreira, B. Kessing, et al. 2011. A molecular phylogeny of living primates. *PLOS Genetics* 7:e1001342. doi:10.1371/journal.pgen.1001342.
Peters, A. 2002. Structural changes in the normally aging cerebral cortex of primates. *Progress in Brain Research* 136:455–65.
Pfefferle, L. W., and G. A. Wray. 2013. Insights from a chimpanzee adipose stromal cell population: opportunities for adult stem cells to expand primate functional genomics. *Genome Biol Evol* 5:1995–2005. doi:10.1093/gbe/evt148.
Promislow, D. E. L., and P. H. Harvey. 1990. Living fast and dying young: a comparative analysis of life-history variation among mammals. *Journal of Zoology* 220:417–37.
Provis, J. M. 2001. Development of the primate retinal vasculature. *Progress in Retinal and Eye Research* 20:799–821. doi:10.1016/S1350-9462(01)00012-X.
Ramsden, C. M., M. B. Powner, A.-J. F. Carr, M. J. K. Smart, L. da Cruz, and P. J. Coffey. 2013. Stem cells in retinal regeneration: past, present and future. *Development* 140:2576–85. doi:10.1242/dev.092270.
Rosenberg, L., A. I. Vinik, G. L. Pittenger, R. Rafaeloff, and W. P. Duguid. 1996. Islet-cell regeneration in the diabetic hamster pancreas with restoration of normoglycaemia can be induced by a local growth factor(s). *Diabetologia* 39:256–62. doi:10.1007/BF00418339.
Ross, C. F. 1988. The intrinsic rate of natural increase and reproductive effort in primates. *Journal of Zoology* 214:199–219.

Ross, C. F. 1998. Primate life histories. *Evolutionary Anthropology* 6:54–63. doi:10.1002/(sici)1520-6505(1998)6:2<54::aid-evan3>3.0.co;2-w.

Ross, C. F., and E. C. Kirk. 2007. Evolution of eye size and shape in primates. *Journal of Human Evolution* 52:294–313. doi:10.1016/j.jhevol.2006.09.006.

Ross, C. F., and R. D. Martin. 2007. The role of vision in the origin and evolution of primates. In *Evolution of Nervous Systems*, ed. Jon Kaas, 59–78. San Diego: Academic Press.

Rossi, D. J., D. Bryder, J. M. Zahn, H. Ahlenius, R. Sonu, A. J. Wagers, and I. L. Weissman. 2005. Cell intrinsic alterations underlie hematopoietic stem cell aging. *PNAS* 102:9194. doi:10.1073/pnas.0503280102.

Rowe, R. G., and G. Q. Daley. 2019. Induced pluripotent stem cells in disease modelling and drug discovery. *Nature Reviews Genetics* 20:377–88. doi:10.1038/s41576-019-0100-z.

Sato, T., J. H. van Es, H. J. Snippert, D. E. Stange, R. G. Vries, M. van den Born, N. Barker, et al. 2011. Paneth cells constitute the niche for Lgr5 stem cells in intestinal crypts. *Nature* 469:415–18. doi:10.1038/nature09637.

Schofield, R. 1978. The relationship between the spleen colony-forming cell and the haemopoietic stem cell. *Blood Cells* 4:7–25.

Schwartz, G. T. 2012. Growth, development, and life history throughout the evolution of *Homo*. *Current Anthropology* 53:S395–S408.

Schwartz, G. T., P. Mahoney, L. R. Godfrey, F. P. Cuozzo, W. L. Jungers, and G. F. N Randria. 2005. Dental development in *Megaladapis edwardsi* (Primates, Lemuriformes): implications for understanding life history variation in subfossil lemurs. *Journal of Human Evolution* 49:702–21.

Shepherd, B. E., H.-P. Kiem, P. M. Lansdorp, C. E. Dunbar, G. Aubert, A. LaRochelle, R. Seggewiss, et al. 2007. Hematopoietic stem-cell behavior in nonhuman primates. *Blood* 110:1806–13. doi:10.1182/blood-2007-02-075382.

Shi, Y., P. Kirwan, J. Smith, H. P. Robinson, and F. J. Livesey. 2012. Human cerebral cortex development from pluripotent stem cells to functional excitatory synapses. *Nat Neurosci* 15:477–86. doi:10.1038/nn.3041.

Sibley, C. G., and J. E. Ahlquist. 1984. The phylogeny of the hominoid primates, as indicated by DNA-DNA hybridization. *Journal of Molecular Evolution* 20:2–15.

Simpson, S. W., J. Quade, N. E. Levin, R. Butler, G. Dupont-Nivet, M. Everett, and S. Semaw. 2008. A female *Homo erectus* pelvis from Gona, Ethiopia. *Science* 322:1089–92.

Skelly, D. A., A. Czechanski, C. Byers, S. Aydin, C. Spruce, C. Olivier, K. Choi, et al. 2020. Mapping the effects of genetic variation on chromatin state and gene expression reveals loci that control ground state pluripotency. *Cell Stem Cell* 27:459–69. doi:10.1016/j.stem.2020.07.005.

Slack, J. M. W. 2000. Stem cells in epithelial tissues. *Science* 287:1431. doi:10.1126/science.287.5457.1431.

Smith, T. M., M. Toussaint, D. J. Reid, A. J. Olejniczak, and J.-J. Hublin. 2007. Rapid dental development in a middle Paleolithic Belgian Neanderthal. *PNAS* 104:20220–5.

Speakman, J. R. 2005. Body size, energy metabolism and lifespan. *Journal of Experimental Biology* 208:1717–30.

Spear, P. D., R. J. Moore, C. B. Kim, J. T. Xue, and N. Tumosa. 1994. Effects of aging on the primate visual system: spatial and temporal processing by lateral geniculate neurons in young adult and old rhesus monkeys. *Journal of Neurophysiology* 72:402–20. doi:10.1152/jn.1994.72.1.402.

Sun, X., E. N. Meyers, M. Lewandoski, and G. R. Martin. 1999. Targeted disruption of Fgf8 causes failure of cell migration in the gastrulating mouse embryo. *Genes & Dev.* 13:1834–46.

Suwińska, A. 2012. Preimplantation mouse embryo: developmental fate and potency of blastomeres. In *Mouse Development: From Oocyte to Stem Cells*, ed. J. Z. Kubiak, 141–63. Berlin, Heidelberg: Springer Berlin Heidelberg.

Takahashi, K., and S. Yamanaka. 2006. Induction of pluripotent stem cells from mouse embryonic and adult fibroblast cultures by defined factors. *Cell* 126:663–76. doi:10.1016/j.cell.2006.07.024.

Tewary, M., N. Shakiba, and P. W. Zandstra. 2018. Stem cell bioengineering: building from stem cell biology. *Nature Reviews Genetics* 19:595–614. doi:10.1038/s41576-018-0040-z.

Theiler, K. 1972. *The House Mouse: Atlas of Embryonic Development*. New York: Springer.

Thompson, D. D., and E. Trinkaus. 1981. Age determination of the Shanidar 3 Neanderthal. *Science* 212:575–7.

Tropepe, V., B. L. K. Coles, B. J. Chiasson, D. J. Horsford, A. J. Elia, R. McInnes, and D. van der Kooy. 2000. Retinal stem cells in the adult mammalian eye. *Science* 287:2032–6. doi:10.1126/science.287.5460.2032.

Voog, J., and D. Jones. 2010. Stem cells and the niche: a dynamic duo. *Cell Stem Cell* 6:103–15. doi:10.1016/j.stem.2010.01.011.

Waterson, R. H., E. S. Lander, R. K. Wilson, Sequencing the chimpanzee, and consortium analysis. 2005. Initial sequence of the chimpanzee genome and comparison with the human genome. *Nature* 437:69–87. doi:10.1038/nature04072.

Watts, E. S. 1990. Evolutionary trends in primate growth and development. In *Primate Life History and Evolution*, ed. C. J. De Rousseau, 89–104. New York: Wiley-Liss.

Weaver, T. D., and J.-J. Hublin. 2009. Neandertal birth canal shape and the evolution of human childbirth. *PNAS* 106:8151–6.

Western, D. 1979. Size, life history and ecology in mammals. *African Journal of Ecology* 17:185–204.

Wootton, J. T. 1987. The effects of body mass, phylogeny, habitat, and trophic level on mammalian age at first reproduction. *Evolution* 41:732–49.

Wunderlich, S., M. Kircher, B. Vieth, A. Haase, S. Merkert, J. Beier, G. Göhring, et al. 2014. Primate iPS cells as tools for evolutionary analyses. *Stem Cell Res* 12:622–9. doi:10.1016/j.scr.2014.02.001.

Xie, T., and A. C. Spradling. 2000. A niche maintaining germ line stem cells in the *Drosophila* ovary. *Science* 290:328–30.

Yamaguchi, T. P. 2001. Heads or tails: WNTS and anterior–posterior patterning. *Curr Biol* 11:R713–24.

Yamaguchi, T. P., and J. Rossant. 1995. Fibroblast growth factors in mammalian development. *Curr Opin Genet Dev* 5:485–91.

Yang, H., F. Qu, R. E. Myers, G. Bao, T. Hyslop, G. Hu, F. Fei, and J. Xing. 2012. Genetic variations in stem cell-related genes and colorectal cancer prognosis. *Journal of Gastrointestinal Cancer* 43:584–93. doi:10.1007/s12029-012-9388-z.

Yuan, Y., Y.-P. P. Chen, S. Ni, A. G. Xu, L. Tang, M. Vingron, M. Somel, and P. Khaitovich. 2011. Development and application of a modified dynamic time warping algorithm (DTW-S) to analyses of primate brain expression time series. *BMC Bioinformatics* 12:347. doi:10.1186/1471-2105-12-347.

Zakrzewski, Wojciech, M. Dobrzyński, M. Szymonowicz, and Z. Rybak. 2019. Stem cells: past, present, and future. *Stem Cell Research & Therapy* 10:68. doi:10.1186/s13287-019-1165-5.

Zhao, G., S. Guo, M. Somel, and P. Khaitovich. 2014. Evolution of human longevity uncoupled from caloric restriction mechanisms. *PLoS One* 9:e84117.

Zhao, H., J. Feng, T.-V. Ho, W. Grimes, M. Urata, and Y. Chai. 2015. The suture provides a niche for mesenchymal stem cells of craniofacial bones. *Nature Cell Biology* 17:386–96. doi:10.1038/ncb3139.

Zhao, W., X. Ji, F. Zhang, L. Li, and L. Ma. 2012. Embryonic stem cell markers. *Molecules* 17:6196–236. doi:10.3390/molecules17066196.

Zhou, B. O., L. Ding, and S. J. Morrison. 2015. Hematopoietic stem and progenitor cells regulate the regeneration of their niche by secreting angiopoietin-1. *eLife* 4:e05521. doi:10.7554/eLife.05521.

Index

Note: Page numbers in *italics* indicate a figure and page numbers in **bold** indicate a table on the corresponding page.

A

adipocytes, 132, *see also* adipose tissue; brown adipose tissue
- beige, 133–135, *134*, 137, 144, 147

adiponectin, 140
adipose tissue
- bone marrow adipose tissue (BMAT), 132
- brown (*see* brown adipose tissue)
- white, 132, 137, 140, 146, 149

adult stem cells, 216–222, 224–227
aging, delay in primates, 225–227
agouti-related peptide (AgRP), 140
agouti-signaling protein (ASIP), *5*, **16**, 38
AIM1, see SLC24A5
air conditioning function of the nasal cavity, 163
albinism, 39, 41–42
Allen's rule, 132, 162
ALX Homeobox 3 (*Alx3*), 15, 65
α-melanocyte stimulating hormone (α-MSH), 38
ameloblasts, 85
AMP-activated protein kinase (AMPK), 139–141
ancient DNA (aDNA), 44–45, 87–88
angiopoietin-like protein 8 (ANGPTL8), 137
anthropoids
- dental formula, 65–71, 74
- eye growth, 198
- life history, 223
- nasal capsule, 198

apoptosis (programmed cell death), in tooth development, 60, 67
ASIP (agouti-signaling protein), *5*, **16**, 38
Asip gene, 13–15, 20, 43
ATAC-seq, 88
attractin (*Attrn*), 13
australopithecines, 67
aye-aye, 73

B

baboons
- DNA methylation in, 92, 95
- locomotion of, 93
- osteoarthritis in, 93–95

Barx1, 64–65
β-defensin 3, 38
beige adipocytes, 133–135, *134*, 137, 144, 147
Bergmann's rule/law, 132, 162
β-catenin, 7, 13, 23, 173
black and white colobus (*Colobus guereza*), 7
black lion tamarin, 43
BMD (bone mineral density), 146
BMP (bone morphogenetic protein), 6–7, 113, 125, 224
bone, *see also* ossification
- appositional, 205
- brown adipose tissue volume and, 146
- effects of environmental temperature on formation, 169–174
- functions of, 85
- loss induced by cold exposure, 132, 146
- modeling, 195, 201–205, *202*
- remodeling, 195
- shifting in space of, 195
- temperature influence on upper airway, 171
- turnover, 86

bone marrow, 85
bone marrow adipose tissue (BMAT), 132
bone marrow biopsies, 86
bone mineral density (BMD), 146
bone morphogenetic protein (BMP), 6–7, 113, 125, 224
bonobos (*Pan paniscus*), stem cells of, 219–220, 222
bottlenecks, population, 20
brown adipose tissue, 132–149, 171, 174
- in adolescence and puberty, 138
- in adulthood and aging, 138–139
- body composition in subadults and, 137–139
- childhood, 138
- chronic stressors and glucocorticoids effect on, 141
- depot regression with age, 133, 138–139
- diet and obesity in UCP1 knockout mice, 144
- diet-induced thermogenesis and, 143
- energetics, 134–135
- exercise upregulation of, 146–147
- experimental effects of energy restriction on, 140
- fatty acid uptake, 135–136, 142–143, 148
- in fetal stage and infancy, 137–138
- future directions for research on, 148–149

glucose uptake, 134–143, 146–149
implications for hominin evolutionary biology, 147–148
lactation and, 145–146
location of, 133, *134*
obesity and, 133–134, 136, 139, 141–144, 147–148
overview of, 132–133
pregnancy and, 144–145
quantification of volume and metabolic activity, 135–137
bushbaby (*Otolemur crassicaudatus*), *204*

C

Callimico goeldii, 67, 73
Callithrix jacchus, 22
camouflage, pigmentation and, 2–3, 13, 20, 43
cartilage
formation of nasal capsule, 200–201
functions of, 85
lack of remodeling and repair, 86
temperature influence on upper airway, 171
catarrhines
dentition, 56, **66**, 67–68, 73
divergence from platyrrhines, 56
nasal capsule, 198
stem cells, 219, 222, 225
cell culture systems, 96–97
cell cycle regulation, 67, 225
Ceropithecus spp., 2, 23
chimpanzee (*Pan* spp.)
DNA methylation, 89–90, 92, 94
induced pluripotent stem cell (iPSC) lines, 96–97
locomotion of, 93
osteoarthritis in, 94
stem cells, 96–97, 219, 222, 224
whole genome sequencing, 20
ChIP-seq, 88
chondrification, 200
chondroblasts, 85, 193, 196, 201
chondroclasts, 201, 205, 206–207
chondrocytes, 85, 193, 196, 206
apoptosis, 206
hypertrophy, 201
chondrogenesis, 200–201, 206
circadian clocks, 173
cleft palate, 64
clock genes, 173
coevolution, 2
cold adaptation/acclimation, 132, 142, 147, 174
cold exposure, 132–149
cold-inducible RNA-binding protein (CIRP), 174
cold stress, 132, 134, 170–171
Colobus guereza (black and white colobus), 7
coloration, *see* pigmentation
condensations
chondrification of, 200
formation of mesenchymal, 200, 206
convergent evolution of pigmentation phenotypes, 46
cranial neural crest cells (CNCCs), 111–114, 120
craniofacial region
cranio-mandibular modularity, 120–123
evolution in primates, SHH and, 123–124
modularity and integration, SHH and, 116–120
shape changes among primates, 111
Sonic Hedgehog signaling and development, 112–114
temperature and morphology, 161–176
crown-rump length (CRL), 196, *197*

D

Daubentonia madagascarensis, 191
DBB1 gene, 46–47
DCT (dopachrome tautomerase), 38, *38*, 41
deciduous teeth, *57*, **58**, 59, 62, 65, **66**, 67, 70–74
Denisovans, DNA methylation in, 89–90
dental combs, *57*, 64–65, 67–68, 73
dental formula, 55–74
deciduous teeth, *57*, **58**, 59, 62, 65, **66**, 67, 70–74
divergent evolution of upper and lower, 70–71
diversity of, *57*, *61*
evo-devo of primate dental lamina, 71–74
of extant primates, **58**, **66**
odontogenic homeobox code, 63–65, 68, 74
phylogeny of living primates and, *57*, *61*
primitive eutherian mammal heterodont, 56
prosimians and anthropoids compared, 65–71
tooth loss, evolutionary, 56, 58, 65, 67–70
dental lamina, 60–65, 67–74
described, 60
evo-devo of primate, 71–74
dentine matrix, 85
Dermoptera (colugos), 56
development
evo-devo dental anthropology, 59–60
gene regulatory processes in the development and evolution of primate skeletal traits, 83–97
nasal airways in primates, 189–207
overview of early tooth development in mammals, 60
primate life history as long, slow development, 223–225, *225*
temperature-sensitive developmental pathways, *172*, 172–174
dichromatic predators, 2–3
diet-induced thermogenesis (DIT), 143–144
Dlx1/2 genes, 69, 71

Dlx5/6 genes, 64
DNA
  ancient, 44–45, 87–88
  obtaining from skeletal tissues, 87
  repair capacity, 42–43
DNA methylation, 87–94, 173
dopachrome tautomerase (DCT), 38, *38*
Down syndrome, mouse model for, 114
dusky langurs (*Trachypithecus obscurus*), 2
dynamic time warp technique, 224–225, *225*

E

ectodysplasin A (*Eda*), 7, 22
ectodysplasin A receptor (*Edar*), 7
embryonic stem cells, *215*, 217–222, 224–227
emperor tamarin (*Saguinus imperator*), 7
enamel, 60, 65
enamel-dentine junction (EDJ), 62
enamel knots, 60, 63
enamel matrix, 85
endochondral bone, 193
endocrine system, temperature effect on, 174
energetics, brown adipose tissue and, 134–135
energy restriction, effect on brown adipose tissue, 140
Engrailed Homeobox 1 (*EN1*), 7
epidermal melanin (EM) unit, 36, *36*
epidermis, pigment cells in, *36*, 36–39
epigenetics
  "memory" of iPSCs, 218
  skeletal traits and, 84, 87–90, 92–95
  skull modularity and, 111
  turbinal lamellae and, 205
epistatic interactions, in skin pigmentation, 45–46
epithelial-mesenchymal interactions (EMI), 5–6
ERK, 173
estrogen, effect on brown adipose tissue, 139
ethical concerns, 22, 86, 95, 218, 227
ethmoturbinals, 166, 190–191, *194*, 195, 198, 200, 203
Eulemur fulvus, 3
eumelanin, 7, *12*, 13, *19*, 37–39, *38*, 41–42, 47
evolution
  craniofacial morphology in humans, 161–176
  gene regulatory processes in the development and evolution of primate skeletal traits, 83–97
  mammalian hair, 1–4
  modularity in mammalian skull, 109–125
  nasal airways in primates, 189–207
  of pigmentation variation in humans, 35–48
  of primate dental form and formula, 55–74
evolutionary developmental biology (evo-devo)
  dental anthropology, 59–60
  of primate dental lamina, 71–74
exercise, BAT upregulation by, 146–147
exome sequencing, 20–22
eyes
  convergence, 190, *194*, 195, 196, 200–201
  development, *194*, 196, *197*, 198
  as functional matrix, 193
  iris pigmentation, *40*, 40–41, 46
  position of, *194*, 195, *197*
  size of, 195, 198–201, 206–207
  timing of growth, 195, 198

F

fatty acids, as preferred substrate for brown adipose tissue, 135–136, 142–143, 148
fibroblast growth factor (FGF), 6–7, 113, 174
folate, UV-induced photolysis of, 43–44
follicular-melanin unit, 39–40
fossil record
  appearance of euprimates in, 55–56
  tooth loss in, 56
frontonasal ectodermal zone (FEZ), 113
functional matrix hypothesis, 111

G

geladas (*Theropithecus gelada*), 7
gene expression
  dynamic time warp technique, 224–225, *225*
  odontogenic homeobox code, 63–65, 68, 74
gene-gene interactions, pigmentation and, 45
Gene ORGANizer, 90
generalized Procrustes analysis (GPA), 115
gene regulation, in development and evolution of primate skeletal traits, 83–97
gene regulatory networks, stem cells and, 219–220
genes, temperature-sensitive, 164, 169–170, 172
genome-wide association studies (GWASs), 35, 47
genomics, use to study hair, 19–22
Genotype-Tissue Expression project, 87
geometric morphometrics, 110
ghrelin, 140
glucocorticoids, suppression of NST, 141
glucose, brown adipose tissue uptake of, 134–143, 146–149
gorilla (*Gorilla gorilla*)
  dental formula, 63
  stem cells, 219–220, 222
  teeth, 63, 70
ground substance, 193
growth plates, 170

H

hair, 1–24
  comparing with vertebrate taxa, 22

density of, 3
evolution of mammalian, 1–4
evolution of pigmentation variation in humans, 46–47
functions of, 2
genes and mechanisms driving growth, 6–7, **8–11**
genomics used to study primate, 19–22
graying with age, 3, 39–40
growth cycle, *5*, 6, 39
kertinization of, 3–7
melanogenesis pathway, 7, *12*, 12–14
pigmentation (*see* hair pigmentation)
primate hair biology, 3–4
research on evolution of, 3–4
variation, 2–4, 14–15, 19
hair cells, melanin production in, 39–40
hair follicle, 4–7, 22
apoptosis, 39
growth cycle, *5*
hair pigmentation, 2–3, 7–19, *19*
camouflage and, 2–3, 13, 20, 43
evolution of variation in humans, 46–47
genes and mechanism underlying variation in primate, 14–15, **16–18**, 19, *19*
patterns, 2, 12, 15, 22–23
haplorrhines
divergence from strepsirrhines, 56
nasal system, 190–191, 196, 200, 203
teeth, *61*, 69, 112
heat dissipation limit theory, 145
heat shock proteins (HSPs), 173
HERC2 gene, 42–43, 45
hibernoma, 141
histone modifications, 87–88, 92
holoprosencephaly (HPE), 113
Homo erectus, expansion into colder climates by, 147
Homo neanderthalensis, *see* Neanderthals
HOXD10 gene, 89
HSPs (heat shock proteins), 173
humans
comparative skeletal tissue studies, 89–91
craniofacial morphology, 161–176
dentition, 56, 62–64, 67–72, 74
larynx evolution, 90–91
molecular data from skeletal tissues, 86–89
osteoarthritis in, 93–94
pigmentation variation in, 35–48
hyaline cartilage, 193
hypodontia, 67–69, 72
hypothermia, 173

I

IGF (insulin-like growth factor), 174
Indian Hedgehog (IHH), 118, 120–121
induced pluripotent stem cells (iPSCs), *215*, 217–222, 224, 226–227
chimpanzee, 96–97
epigenetic "memory" of, 218
transcription factors necessary to produce, 218
use in comparative stem cell biology in primates, 218–222
infrared skin temperature measurement, *136*, 136–137
I-NOS (NOS-2) enzymes, 169
insectivorous dentition, 56
insulin-like growth factor (IGF), 7, 137, 174
IRF4 gene, 47
iris
evolution of pigmentation variation in humans, 46–47
melanin in, *40*, 40–41
irisin, 146–147, 148
iris pigmented epithelium (IPE), 40, *40*

J

jaw, *see also* mandible
evolution of hinged, 58
length, 63

K

keratinocyte, 4–6, 12–13, *36*, 36–37, 39–40, 48

L

lactation, brown adipose tissue and, 145–146
larynx, 90–91
lemur
adaptive radiation, 73
dentition, 56, **58**, *61*, 65, **66**, 67–68, 70, 74
divergence from haplorhines, 56
eye color, 43
hair pigmentation, 23
nasal region of, 191, *194*, 198, *202*, *204*
teeth, 56, **58**, *61*, 65, **66**, 67–68, 70, 73–74
leptin, 174
life history
primates as long, slow development and delayed aging, 223–227, *225*
theory, 134
life spans, primate, 224
limb shortening, in response to a decrease in temperature, 170
Linchpin hypothesis, 71–73
locomotion, gene regulation patterns in skeletal tissues affecting, 92–93
lorises, 56, 68, 191, 198, *202*, 203

M

Macaca fascicularis, 22, 219–220, 222
macaques
  development of nasal capsule, 198
  DNA methylation in, 89, 92
  eye volume, 198–199
  gene expression during gestation, *225*
  hair, 22, 43
  locomotion of, 93
  skeletal morphology, 92–93
  stem cells, 219–220, 224
mandible, 110
  cranio-mandibular modularity, 120–123
  dentition and, 59, 64, 70–71
  evolution of hinged jaw, 58
  formation of, 113
  hypoplasia, 114
marmosets, 22
  DNA methylation in, 92
  locomotion of, 93
Marshall-Smith syndrome, 91
MATP, *see SLC45A2*
maxilloturbinal, 191, *199*, *202*
Mc1r gene, *see* melanocortin-1-receptor (*Mc1r*) gene/protein
Meckel's cartilage, 113–114, 120
median nasal processes (MNP), 113
melanin, 35–48
  as natural sunscreen, 39
  production by Raper-Mason pathway, 37–38, *38*
  production in epidermis, 37–39
  production in hair cells, 39–40
  production within melanosomes, 37–38
melanocortin-1-receptor (*Mc1r*) gene/protein, 13–14, 20, 38, 42–43, 46
melanocyte, *5*, 7, 12, *12*, 14, *19*, 35–37, *36*, 39–40, *40*, 48
melanogenesis pathway, in hair, 7, *12*, 12–14
melanoma, 42–43
melanosomes, 12–13, *36*, 36–42, 47–48
methylation, DNA, 87–94
MFSD12 gene, 46–47
Microcebus, dental formula of, 63
micrognathia, 114
microphthalmia-associated transcription factor (MITF), 12, 15, 37–38
MNP (median nasal processes), 113
modularity in the mammalian skull, 109–125
molecular clock, 56
morphological integration, 109–112
mouse lemur (*Microcebus murinus*), *194*, 196, 198
Msx1, 65, 69
mucous membrane, nasal cavity, 165–166
multipotent stem cells, 215, *215*
muscle contractions, as a source of heat, 132

N

nasal capsule, 193, *194*, 195
  formation of nasal capsular cartilage, 200–201
  perinatal bone modeling patterns and fate of the posterior, 201–205, *202*, *204*
  perinatal development of, 198–200, *199*
nasal cavity
  air conditioning function and climatic adaptation, 163
  climate-related variation of nasal structures, 166–169
  overview of, 190
  reduction trend in primates, 190
  structural elements of the incoming nasal airflow pathway, 165, *165*
  surface area to volume (SA/V) ratio, 167
  temperature-related variation, 163
nasal mucosae, 193–194
nasal region
  anatomy and physiology, 164–166
  functional matrices in, 193–196
  influence of soft tissue growth on development of, 193–200
  postnatal factors influencing nasal complexity, 205
  skeletal tissues, 193
  tissue-level changes in development of, 200–205
nasal skeleton
  bone modeling patterns, 201, *202*, 203
  variation in living primates, 191–193
natural selection, geographic distribution of human skin pigmentation variation and, 43–45
Neanderthals, 13
  craniofacial morphology, 162
  DNA methylation, 89–91
  facial shape, 90–91
  larynx, 90
  nasal cavity, 163
  pigmentation-associated mutations, 44–45
neural crest cells, 36, 86, 97
neuropeptide Y (NPY), 140
NFIX gene, 90–91
nitric oxide, 169
nonshivering thermogenesis (NST), 132–137, 139–147, 174, *see also* brown adipose tissue
  chronic stressors and glucocorticoids effect on, 141
  energy restriction, effect of, 140
  in pregnancy and lactation, 144–146
  uncoupling of cellular respiration, 132–133, *133*
NPY (neuropeptide Y), 140

Nycticebus coucang, 195
Nycticebus pygamaeus (pygmy slow loris), 191

O

obcsity
 BAT and, 133–134, 136, 139, 141–144, 147–148
 brown adipose tissue (BAT) and, 134, 139, 141–144
OCA2 gene, 13, 15, 42–43, 45
oculocutaneous albinism, 39, 41
oculocutaneous albinism II melanosomal transmembrane protein (*Oca2*), 13, 15, 42–43, 45
odontoblasts, 85
odontogenic homeobox code, 63–65, 68, 74
olfaction reduction in primates, 190
olfactory bulb, 207
oligodontia, 68
orangutan, coat pigmentation in, 43
ossification
 endochondral, 85, 121, 170, 193, 201, 205
 intramembranous, 85–86
 membranous, 193
osteoarthritis (OA), 93–95
osteoblasts, 85–86, 173, 193–194, 201
osteocalcin, 173
osteoclasts, 86, 173, 193–194, 201
osteocytes, 85
osteoid, 86
osteopontin, 173

P

Pan paniscus (bonobo), stem cells of, 219–220, 222
Pan troglodytes, *see* chimpanzee
paranasal sinuses, 164, 167–169, *168*
Paranthropus, 63, 67
PAX9, 69
pelage, 1–24, *see also* hair
 primate hair biology, 2–3
 variation in, 2
phaeomelanin, 7, *12*, 13, *19*, 37–39, *38*, 41–42, 47
phenotype
 modular elements of, 110
 stem cell, 216–218
phenotypic plasticity, 171
pigmentation
 development as a two-stage process, *19*
 genetics of, 41–43
 pelage, 2–3, 7–19
 skin (*see* skin pigmentation)
 thermoregulation and, 2–3
pigmentation in humans
 ancient DNA, evidence from, 44–45
 convergent evolution and, 46
 convergent evolution of phenotypes, 46
 epidermis, *36*, 36–39
 Eurocentric bias in pigmentation studies, 47
 evolution of variation, 35–48
 genetics of, 41–43
 hair cells, 39–40
 iris, *40*, 40–41
 skin pigmentation variation, 43–46
 ultraviolet radiation and, 35, 39, 43–44, 46, 48
Pitx1, 64
platyrrhines
 dentition, **66**, 67–68, 70
 divergence from catarrhines, 56
 locomotion of, 93
 pigmentation, 2
pluripotent stem cells, 214–215, *215*
Pomc gene, *12*, 12–13
pregnancy, brown adipose tissue and, 144–145
primate evolution
 pelage growth and pigmentation, 1–24
 stem cells in, 213–227
primates
 associations between gene regulation and trait variation in primate skeletons, 91–95
 comparative skeletal tissue studies, 89–91
 craniofacial evolution and Sonic Hedgehog signaling, 123–124
 evolution of dental form and formula, 55–74
 gene regulatory processes in the development and evolution of primate skeletal traits, 83–97
 life history as long, slow development and delayed aging, 223–227, *225*
 molecular data from skeletal tissues, 86–89
 nasal airways, 189–207
 postcranial skeletal morphologies, 92–93
 postcranial skeletal pathologies, 93–95
programmed cell death (apoptosis), in tooth development, 60, 67
proopiomelanocortin (POMC) protein, 38
Propithecus diadema, 23
prosimians, dentition of, 56, 59, 65–71, 73–74

R

RANKL/RANK pathway, 173
Raper-Mason pathway, 37–38, *38*
rhesus macaques, 198–199, 224
Rhinopithecus spp. (snub-nosed monkeys), 20
RNA, measuring expression levels of, 88
RNA editing, 173
RNAseq, 21
Rousettus leschenaulti, *202*
RUNX1 gene, 94
RUNX2, 72, 174

S

SAG, 114, 116–119, *117*, *119–122*, 121–122, 124
Saguinus spp., 7, *202*
Saimiri spp.
  eye size, 200, 203
  orbitonasal development, *199*, 199–200
  S. boliviensis, *199*
Scandentia (tree shrews), 56
selection
  directional, 44
  on hair genes, 20
  keratinization of hair and, 3
  natural, 2
  pigmentation and, 42–47
  positive, 4
  sexual, 2, 43, 47, 64
  underlying human hair phenotypes, 3–4
selection scans, 35
selective brain cooling (SBC), 171
self-renewal, stem cells and, 214–216, 220, 226–227
sexual selection, 43, 47, 64
Shh, *see* Sonic Hedgehog
sinus, *see* paranasal sinuses
skeletogenesis, 195, 200, 206
skeleton
  associations between gene regulation and trait variation in primate skeletons, 91–95
  cell types, 85–86
  comparative anatomy, 84
  comparative studies in human and nonhuman primates, 89–91
  craniofacial morphology, 161–176
  functions of, 83–84
  gene regulatory processes in the development and evolution of primate skeletal traits, 83–97
  nasal, 189–207, 193
  obtaining molecular data from human and nonhuman primate skeletal tissues, 86–89
  primate postcranial skeletal morphologies, 92–93
  primate postcranial skeletal pathologies, 93–95
skin cancers, 43
skin pigmentation
  ancient DNA, evidence from, 44–45
  convergent evolution of phenotypes, 46
  epidermis pigment cells, *36*, 36–39
  epistatic interactions, 45–46
  evolution of pigmentation variation in, 43–46
  ultraviolet radiation and, 35, 39, 43–44
skin surface temperature, brown adipose tissue and, *136*, 136–137
skull
  developmental/functional modularity and integration in, 111–112
  modularity in mammalian, 109–125
  temperature and craniofacial morphology, 161–175
SLC24A5, 13–15, 41–42, 44–46
SLC45A2, 41, 44–45
Smoothened (SMO), 114
snub-nosed monkeys (*Rhinopithecus* spp.), 20
Sonic Hedgehog (*Shh*), 7, 109–125
  case study of, 114–123, **115**, *116–117*, *119–122*
  craniofacial development and, 112–114
  craniofacial evolution in primates and, 123–124
  craniofacial modularity and integration, 116–120
  cranio-mandibular modularity and, 120–123
  SAG (Smoothened agonist), 114, 116–119, *117*, *119–122*, 121–122, 124
  skull development and, 109–125
SPR2/4, 69
squirrel monkey (*Saimiri sciureus*), *199*
stem cells
  adult, 216–222, 224–227
  aging delay in primates, 225–227
  classifications of, 214–215, *215*
  comparative biology in primates, 218–222, *221*
  conservation in primate stem cells biology, 219–220
  definition, 214
  developmental length increase in primates, 224–225, *225*
  embryonic, *215*, 217–222, 224–227
  evolution of primate characteristics, 222–227
  hair follicles and, 4
  induced pluripotent stem cells (iPSCs), *215*, 217–222, 224, 226–227
  location of, 215–216
  phenotype, 216–218
  in primate evolution, 213–227
  self-renewal, 214–216, 220, 226–227
  temperature effect on, 173
  variation in primate stem cell biology, 220–222
strepsirrhines
  divergence from haplorhines, 56
  exome sequencing, 21
  eyes, 198
  life span, 223
  nasal system, 190–191, 198, 203–204, 206
  teeth, *61*, 69–70, 112
stress
  BAT response to, 134, 139–141, 148
  cold, 132, 134, 170–171

Sulawesi macaque, 43
supernumerary teeth, 67, 69–71
sympathetic nervous systems, BAT and, 134–135, 139–141, 144, 146, 148

T

tarsiers, 56, 68, 195, 196, *197*, 198, *202*, 206, 219
teeth, *see also* dental formula; tooth morphogenesis
    crown morphology, 64–65
    deciduous, *57*, **58**, 59, 62, 65, **66**, 67, 70–74
    dental combs, *57*, 64–65, 67–68, 73
    evo-devo dental anthropology, 59–60
    evo-devo of the primate dental lamina, 71–74
    evolution of, 58
    evolution of dental form and formula, 55–74
    loss in evolution, 56, 58, 65, 67–70
    odontogenic homeobox code, 63–65, 68, 74
    overview of early tooth development in mammals, 60
    supernumerary, 67, 69–71
temperature
    Allen's rule, 162
    Bergmann's rule, 162
    bone formation, effects on, 169–174
    craniofacial morphology and, 161–176
    developmental pathways temperature-sensitive, 172–174
    mechanisms of influence on upper airway bone and cartilage, 171–172
    temperature-sensitive developmental pathways, *172*, 172–174
tenrec (*Tenrec ecaudatus*), *194*
terminally differentiated cells, *215*, 215–219, 221
testosterone, effect on brown adipose tissue, 139
thermogenesis, 132
    diet-induced, 143–144
    NST (*see* nonshivering thermogenesis)
thermoregulation, 162, 166, 169
    in extreme climatic environments, 170–171
    hair pigmentation and, 2–3
Theropithecus gelada (geladas), 7
thyroid hormones, 174
Tn5 transposase, 88
tooth morphogenesis, 56
    evolutionary adaptability of, 64
    genetics of, 67–68, 70
    stages of, 60
totipotent stem cells, 214, *215*
Tp63-null mouse, 59, 67, 70, 74
Trachypithecus obscurus (dusky langurs), 2
transcriptome profiling, 21
Tupaia spp., 201, *202*, 203, *204*, 207
turbinals, 190–191, 200–201, 203, 205–206
turbinates, *165*, 165–166, 171
TYR gene, 37–38, 41, 46
tyrosinase, *37*, 37–39, 40, 41–42
tyrosinase-related protein 1 (*Tyrp1*), 13–14, 38, *38*, 41, 44

U

ultraviolet radiation (UVR)
    intensity, 35, 39, 43–44, 46, 48
    photolysis of folate by, 43–44
    vitamin D_3 synthesis and, 43–44
uncoupling protein-1 (UCP1), 132–133, *133*, 140–141, 143–144, 146–147, 174
unipotent stem cells, 215, *215*
upper airway
    morphology, 164–169
    temperature influence on bone and cartilage, 171–172

V

Varecia spp., 23, 205
vascular endothelial growth factor (VEGF), 173
vervets
    DNA methylation in, 92
    locomotion of, 93
vision degradation with age, 226
vitamin D_3, UV-mediated synthesis of, 44

W

white adipose tissue (WAT), 132, 137, 140, 146, 149
whole-genome sequencing, 20–21
Wnt, 6–7, 72, 113, 173–174, 224